Chemistry of Biomolecules: An Introduction

Richard J. Simmonds
Department of Biochemistry, University College of Wales, Aberystwyth

ISBN 0-85186-883-5

A catalogue record for this book is available from the British Library

Published by The Royal Society of Chemistry,
Thomas Graham House, Cambridge CB4 4WF

Phototypeset by Computape (Pickering) Ltd, North Yorkshire
Printed in Great Britain by Billing & Sons Ltd, Worcester

Preface

The aim of this book is to provide a readily accessible source of information on the chemistry of biomolecules and to convey some of the fascination of the chemistry responsible for sustaining life. In addition to descriptions of syntheses and chemical properties of biomolecules, emphasis is given to how events at the molecular level (*i.e.* chemistry) explain biological properties. This distinguishes the book from standard texts in organic chemistry, which tend to give this relationship scant treatment and from those in biochemistry, which normally neglect much of the basic chemistry of biomolecules.

The book covers the chemistry of those groups of compounds of biological importance which are included in most undergraduate chemistry courses, namely carbohydrates, proteins, nucleic acids, and steroids, with two areas of pharmaceutical interest—prostaglandins and β-lactam antibiotics. The coverage can not be comprehensive and the rather diverse groups of the alkaloids and the lipids receive only passing mention. The material is presented in a lively style and includes sufficient detail to relieve most students of the (often neglected) need to consult texts specializing in the various types of biomolecule.

No knowledge of biology is assumed and the main target readership is students on advanced chemistry courses such as second and third year undergraduates in the UK and post-graduate course students in the USA. In addition, it should be a useful text throughout courses in biological or medicinal chemistry and for advanced courses in pharmaceutical sciences.

Contents

Acknowledgements

I am indebted to those who generously gave their time in reading and commenting on sections of this book. These were Professor R.J. Stoodley (UMIST), Drs. S. Evans, D.J. Hopper, M.G. Jones, E.I. Mercer, A.H. Price, and A.J. Smith (colleagues from this department). Their efforts resulted in a clearer presentation, the correction of many errors, and the inclusion of more up to date material. Those errors that remain are entirely my own responsibility and I shall welcome any suggestions for improvement of the text.

CHAPTER 1

Introduction

The term 'biomolecules'—the chemicals of life—is valuable for its implication that life has a chemical basis. In the early nineteenth century scientists believed that the chemicals found in plants and animals were fundamentally different from those found in minerals. The separate branches of organic chemistry and inorganic chemistry thus developed but gradually chemists realized that the division was artificial and organic chemistry broadened to the study of carbon compounds in general. The illusory distinction between the chemicals of life and other materials is still perceived by most non-chemists and there is no widespread realization that living material consists of chemicals. Still more dangerous is the prevalent believe that all synthetic substances are automatically environmentally dangerous 'chemicals' and totally apart from hazard-free 'organic' matter in life forms. In addition to providing information on the chemistry of biomolecules, it is an aim of this book to give the reader some idea of how chemical interactions of biomolecules may account for life.

Increasing use of the term biomolecules has coincided with a renewed interest in the overlap between chemistry and biochemistry. For most of the twentieth century biochemists offered largely descriptive explanations of cellular processes rather than attempting to specify the interactions of biomolecules involved. This approach has been highly successful in explaining most basic biological functions of cells including their regulation, reproduction, energy production, nutrition and differentiation. Recently, detailed structures of complex biomolecules such as enzymes and nucleic acids have been determined and consideration of their exact (atom by atom!) structures has begun to reveal the molecular basis for their mechanisms of reaction and modes of interaction. The (still incomplete) understanding of genetics at the molecular level, for example, has led to the production of 'human' insulin from micro-organisms through genetic engineering. We have entered an era when drugs, agrochemicals, synthetic vaccines, and biologically compatible materials may be rationally designed and a more thorough understanding of molecular recognition between biomolecules promises further great advances in such areas.

The remaining pages of this chapter provide an overview of the how biomolecules control the processes necessary to sustain life and describe the main types of non-covalent interaction between molecules which are responsible for molecular recognition.

1.1 FUNCTIONS OF BIOMOLECULES

1.1.1 Nucleic Acids

The most fundamental feature of life is the ability to reproduce. All the information necessary to beget progeny true to type must be carried by parents and passed on to their offspring. Although life on Earth is immensely variable, the storage and reproduction of all the characteristics of every known life form is undertaken by nucleic acid chains composed of four different nucleotide building blocks and just five different atoms. Although there are several different types of nucleic acid each appears to be concerned principally with the storage, transmission, or expression of genetic information. All organisms except bacteria and blue–green algae concentrate their nucleic acid genetic material into a cell nucleus. The first process by which this information is converted into recognizable plant or animal characteristics is its use as a blueprint for specific structures of the other ubiquitous constituent of living matter—proteins. Every protein is synthesized to the pattern set by the nucleotide sequence of its gene. Each gene consists of only a small proportion of the length of an organism's nucleic acid chains and even viruses, the simplest type of life form, contain between three and three hundred genes. A complete set of genes is present in almost every living cell (red blood cells are an exception in humans). This poses the question of how cells from the same organism can appear so different and perform specialist functions. The answer is that only a few genes in a cell are functioning to produce their specific protein at any one time. The mechanisms by which genes are switched on and off are still being unravelled. In some cases the gene is inactivated by a repressor protein but is actuated when an inducer molecule complexes with the repressor.

1.1.2 Proteins

Proteins form the bulk of many tissues, but probably the most crucial proteins for sustaining life are those which function as enzymes, catalysing the (bio)chemical reactions of life. Almost every biological bond-breaking and bond-forming reaction, including breakdown of foodstuffs and their reconstruction into other biomolecules, is catalysed by one or more enzymes. These chemical reactions occur mainly in the cells' cytoplasm which surrounds the nucleus and is in turn bounded by the cell membrane or plasma membrane.

Cellular levels of metabolites (*i.e.* biologically produced chemicals) are kept within narrow, but controllable, limits by enzymic control. The initial abundances of enzymes are usually governed by other proteins which induce or repress their synthesis. Additional regulation, known as feed-back control, occurs when the products of an enzyme reaction inhibit enzyme synthesis or inhibit the enzyme reaction itself. Regulation of enzyme activity not only allows the efficient use of resources within individual living cells but also provides one mechanism by which cellular activity can be influenced by external conditions such as other cells in multicellular organisms.

1.1.3 Hormones

Differentiation of cells into different types with specialist functions is a feature of all but the smallest of microscopic life forms. Multicellular organisms are capable of great diversity of function but communication between the different cell types is essential. Such signalling is usually by release of a chemical by one cell which will alter the functioning of another cell. Insulin is an example of such a chemical messenger or hormone. It is a small protein (peptide) which is released by specialized cells in the pancreas and acts on nearby liver cells to suppress the breakdown of glycogen to glucose. Signalling between cells does not rely on their being touching or even close together; follicle-stimulating hormone, for example, is a protein released from the brain (pituitary gland) of vertebrates which stimulates the release of steroidal hormones from the ovary or testis.

For a cell in one part of an organism to influence a remote cell would seemingly require a large amount of the hormone to be produced. This requirement is avoided, however, by provision of receptor sites on (or in) sensitive cells to trap the hormone. Although all cells will encounter hormones most are unaffected and do not retain the hormone, leaving it free to proceed to its target tissue. Hormone receptors are usually proteins and they are extremely specific in their molecular recognition—an alteration of a single atom in certain hormones has been found to change their sites (and thus type) of biological action.

Steroids exert their effect in the nucleus. Steroids pass from their source tissue to target cells where they diffuse through the cell membrane and bind to receptor proteins in the cytoplasm*. This complex migrates to the nucleus and switches on a gene by binding to the repressor protein already bound to the gene.

1.1.4 Carbohydrates

Carbohydrates occur in large quantities in plants and are important for energy provision and in load-bearing tissues for both plants and animals. Such carbohydrates have simple repetitive structures. Carbohydrates showing great complexity are found in much smaller amounts, often on cell surfaces, and are used mainly to govern a cell's behaviour when it encounters other cells.

The surfaces of the cells of vertebrates carry many such short but complex carbohydrates which mark the type of cell. One of their functions is to control the production of different cell types (differentiation) in the developing embryo. Cells with the same function bear similar carbohydrate markers causing them to assemble and often stick together while being repelled from dissimilar cells. Many other key events in development of healthy cell populations are controlled by the presence of cell-surface carbohydrates and they are also crucial in the functioning of the immune response. Dangerous invaders such as bacteria and cancerous cells may be recognized by their cell-surface carbohydrates and targeted for destruction. In vertebrates, proteins known as antibodies become non-covalently attached to such parasites. This attachment kills or immobilizes them, or makes them susceptible to engulfment by phagocyte cells.

* The cell contents excluding the nucleus.

1.2 MOLECULAR RECOGNITION

Most of the continuous processes responsible for maintaining life forms rely on non-covalent interactions of molecules. A molecule released by one cell may then pass by countless cells to encounter a specific receptor only on the type of cell which it is designed to influence. Such signalling can also occur between the various parts of single cells and requires very little energy. How non-covalent interactions between molecules can be very specific and strong is incompletely understood but it is certain that both steric and polar effects are involved.

Molecular recognition in enzyme action was the first to receive attention and the analogy of a lock and key was introduced. By this widely used theory the specificity of biomolecular interactions relies on the shape of a recognition site (the lock) being such that only a certain molecular shape (the key) will fit it.

The question remaining is what attracts the key to the lock; not only must the key reach the lock but there must then be enough energy liberated to turn the lock and begin desired alterations. Interaction of biomolecules often results in a shift in conformation of the receptor which effects further change. In enzymes the important change will be at their active sites which may change shape to accommodate the incoming substrate (an induced fit). The substrate will often undergo strain during binding and this brings it closer to the required transition state geometry.

Non-covalent bonding is usually weak (bonding energies 1–40 kJ mol^{-1}) in comparison to hydrogen bonding (up to 100 kJ mol^{-1}) and covalent bonds (200–500 kJ mol^{-1}) but many such bonds form between interacting biomolecules. The bonding arises mainly from electostatic forces, both attractive and repulsive, which may be split for convenience into steric, electrostatic, dispersive, and hydrogen bond forces.

1.2.1 Non-polar Interactions—Steric Repulsion and London Dispersion Attraction

Steric repulsion results from filled molecular orbitals (*i.e.* electrons) trying to occupy the same space. As the atoms get closer, the repulsion between them increases with such rapidity that consideration of the atoms as solid spheres is reasonable.

Weakly attractive van der Waals* (mainly London dispersive) forces attract non-polar molecules together. Non-polar molecules are those, like hydrocarbons, in which the electrons are fairly evenly distributed between the various component atoms; having little charge separation, electostatic forces should not be important between such molecules. However, instantaneous charge density fluctuations on one molecule induce similar but opposite charges on the other molecule resulting in net attractive forces between them. These attractions, or London dispersion forces, only become significant when the atoms are virtually

* van der Waals forces include dipole–dipole and dipole–induced dipole as well as London dispersion forces.

touching (at their van der Waals radii) since the interaction energy is proportional to $1/r^6$ (r being the distance between interacting groups).

1.2.2 Polar Interactions

Inclusion of atoms of dissimilar electronegativity in a molecule causes a build up of negative charge on the more electronegative atom(s) and positive charge on less electronegative one(s) thus forming dipoles. Such a molecule is said to be polar and its dipole(s) attract oppositely orientated dipoles on other molecules. The interaction energy may be repulsive or attractive but, once the dipoles are correctly aligned, it is proportional to the product of the charges and to $1/r^3$. An example of dipole-dipole attraction between two carbonyl groups is shown in Figure 1.1.

1.2.2.1. Hydrogen Bonding. Interaction energies between dipoles involving hydrogen, such as O—H with NR_3, can be fairly large (up to 100 kJ mol^{-1}). In this case it is not merely electrostatic attraction; the strong binding arises from an across-space donation of electrons to the hydrogen. A hydrogen bond is said to be formed and is frequently represented by a dotted line, as in O—H···NR_3. In principle any atom with lone pairs of electrons could donate them to a hydrogen bond but O and N are the only important atoms; halogen (X), and π-electron systems (benzene rings, acetylenes) might act as **electron** donors for some hydrogen bonds but the evidence is inconclusive. **Hydrogen** donors are usually O—H or N—H and occasionally S—H, X—H, or P—H; C—H is only involved in rare cases such as HCN where the H is unusually acidic.

An important example of hydrogen bonding is that shown by liquid water which has a remarkably complex intermolecular structure. Large clusters of molecules form in which each internal H_2O is linked to four other molecules (Figure 1.2), but molecules with one, two, or three hydrogen bonds are present anywhere. Water molecules will readily join or leave clusters giving a structure which changes rapidly and continually. The energy from the hydrogen bonding makes the overall structure very favourable. Insertion of molecules will remove some hydrogen bonding and increase the system's energy unless there is alternative strong bonding to water; consequently non-polar molecules can not enter

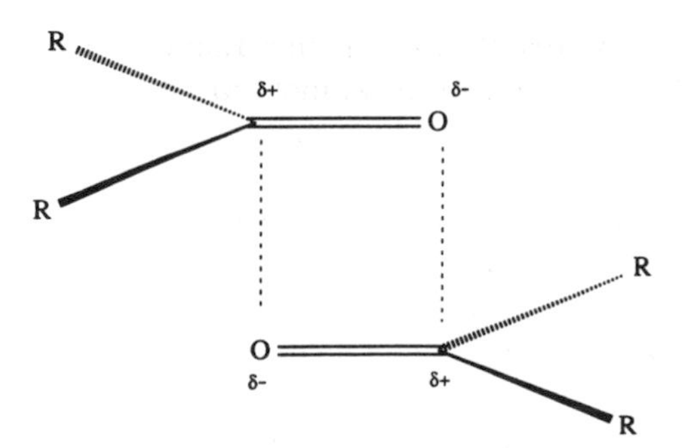

Figure 1.1 *Dipole–dipole interaction between two carbonyl groups*

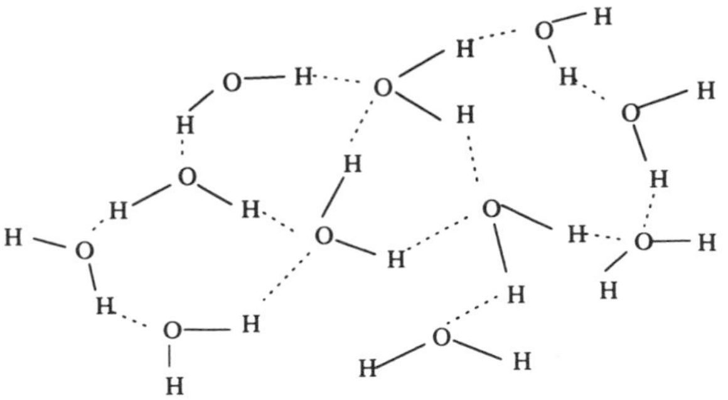

Figure 1.2 *A cluster of water molecules (two-dimensional sketch)*

the structure and are water-insoluble. It is often said that such molecules are hydrophobic, water hating. This is useful terminology but not strictly correct since water dipoles induce polarization on non-polar molecules resulting in favourable interactions between the two. However, such interactions will not be stable if they prevent more advantageous water–water contacts.

Hydrogen bonding is less tolerant of variations in geometry than most non-covalent bonding. Firstly the angle between atoms on either side of the hydrogen-bonded hydrogen should be close to 180° for strong hydrogen bonding. Secondly, the distance between the two electronegative atoms on each side of the hydrogen should not be more than the sum of their van der Waals radii (*ca.* 2.9 Å for O—H···O).

1.2.2.2 Ionic Bonding. One of the strongest non-covalent bonding forces and that extending farthest is electrostatic interaction between ions which declines with the distance (r) between them such that the interaction energy is proportional to $1/r$. The bonding is the same as in ionic crystals but when the ion is on a large organic molecule it is not usually possible to form arrays of matched ions but only to bring single positive and negative charges together. Ions may also interact attractively or repulsively with dipoles, such as $^{\delta+}C{=}O^{\delta-}$, and since the interaction energy varies as $1/r^2$ such alignments have considerable importance.

In conclusion non-covalent interactions of biomolecules are not simple to predict since they are made up of many components, both attractive and repulsive. Where possibilities for strong bonding (ionic or hydrogen bonding) exist a good match can often be recognized by consideration of molecular models. More usually, reasonable predictions of biomolecular interactions require the use of sophisticated computer programs.

CHAPTER 2

Monosaccharides

Carbohydrates are the most abundant group of biomolecules and important biological functions are associated with their formation, their properties either alone or in combination with other biomolecules, and even with their breakdown. Their biosynthesis is of fundamental importance since the carbohydrates formed by plants from carbon dioxide during photosynthesis provide almost all the carbon for living organisms. The sun's energy used in photosynthesis may later be recovered for use in plants and animals by metabolic break-down of carbohydrates, for example in glycolysis. The simple sugars may alternatively serve as units in more complex biomolecules and merely inter-linking one type of sugar unit may give several fundamentally different polysaccharides. Glucose, for example, gives cellulose, the biopolymer providing rigidity in plants and holding over half the world's organic carbon, glycogen which acts as a readily mobilized energy reserve in animals, and starch which serves as the nutritional reservoir in plants. Often non-carbohydrate monomers are incorporated into such polysaccharides; of particular importance is the copolymerization of D-ribose or 2-deoxy-D-ribose with phosphoric acid to form the back-bones of nucleic acids responsible for storage and replication of genetic information. Carbohydrates are thus used extensively as structural building blocks and fuel reserves but their great structural diversity suggests more complex biological functions and recently it has become apparent that they also have crucial roles in many biomolecular recognition processes such as immunological defence.

Carbohydrates also have great commercial importance in the manufacture of sucrose (sugar), paints, paper, textiles, foodstuffs, and certain pharmaceuticals. Such uses normally retain the carbohydrate structure but since carbohydrates are a continually renewing cheap source of organic chemicals it is probable that they will eventually replace oil as the major feedstock of the chemical industry.

2.1 MONOSACCHARIDE STRUCTURE AND NOMENCLATURE

Historically the name 'carbohydrate' derives from the observation that all the then known sugars had the general formula $C_m(H_2O)_n$. Nowadays the term is applied to the wide range of compounds which are structurally similar, or may be hydrolysed, to monosaccharides (the smallest carbohydrate building blocks). The term saccharide, meaning sugar-like, is gradually replacing the ambiguous term 'sugar', used by chemists to mean mono- or oligo-saccharides but by others to mean glucose or sucrose. Oligosaccharides are sugars composed of two to ten

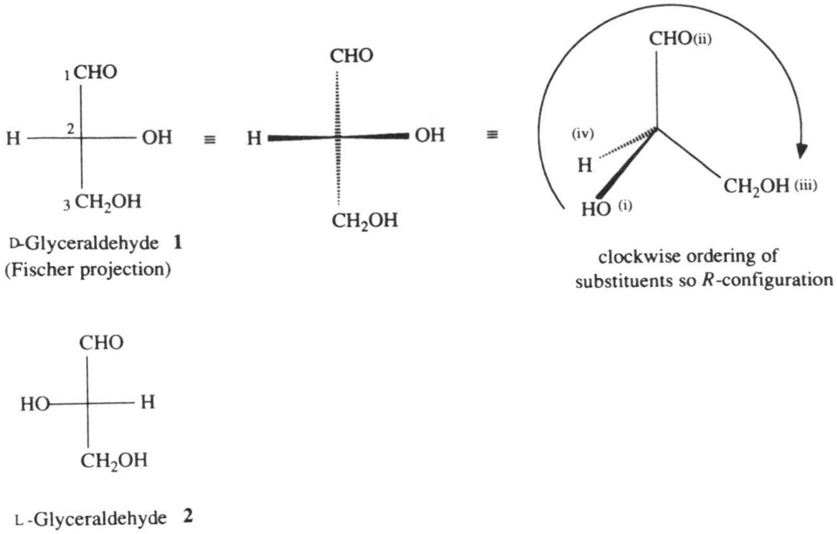

D-Glyceraldehyde **1**
(Fischer projection)

clockwise ordering of substituents so *R*-configuration

L-Glyceraldehyde **2**

Figure 2.1 *Absolute configuration of glyceraldehyde*

monosaccharide units linked together, and their chemistry is described mainly in the next chapter with that of the longer chain polysaccharides.

Monosaccharides contain several hydroxyl groups and a carbonyl group and are known as trioses, tetroses, pentoses, hexoses, *etc.*, depending on the number (generally three to eleven) of carbon atoms they contain. They may be either aldehydes or ketones (known as aldoses and ketoses respectively) and the carbon at the end of the chain which is nearest to the carbonyl group is numbered 1 and placed uppermost in Fischer projections (*e.g.* **1**). The simplest aldose, glyceraldehyde (**1**), has one asymmetric carbon and thus exists in two enantiomeric forms; that with the OH group on the right in the Fischer projection is D-glyceraldehyde and that with the OH group on the left is L-glyceraldehyde (**2**). Remembering that horizontal bonds in the Fischer projection point towards us and the vertical ones away, it is easily ascertained (Figure 2.1) that D-glyceraldehyde has the *R*-configuration. The aldotetroses erythrose and threose (Scheme 2.1) have an extra CH(OH) in their chains and since either the *R*- or the *S*-configuration is now possible at two carbons there are two diastereoisomers (*RR* and *SR*), each having a mirror image form or enantiomer (*SS* and *RS*) giving a total of four stereoisomers. Similarly there are eight aldopentoses, sixteen aldohexoses and, in general, 2^n stereoisomers for an aldose having *n* chiral carbon atoms.

Each monosaccharide which has the OH group on its highest-numbered asymmetric carbon atom to the right in its Fischer projection is assigned a different name prefixed by D-; their enantiomers, having this OH to the left, are given the same names but with the prefix L-. The Fischer structures of all D-monosaccharides up to the hexoses are shown in a chart (Scheme 2.1) in which the sugars are vertically positioned according to their sizes. Starting at D-glyceraldehyde, insertion of a CH(OH) immediately below its aldehydo group gives a

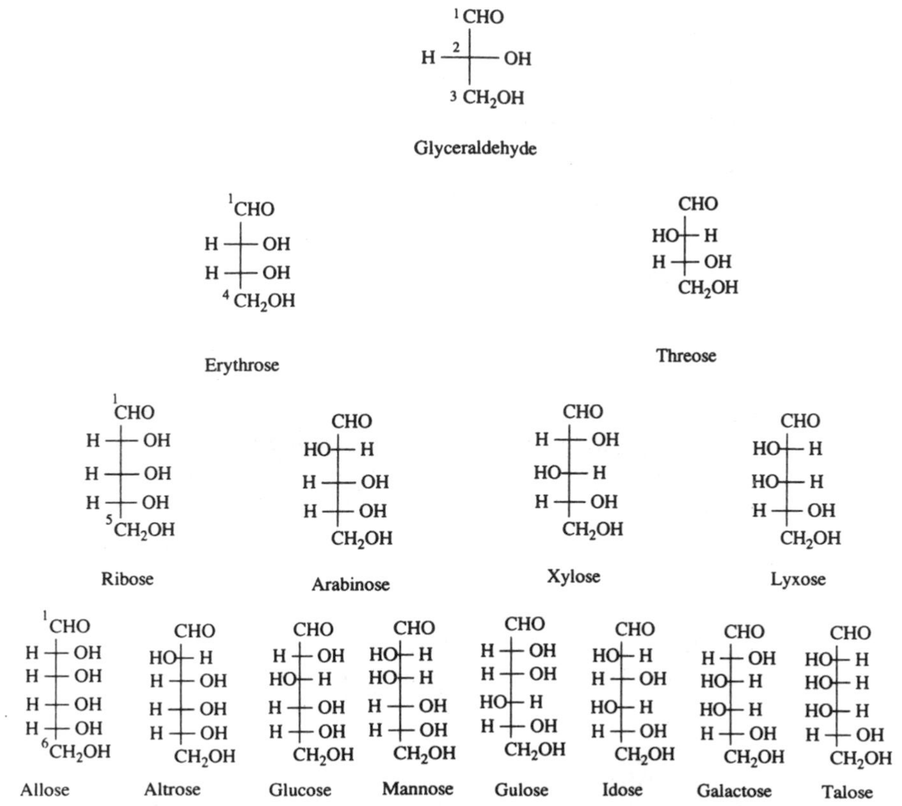

[1]CHO
H–[2]–OH
[3]CH_2OH

Glyceraldehyde

[1]CHO
H–OH
H–OH
[4]CH_2OH

Erythrose

CHO
HO–H
H–OH
CH_2OH

Threose

[1]CHO
H–OH
H–OH
H–OH
[5]CH_2OH

Ribose

CHO
HO–H
H–OH
H–OH
CH_2OH

Arabinose

CHO
H–OH
HO–H
H–OH
CH_2OH

Xylose

CHO
HO–H
HO–H
H–OH
CH_2OH

Lyxose

[1]CHO
H–OH
H–OH
H–OH
H–OH
[6]CH_2OH

Allose

CHO
HO–H
H–OH
H–OH
H–OH
CH_2OH

Altrose

CHO
H–OH
HO–H
H–OH
H–OH
CH_2OH

Glucose

CHO
HO–H
HO–H
H–OH
H–OH
CH_2OH

Mannose

CHO
H–OH
H–OH
HO–H
H–OH
CH_2OH

Gulose

CHO
HO–H
H–OH
HO–H
H–OH
CH_2OH

Idose

CHO
H–OH
HO–H
HO–H
H–OH
CH_2OH

Galactose

CHO
HO–H
HO–H
HO–H
H–OH
CH_2OH

Talose

Scheme 2.1 *The D-aldoses in their acyclic forms*

pair of diastereomeric D-tetroses to be placed below glyceraldehyde; that tetrose (erythrose) having the new OH on the same side as the original is positioned to the left and the other to the right. Repetition of this procedure gives the four D-pentoses, then the eight D-hexoses and the names of the latter may be assigned to this sequence of structures using the mnemonic **all altr**uists **gl**adly **ma**ke **gu**m **i**n **gal**lon **ta**nks. Most naturally occurring sugars are pentoses or hexoses having the D-configuration as in Scheme 2.1 but the structures of L-sugars are easily obtained by drawing the mirror image of these D-forms; L-arabinose thus has structure **3** in which the configuration of each asymmetric centre is the opposite of that in D-arabinose.

Only a few ketoses are common enough to have established trivial names and so most are named systematically as uloses prefixed by an indication of the number of carbons (*e.g.* hex-), the position of the carbonyl group, and a configurational description (such as D-*gluco*-) to indicate the configuration of each carbon. This nomenclature is illustrated for D-fructose (**4**) and D-*ribo*-3-hexulose (**5**). 2-Ketoses are the most common and their names are often abbreviated by dropping the '2' as in D-*arabino*-hexulose for D-fructose.

Some naturally occurring sugars do not have a carbonyl group at all and are

CHO
H—OH
HO—H
HO—H
CH_2OH

L-Arabinose
3

CH_2OH
=O
HO—H
H—OH
H—OH
CH_2OH

Fructose
D-Arabino-2-hexulose
4

CH_2OH
H—OH
=O
H—OH
H—OH
CH_2OH

D-Ribo-3-hexulose
5

CH_2OH
H—OH
HO—H
H—OH
H—OH
CH_2OH

D-Glucitol
6

known as alditols. They are named by replacing the 'ose' in the aldose of which they are a reduction product by 'itol'. D-Glucitol (**6**, trivially sorbitol), for example, which is found in apples and plums and used widely in confectionery and pharmaceuticals for its slightly hygroscopic and sweet properties, is the reduction product of D-glucose (or L-gulose).

A few biologically important sugars have widely used trivial names although they are modified from the basic monosaccharides by the lack of an OH group on one of their carbons (deoxy-sugars) or by possession of carboxylate or amino groups. Examples of deoxy sugars include 6-deoxy-L-mannose (L-rhamnose, **7**) which is found in aminoglycoside antibiotics, plant polysaccharides, and bacterial polysaccharides; and 6-deoxy-L-galactose (L-fucose, **8**) which occurs in the cell-surface oligosaccharides responsible for the antigenicity of blood group substances. Strangely, although 2-deoxy-D-ribose forms the backbone of DNA, it has not acquired an accepted trivial name. Neuraminic acid (**9**, 5-amino-3,5-dideoxy-D-*glycero*-D-*galacto*-2-nonulonic acid) and *N*-acetylneuraminic acid are normal constituents of gangliosides, the major polymeric materials in brain grey matter, and 2-amino-2-deoxy-sugars such as glucosamine (**10**) are found in many polysaccharides and aminoglycoside antibiotics.

CHO
H—OH
H—OH
HO—H
HO—H
CH_3

L-Rhamnose
7

CHO
HO—H
H—OH
H—OH
HO—H
CH_3

L-Fucose
8

CO_2H
=O
H—H
H—OH
H_2N—H
HO—H
H—OH
H—OH
CH_2OH

Neuraminic acid
9

CHO
H—NH_2
HO—H
H—OH
H—OH
CH_2OH

D-Glucosamine
10

2.1.1 Cyclic Forms of Monosaccharides

The structures of the monosaccharides were first established unambiguously by chemical interconversions and analysis and even then it was clear that the acyclic

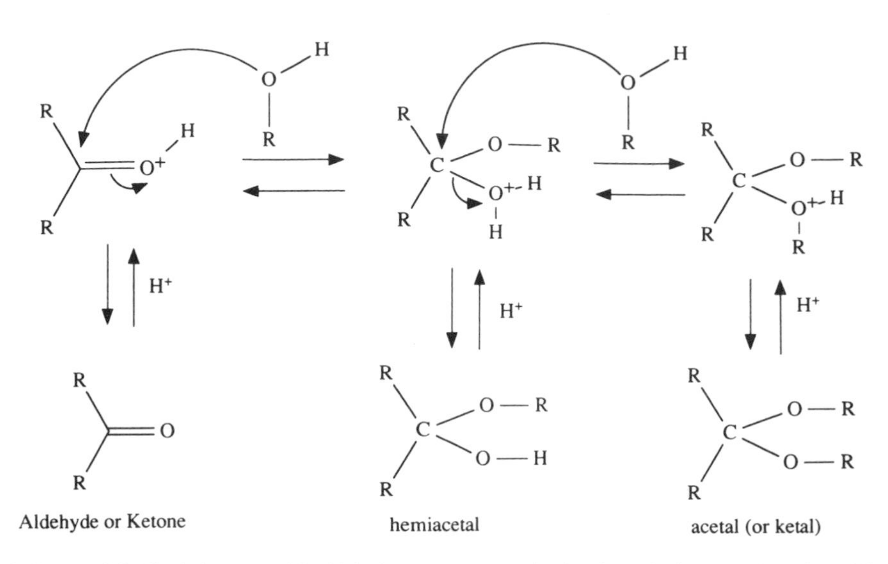

Scheme 2.2 *Alcohols react with aldehydes or ketones to give hemiacetals then acetals under acidic catalysis*

structures could not account for all their properties. D-Glucose for example exists in two crystalline forms (α and β); these show different specific rotations, $[\alpha]_D$, when first dissolved but after some hours in aqueous solution the rotations converge to a single value. This change in optical rotation is known as mutarotation and indicates an isomerization of the pure α- or β-forms to an equilibrium mixture:

	α-D-glucose	⇄ equilibrium mixture ⇄	β-D-glucose
$[\alpha]_D$	+112°	+52.7°	+19°

All aldoses, except glyceraldehyde, show no peaks attributable to C=O stretching frequencies near 1700 cm^{-1} in their infrared spectra nor any peaks near δ 10 indicative of aldehydo protons, HC=O in their ^{1}H NMR spectra. They can not, therefore, exist to any significant extent in the acyclic forms shown in Scheme 2.1. The explanation is that aldehydes and ketones react with OH groups to give hemiacetals then acetals (Scheme 2.2) and in monosaccharides this occurs intramolecularly to give cyclic hemiacetals. Five-membered rings usually form quickly but six-membered rings are generally more stable and predominate in equilibrium mixtures of monosaccharides. This process is shown for D-glucose (Scheme 2.3) together with drawings indicating the relationship between the acyclic Fischer projection formula and the cyclic forms of glucose. Only five- and six-membered rings are found in natural aldoses since these are strongly thermodynamically favoured over both smaller and larger rings; three- and four-membered rings are highly angle-strained while seven-membered and larger rings suffer transannular steric clashes.

The five-membered cyclic acetals (**11** and **12**) are known as furanoses (after furan, **15**) while the six-membered rings (**13** and **14**) are pyranoses (after pyran,

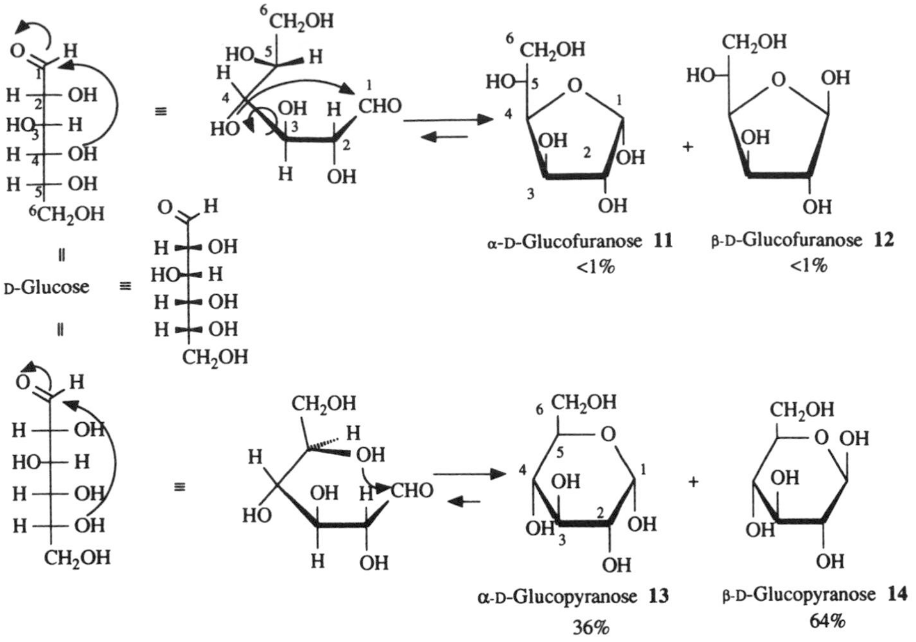

D-Glucose

α-D-Glucofuranose **11**
<1%

β-D-Glucofuranose **12**
<1%

α-D-Glucopyranose **13**
36%

β-D-Glucopyranose **14**
64%

Percentages indicate approximate composition of an equilibrated aqueous solution.

Scheme 2.3 *Fischer projections (acyclic) and Haworth perspective formulae (cyclic) for* D*-glucose*

16). By convention the Haworth perspective formulae are drawn such that numbering is clockwise when viewed from above—usually with the ring oxygen to the rear and carbon-1 furthest right—and in this orientation D-hexopyranoses have the CH_2OH group pointing up whilst L-hexopyranoses have it down. Parts of the carbon chain remaining acyclic, such as C5 and C6 in α-D-glucofuranose (**11**), are shown as a Fischer-type projections (although they may be upside down to conventional Fischer projections).

Furan

15

Pyran

16

Cyclization of a monosaccharide produces an additional chiral centre at position-1 and it is isomerism at this **anomeric** carbon of the pyranose form which is responsible for the two isolatable isomers (anomers) of glucose. The anomeric configuration is α if it is different from the configuration of the highest numbered carbon but β if these two configurations are identical (both *R* or both *S*). This is generally translated to the more readily applicable *α-D- and β-L-anomers have the anomeric OH group down in their Haworth perspective formulae.* The

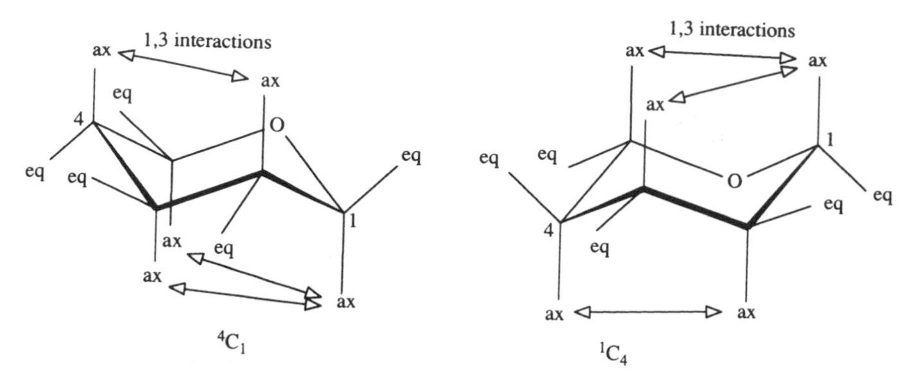

Figure 2.2 *Chair conformations of pyranoses*

crystalline form of D-glucose (**13**) which has the more positive optical rotation is thus named α-D-glucopyranose and the other crystalline form is β-D-glucopyranose (**14**).

Haworth perspective formulae as shown in Scheme 2.3 make it obvious whether OH groups are *cis* or *trans* to each other but they do not show which are axial and which equatorial. The preferred conformations of monosaccharides have been determined and it is now common for pyranoses to be drawn according to the rules for Haworth formulae but as their most stable (generally chair) forms so providing this extra information. Every pyranose may exist in two distinct chair forms which interconvert by ring flips, although one of these usually predominates. The two different chairs (Figure 2.2) are called $^1C_4$ and $^4C_1$ where the C denotes a chair, and the superscript and subscript refer to the ring atoms above and below the plane of the pyranose ring respectively. This convention has been extended to name the much less common skew (S), boat (B), and half-chair (H) conformations and also envelope (E) and twist (T) conformations of furanoses.

The most stable conformation of a saccharide is chiefly determined by minimization of steric hindrance. In chair conformations there is a strong preference for large substituents to occupy equatorial positions to avoid the 1,3-diaxial interactions (see Figure 2.2) which sterically hinder axial positions. In pyranoses this preference is accentuated for substituents on each side of the ring oxygen since C—O bonds are slightly shorter than C—C bonds so bringing axial substituents at these positions closer to their 1,3-related substituents and increasing the steric hindrance. In D-hexopyranoses, such as α-D-glucopyranose (**13** and **17**) and β-D-galactose (**18**), the CH_2OH group is the largest substituent and this usually makes the $^4C_1$ conformation, in which this group is equatorial, the more stable. The only D-hexoses having significant proportions of the alternative $^1C_4$ conformation are α-D-idose (**19**) and α-D-altrose (**20**). Since L-hexopyranoses have the opposite configuration at position-5 they generally adopt the $^1C_4$ conformation as shown for α-L-glucopyranose (**17b**). These different conformational descriptions for α-L- and α-D-glucopyranoses, which are mirror images of each other, arise from the convention of drawing pyranoses such that numbering is clockwise when viewed from above.

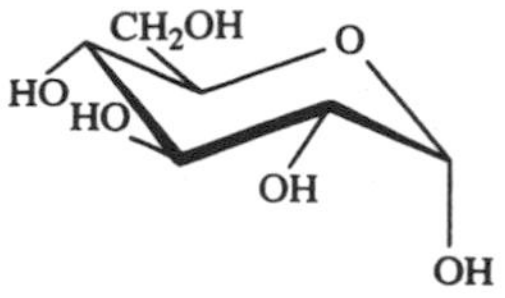

α-D-Glucopyranose
17

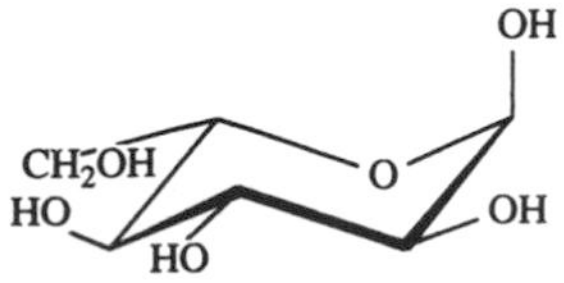

α-L-Glucopyranose
17b

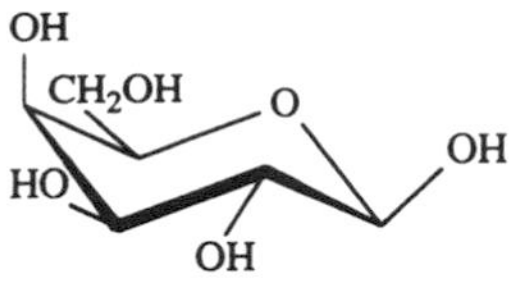

β-D-Galactose
18

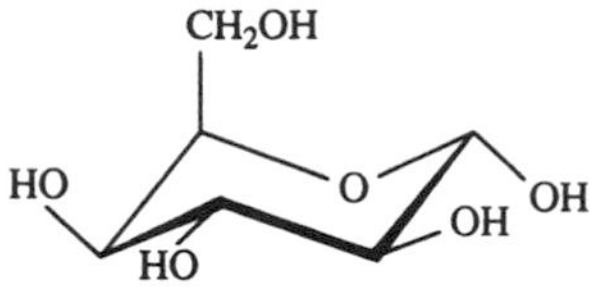

α-D-Idopyranose
19

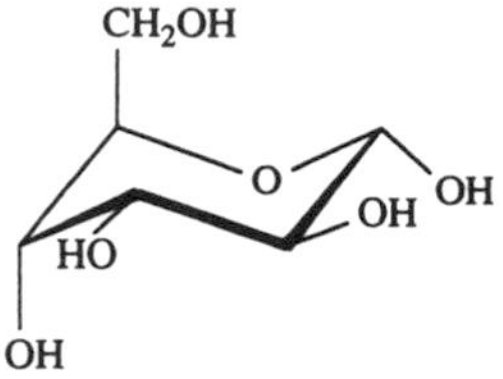

α-D-Altropyranose
20

2.1.2 The Anomeric Effect

So far we have been able to explain preferred conformations of monosaccharides solely in terms of steric effects (*i.e.* the bulkiness of groups) but stereoelectronic effects (*i.e.* the direction of bonded or non-bonded electron pairs) can also be important. A simple example is equilibration of methyl-D-glucopyranosides by treatment with HCl in methanol which produces twice as much of the α-anomer (having a hindered axial 1-OMe) as the β-anomer (Scheme 2.4). This is due to the anomeric effect—*electronegative substituents at the anomeric position prefer the axial orientation*. The anomeric effect may be explained either on the basis of the equatorial anomer being destabilized (Figure 2.3) by electrostatic interactions or on the basis of the axial anomer being stabilized (Figure 2.4) by overlap of correctly positioned orbitals. Both provide explanations for the experimental data; most students find electrostatic interactions easier to understand but the orbital overlap explanation is of wider application.

The electrostatic destabilization of equatorial anomers (Figure 2.3) is due to a repulsive interaction between the local dipoles involving the ring oxygen and the anomeric oxygen. Differences in electronegativity give a partial negative charge on the ring oxygen with partial positive charges on the two attached carbons resulting in a dipole having the direction indicated [Figure 2.3(b)]. An electronegative element at the anomeric position will also bear a partial negative charge, and its associated dipole, being in roughly the same direction as that involving the ring oxygen, produces a repulsive interaction similar to that from two bar magnets with like poles aligned. The alternative axial anomer has these two dipoles in roughly opposite directions and is therefore more stable. Another, almost equivalent, explanation is that the lone pairs on the ring oxygen and on the electronegative anomeric substituent are directed more closely together in the equatorial anomer giving an unfavourable electrostatic interaction [Figure 2.3(a)].

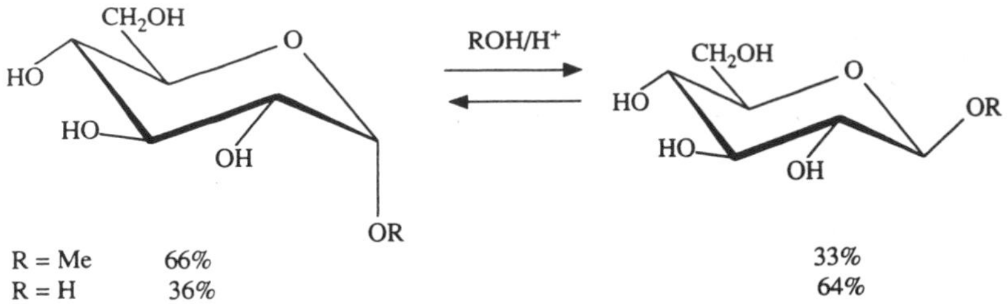

R = Me	66%	33%
R = H	36%	64%

Scheme 2.4 *Equilibration of α-D-glucopyranoses in methanol and water*

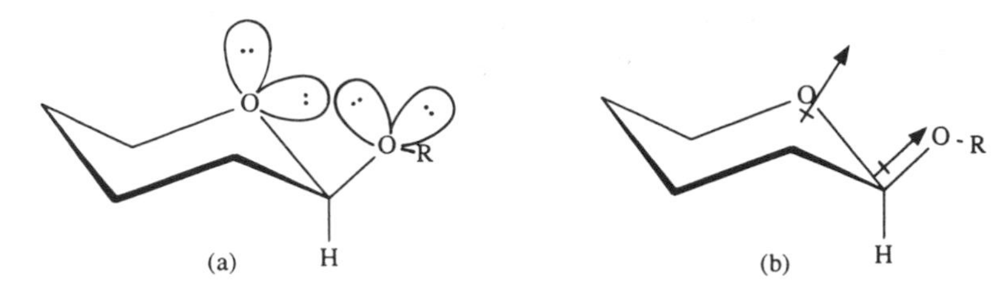

Figure 2.3 *Destabilization of equatorial anomers:* (a) *repulsion between lone pairs and* (b) *repulsion between aligned dipoles*

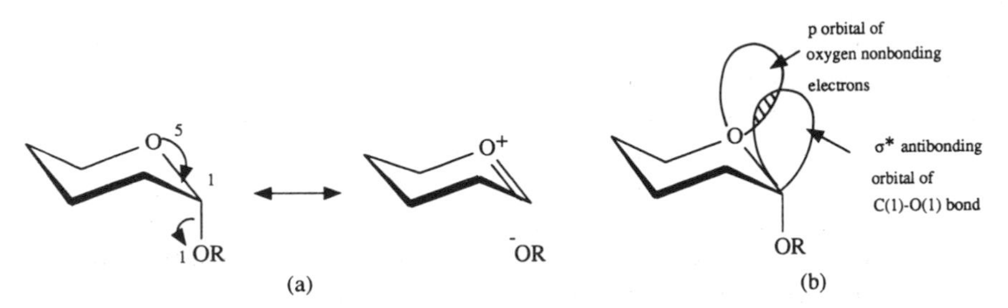

Figure 2.4 *Stabilization of axial electronegative substituents on C-1 (anomeric effect) by* (a) *resonance or* (b) *orbital overlap*

The anomeric effect viewed (Figure 2.4) as a stabilization of the axial anomer results from resonance, similar to hyperconjugation, between the two canonical forms [Figure 2.4(a)]. This resonance involves 'elimination' of the anomeric group and only proceeds well if the attacking lone pair is antiperiplanar to the departing group as in the axial anomer. A more modern expression of this idea is that the highest occupied molecular orbital of the ring oxygen overlaps with the lowest unoccupied orbital (LUMO) of the C1—O1 bond and this LUMO is correctly orientated for overlap only in the axial anomer [Figure 2.4(b)]. Such resonance or orbital overlap causes donation of electron density to the axial anomeric substituent so explaining the preference of electronegative atoms for the axial position.

The axial preference of electronegative groups at the anomeric position is solvent dependent, amounting to *ca.* 3 kJ mol^{-1} in non-polar solvents but being so diminished in polar solvents that an energetic preference for the equatorial

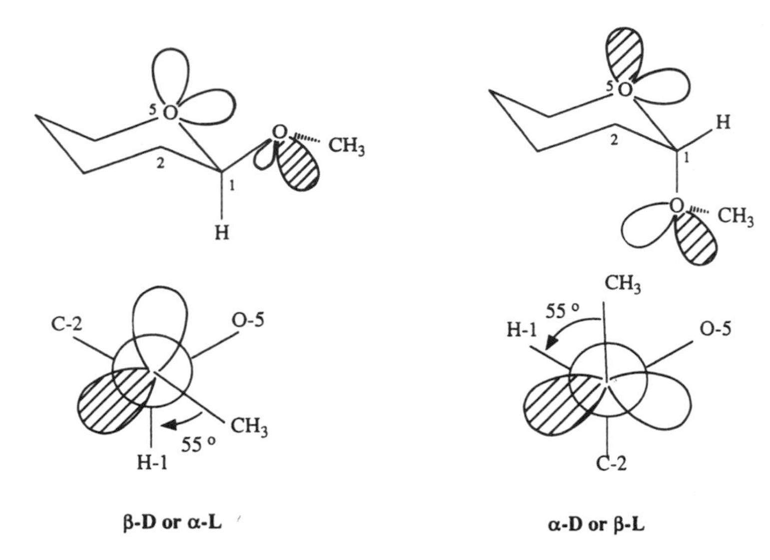

Figure 2.5 *The* exo-*anomeric effect: preferred positions of methyl groups of methyl glycosides (hatched orbitals are antiperiplanar to a C—O bond)*

anomeric position is sometimes seen. Aqueous glucose solution, for example, contains 64% β-D-glucopyranose (equatorial 1-OH) at equilibrium (Scheme 2.4). The explanation is that in water and other polar solvents steric hindrance is increased by the additional bulk of highly solvated polar groups such as OH, but electrostatic interactions are reduced because of the high dielectric constants of polar solvents.

The orientation of the alkyl group of alkyl glycosides is controlled by a stereoelectronic effect very similar to the anomeric effect. The least sterically hindered position for the alkyl group is antiperiplanar to the C1—O5 bond (Figure 2.5) but this position is taken by an oxygen *p*-orbital to allow its overlap with a C1—O5 antibonding orbital so placing the alkyl group in the more hindered synclinal position. This stereoelectronic effect is known as the *exo*-anomeric effect and since it amounts to *ca.* 6 kJ mol^{-1} in aqueous solution almost all glycosides and oligosaccharides have synclinal glycosidic linkages.

2.2 SYNTHESIS OF MONOSACCHARIDES

Most D-sugars up to hexoses are readily obtained from natural sources but their syntheses, especially from other sugars, were important during their structure determinations. Synthesis remains important for the preparation of less common sugars and as a challenge to the synthetic chemist. It is fairly simple to make a polyhydroxylated aldehyde but to achieve the correct stereochemistry at each chiral centre was, until recently, very difficult.

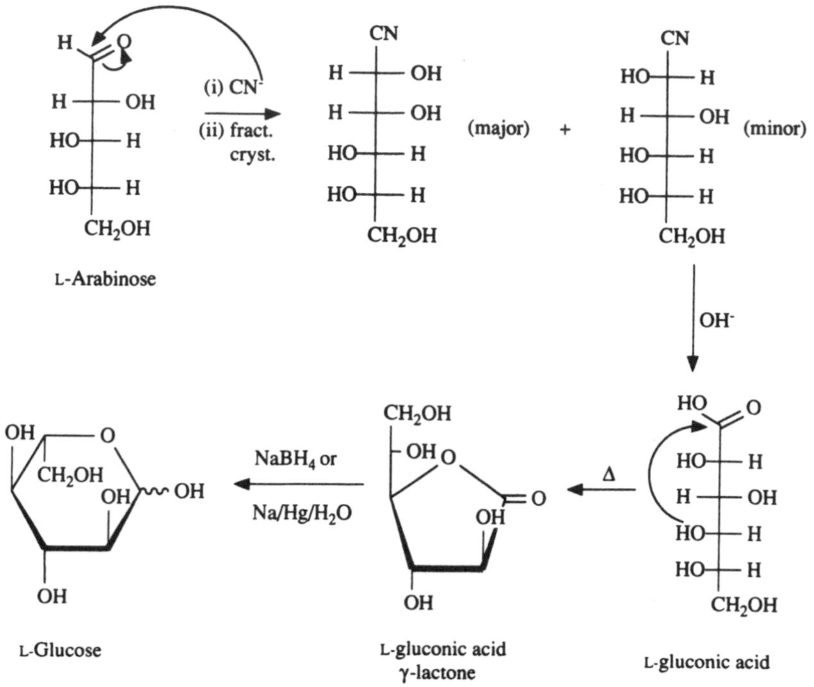

Scheme 2.5 *The Fischer–Kiliani synthesis*

2.2.1 Chain Extension Reactions

The original configurational assignments for sugars relied heavily on structural relationships established between sugars when aldose chains were extended by addition of cyanide to their aldehyde groups (masked as cyclic hemiacetals). The epimeric pair of cyanohydrins formed from an aldose may be hydrolysed to acids which cyclize to lactones and may then be reduced to a pair of aldoses with the same configuration as the starting aldose except for the extra CH(OH) group at C2. This sequence is known as the Fischer–Kiliani synthesis (Scheme 2.5) and was used to convert L-arabinose into L-mannose and L-glucose. The production of much more *manno*- than *gluco*-cyanohydrin from arabinose prompted the first realization that synthesis with asymmetric systems proceeds in an asymmetric manner. The Fischer–Kiliani synthesis is still used, especially for the preparation of isotopically labelled sugars such as the doubly labelled ribose (**21**, Scheme 2.6), and its modern version employs catalytic hydrogenation of the cyanohydrins to give the new sugars directly, thus ascending the series in just two synthetic steps.

Carbon nucleophiles other than CN^- have also been used in chain extension of aldoses. Nitromethane in base, for example, gives $^-CH_2NO_2$ which will add to the carbonyl group; subsequent removal of the nitro group is accomplished by treatment with base then acid (Nef reaction). This gives a viable procedure for the preparation of the uncommon sugars L-glucose and L-mannose from natural L-arabinose (Scheme 2.7).

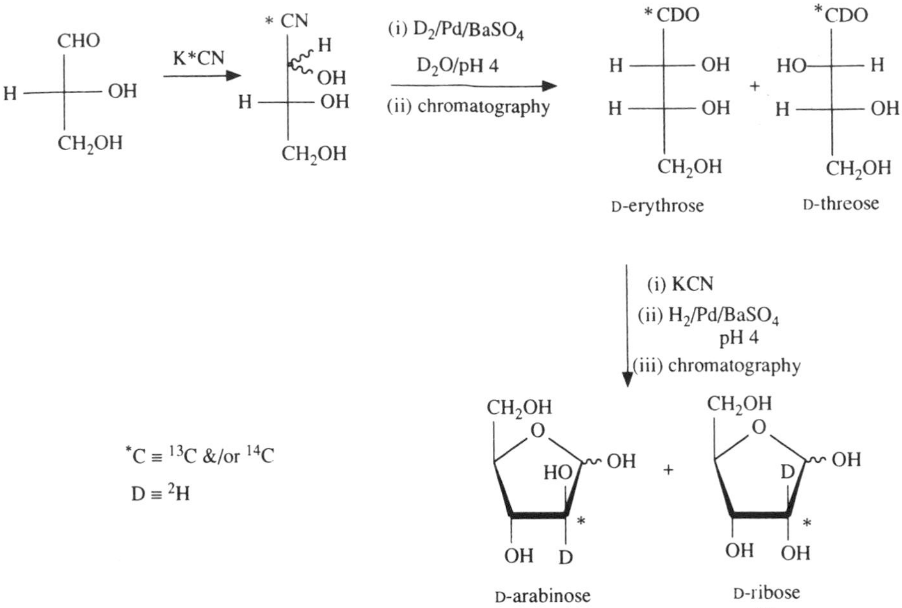

*C ≡ ^{13}C &/or ^{14}C

D ≡ ^2H

Scheme 2.6 *Synthesis of radiolabelled monosaccharides using a modified Kiliani synthesis*

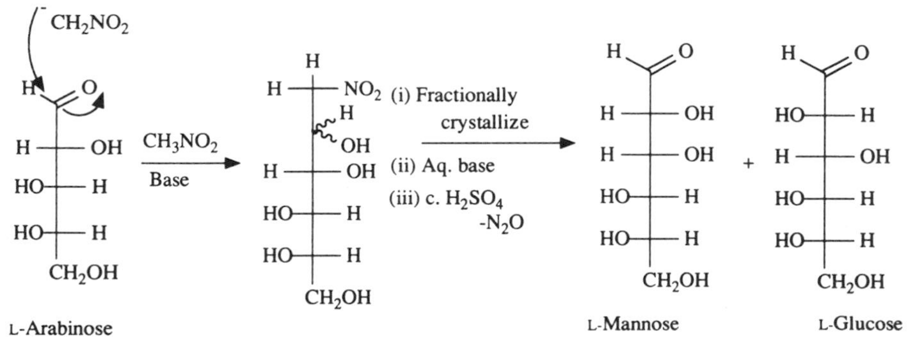

Scheme 2.7 *Synthesis of L-mannose from L-arabinose*

2.2.2 Isomerization Reactions

Reactions inverting the stereochemistry at one or more chiral carbon atoms are very common in carbohydrate chemistry and are used for preparing rare sugars. A simple example (Scheme 2.8) is the conversion of D-galactose into the relatively inaccessible D-talose (**23**) via the unsaturated sugar D-galactal (**22**). The intermediate epoxide is produced *cis* to the 3-OH since hydrogen bonding of this group stabilizes the transition state for epoxidation. Epoxide opening gives diaxial products and is initiated here by nucleophilic attack at the more reactive position, C1, giving a net inversion of configuration at C2.

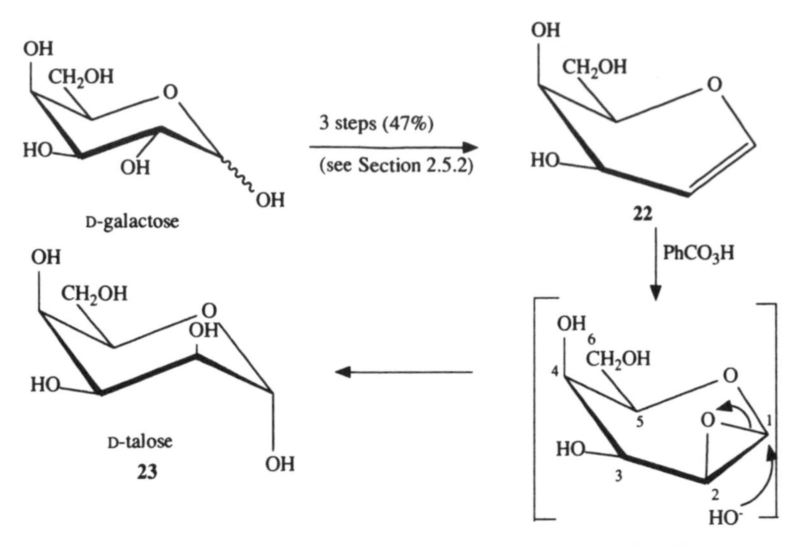

Scheme 2.8 *Synthesis of* D*-talose by inversion of configuration of C2 of* D*-galactose*

2.2.3 Total synthesis

Total synthesis of sugars from non-carbohydrate starting materials has generally used hydroxylation of alkenes to give either *cis*- or *trans*-diols depending on the conditions. Thus the mixture of *cis*-alkenes (**24**) may be *cis*-hydroxylated using potassium permanganate or *trans*-hydroxylated via the epoxide. Removal of the diethylacetal and tetrahydropyranyl protective groups by acidic hydrolysis gives

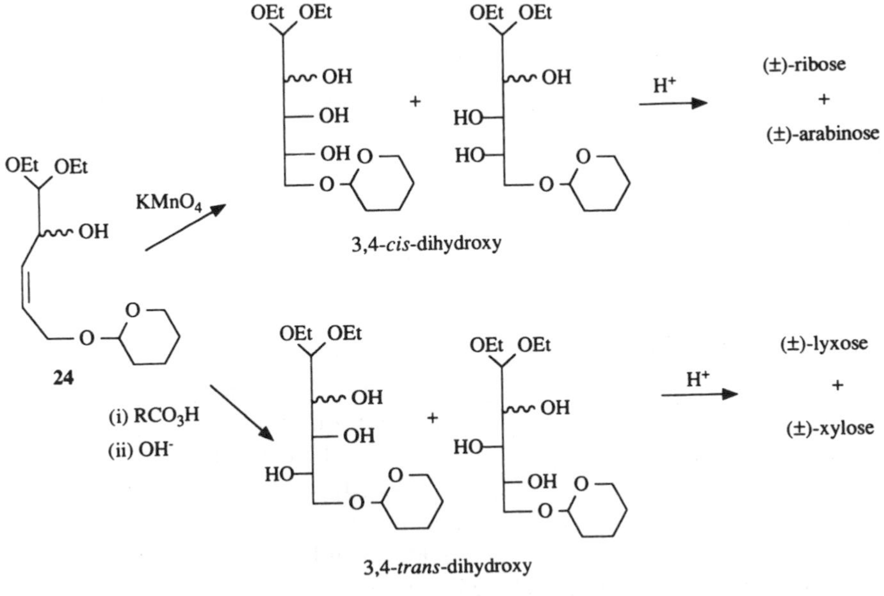

Scheme 2.9 *Total synthesis of sugars using* cis- *or* trans-*hydroxylation of alkenes*

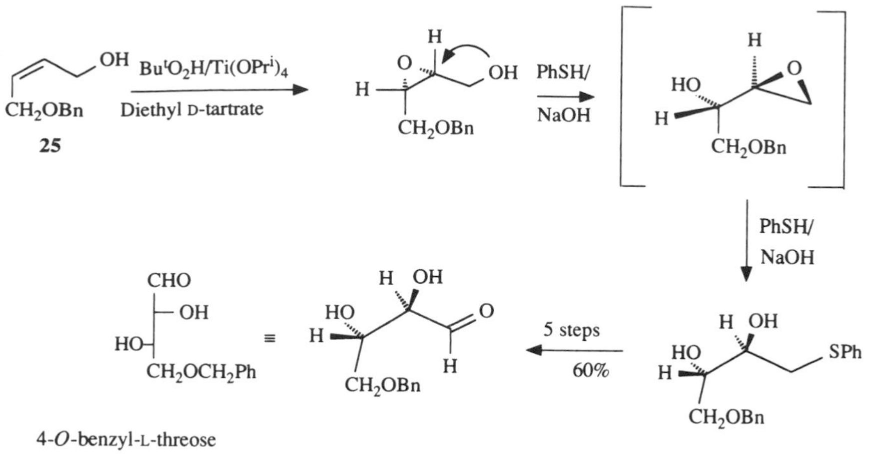

Scheme 2.10 *Asymmetric synthesis of an aldose*

a mixture of (±)-ribose with (±)-arabinose from the *cis*-hydroxylated product and (±)-xylose and (±)-lyxose from the *trans*-hydroxylated product (Scheme 2.9).

During the 1980s it was discovered that epoxidation of alkenes bearing a hydroxymethyl group may give essentially one enantiomer when catalysed by titanium tetraisopropoxide asymmetrically chelated with diethyl tartrate. This made possible the synthesis of chiral sugars as illustrated in the total synthesis of L-threose from the achiral butene (**25**, Scheme 2.10). As yet few sugars have been made by total synthesis but the methodology now exists to make the total synthesis of rare saccharides an attractive alternative to partial synthesis from natural sugars.

2.3 REACTIONS OF MONOSACCHARIDES

A cursory glance at simple monosaccharides may give the erroneous impression that they contain only one type of reactive group, the hydroxyl (OH) group. The anomeric centre (C1 of aldoses) is the most deceptive; both the ring oxygen and an OH are attached to the same carbon thus giving a hemiacetal which not only has a more reactive hydroxyl than a typical alcohol but also gives many of the reactions of an aldehyde. We have already (Section 2.2.2) seen that carbon nucleophiles like cyanide add to the anomeric carbon as though it were an aldehyde and this section deals primarily with other addition reactions at the anomeric centre. Hydroxyls other than that at the anomeric centre show considerable variation in reactivity since they may be primary or secondary, hindered or accessible, axial or equatorial, and *cis* or *trans* to each other but most reactions at non-anomeric hydroxyls make use of protective groups and will be covered in the next section (2.4).

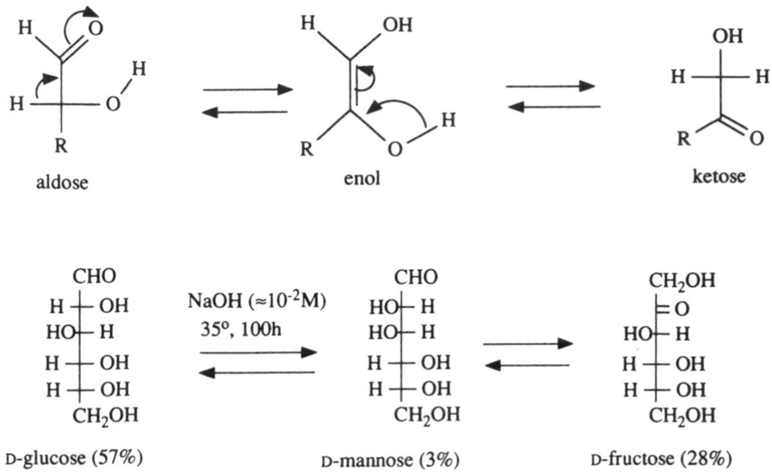

aldose ⇌ enol ⇌ ketose

NaOH (≈10^{-2}M) 35°, 100h

D-glucose (57%) ⇌ D-mannose (3%) ⇌ D-fructose (28%)

Scheme 2.11 *Keto–enol tautomerism of aldoses and ketoses*

2.3.1 Isomerization in Mild Base

Both aldoses and ketoses are isomerized in weak aqueous base. Base abstracts protons α to carbonyl groups thus accelerating the interconversion of aldehyde or keto forms of monosaccharides with their enols (Scheme 2.11). Tautomerism back to a carbonyl can involve either OH attached to the C—C and so a mixture of isomers results.

Thus treatment of D-glucose with 0.035% aqueous sodium hydroxide for 100 hours at 35 °C gives an equilibrium mixture of D-glucose (57%), D-mannose (3%), and D-fructose (28%). A similar mixture is to be expected no matter which of these three sugars is used as starting material and isomerizations of other aldoses have been used to prepare 2-ketoses. With more stringent reaction conditions the epimerization continues to give 3- then 4-epimers (epimers are stereoisomers differing in configuration at one carbon only).

Mild base also reversibly removes the anomeric proton (H1) thus promoting the interconversion of cyclic and acyclic forms. The chief result is accelerated equilibration of α- and β-forms (usually pyranose) also known as anomerization or mutarotation.

2.3.2 Saccharinic Acid Formation or Fragmentation in Strong Base

In stronger base (such as 1% calcium hydroxide) the initially formed enol may dehydrate giving a 1,2-dicarbonyl derivative which undergoes a benzilic acid type rearrangement yielding a 3-deoxyhexonic acid (metasaccharinic acid) (**28**, Scheme 2.12). Alternative enolization and dehydration may give a 2,3-dicarbonyl compound whose subsequent benzilic acid rearrangement yields saccharinic acids (**27**) or isosaccharinic acids (**26**). The proportion of each saccharinic acid is dependent on the structure of the original sugar and the reaction conditions.

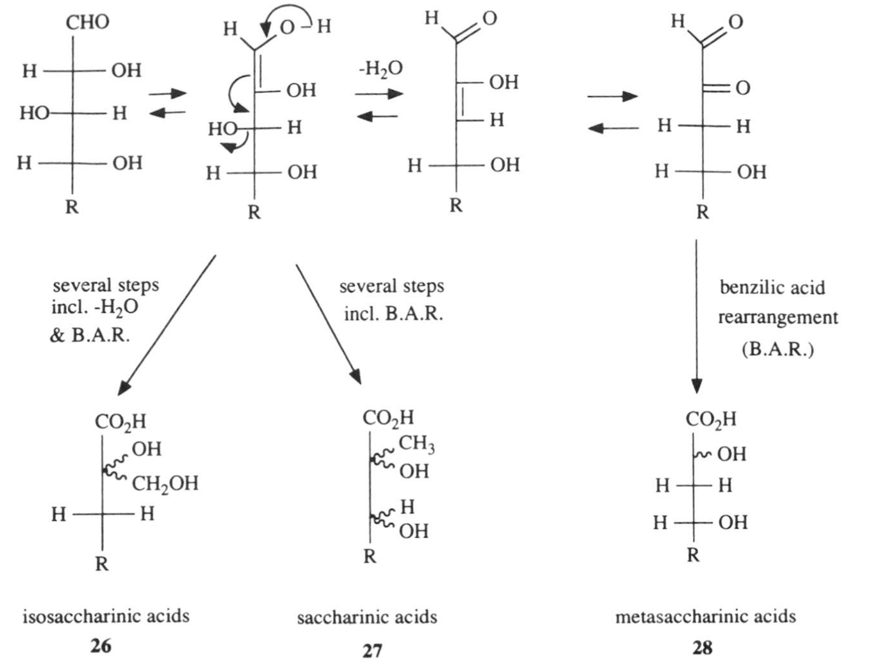

Scheme 2.12 *Monosaccharides treated with 1% calcium hydroxide give various saccharinic acids*

Sugars generally have a hydroxyl β to their carbonyl groups and so could be formed by an aldol reaction. Aldol reactions are reversible and hexoses treated with moderately concentrated aqueous alkali undergo a retroaldol reaction, fragmenting to three-carbon products—initially glyceraldehyde and 1,3-dihydroxypropanone (Scheme 2.13). Recombination of the three-carbon fragments in further aldol reactions leads to 3- and 4-epimers of the original sugar.

Many products are thus possible when sugars, unprotected at the anomeric position, encounter base; the prudent carbohydrate chemist generally does not permit aqueous solutions of unprotected sugars to become basic.

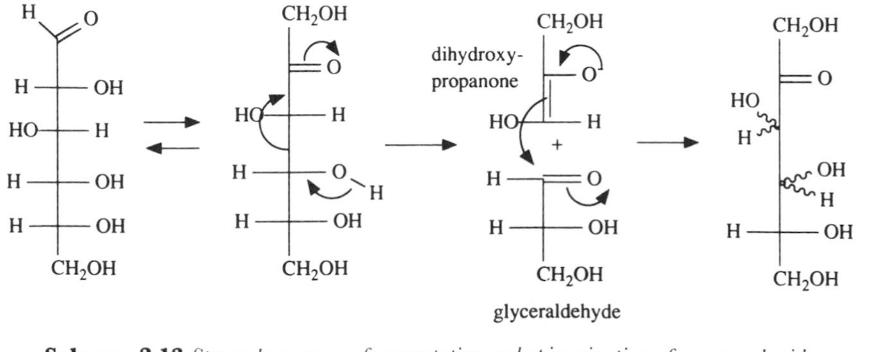

Scheme 2.13 *Strong base causes fragmentation and epimerization of monosaccharides*

2.3.3 Isomerization and Dehydration in Aqueous Acid

Free sugars are much more stable in acid than in base and it requires concentrated mineral acid before any fundamental change occurs—pentoses and hexoses dehydrate to the furan derivatives **29** and **30** respectively.

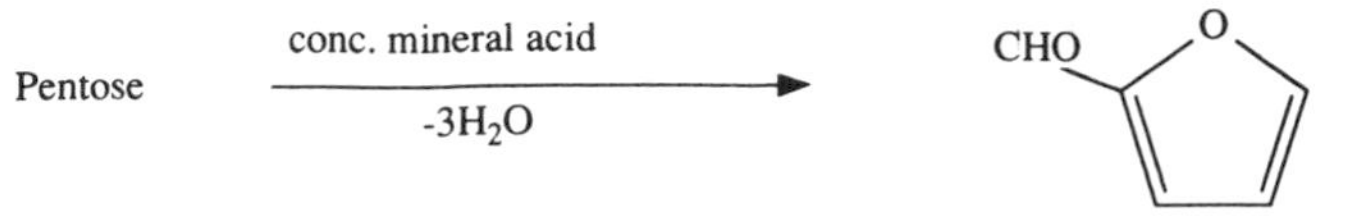

Pentose —conc. mineral acid, $-3H_2O$→ Furfuraldehyde **29**

Hexose —conc. mineral acid, $-3H_2O$→ **30**

However, reactions at the anomeric centre are catalysed by dilute acid and several reversible changes occur readily. The rate of anomerization of aldoses is thus increased and the mutarotation of α-D-glucose accelerates dramatically as the pH falls below 2. The catalysis is due to protonation of the ring oxygen which promotes cleavage of the bond between the ring oxygen and C1 (Scheme 2.14) to give an acyclic intermediate which may recyclize to alternative structures.

The cyclic forms of sugars are hemiacetals and when protonated they may react with alcohols to give acetals (Scheme 2.2). For sugars in acidic solution this means that the anomeric centre may react with other OH groups with elimination of water. This reaction is reversible and in dilute hydrochloric acid glucose gives an equilibrium mixture containing oligosaccharides from which the 1,6-linked disaccharides gentiobiose (**31**) and isomaltose (**32**) may be isolated in 7% and 8% yields respectively. This process is called acid reversion and can be a problem during acidic hydrolysis of polysaccharides to monosaccharides. Glucose may also undergo intramolecular dehydration (Scheme 2.15) to give

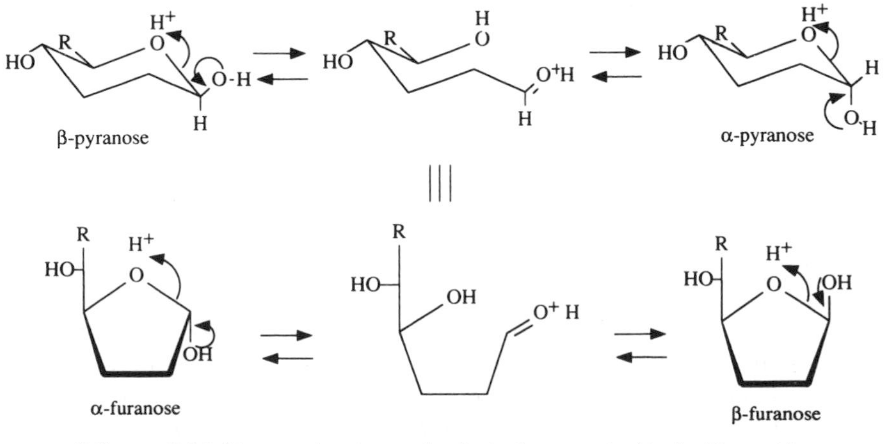

β-pyranose ⇌ α-pyranose ; α-furanose ⇌ β-furanose

Scheme 2.14 *Mutarotation (anomerization) of monosaccharides in dilute acid*

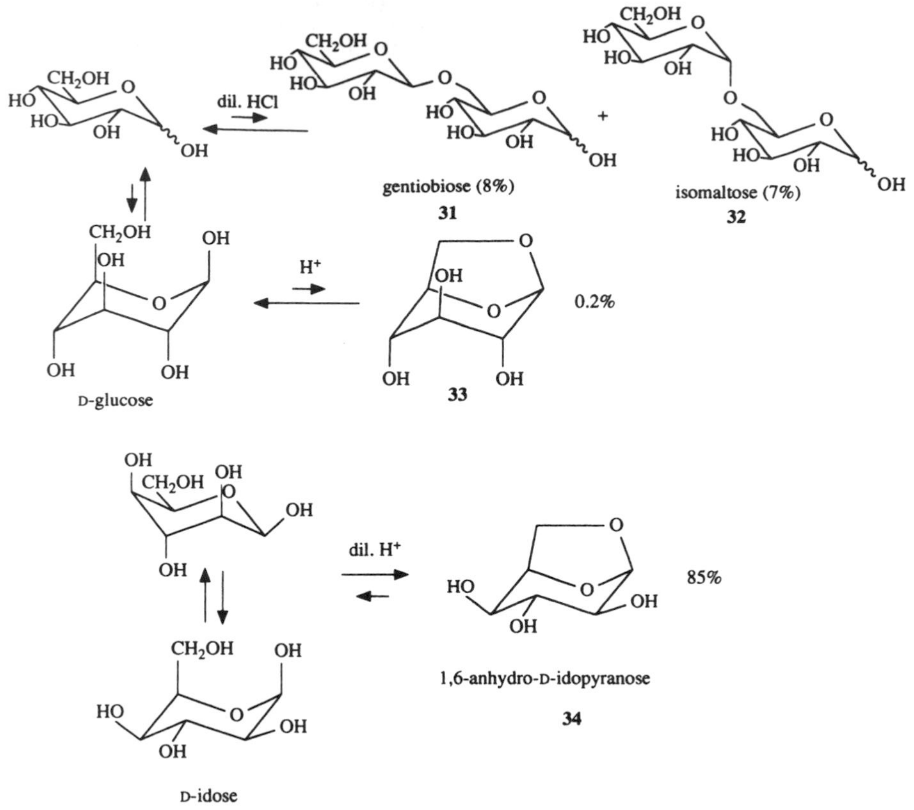

Scheme 2.15 *Acid reversion and anhydro-sugar formation in acid*

1,6-anhydro-D-glucopyranose (**33**) but the anhydroglucose comprises only 0.2% of the equilibrium concentration since it must be formed via the unstable $^1C_4$ conformation of β-D-glucopyranoside which has each substituent axial. In contrast the $^1C_4$ conformation of D-idose has only two axial substituents and gives about 85% of 1,6-anhydro-D-idopyranose (**34**) on equilibration in aqueous acid.

2.3.4 Glycosides by Reaction of Aldoses with Alcohols

The cyclic forms of aldoses are hemiacetals and so they react with alcohols in the presence of acidic catalysts to give acetals in the expected manner (see Scheme 2.2 and the previous section, 2.3.3). In essence the anomeric OH is replaced by an alkoxy group OR derived from the alcohol ROH but since the products, known as glycosides, can no longer interconvert with their acyclic or other cyclic forms in neutral solutions, the isolation of at least four different glycosides, namely α- and β-furanosides and α- and β-pyranosides as shown for glucose reacting with acidic methanol (Scheme 2.16), is possible in principle.

Usually dry hydrogen chloride is bubbled through a mixture of the sugar in

methyl α-D-glucopyranoside methyl β-D-glucopyranoside methyl α-D-glucofuranoside methyl β-D-glucofuranoside

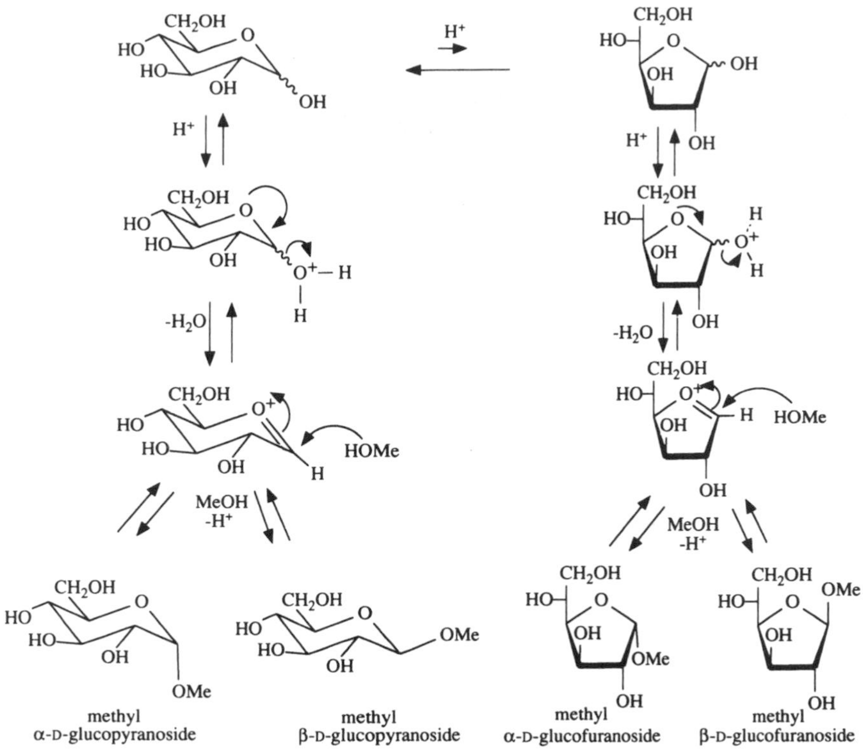

Scheme 2.16 *Methyl glucoside formation*

the alcohol and reaction continued until equilibrium is reached. These acetalation reactions are called Fischer glycosidations and a convenient modification is to use an acidic ion exchange resin in place of the hydrogen chloride as in the preparation (Scheme 2.17) of the four methyl-D-mannosides which may be separated by column chromatography on cellulose powder. The thermodynamically most stable glycoside will be the major product in Fischer glycosidations and this is controlled by factors similar to those governing the favoured isomer of free sugars (Sections 2.1.1 and 2.1.2). Reactions in methanol give mainly the methyl α-pyranoside in the majority of cases because furanose rings are destabilized by unfavourable interactions between *cis* vicinal substituents, and the anomeric effect favours the α-pyranoside (O1 normally axial) over the β-pyranoside. On the other hand ribose and arabinose have unfavourable 1,3-diaxial interactions in their α-pyranosides and usually give the methyl β-pyranosides as major products. Being pentoses, their ${}^1C_4$ conformations are not destabilized by an axial CH_2OH and so the β-pyranosides adopt this conformation to benefit from an anomeric effect.

Furanosides, such as methyl α-D-arabinofuranoside (**35**), may be isolated in reasonable yields by glycosidation of sugars under kinetically controlled conditions (Scheme 2.18)—that is using a short reaction period. In the above case this was achieved by quenching the reaction mixture with pyridine as soon as

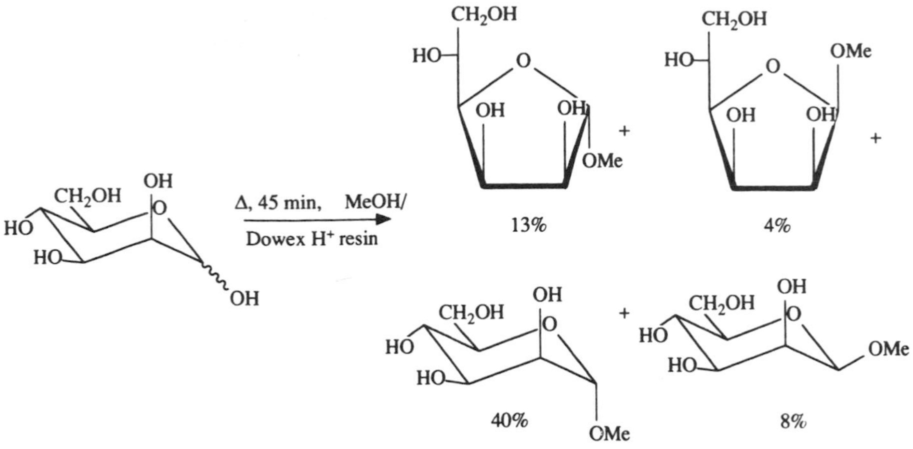

Scheme 2.17 *Fischer glycosidation of* D*-mannose using acidic ion exchange resin*

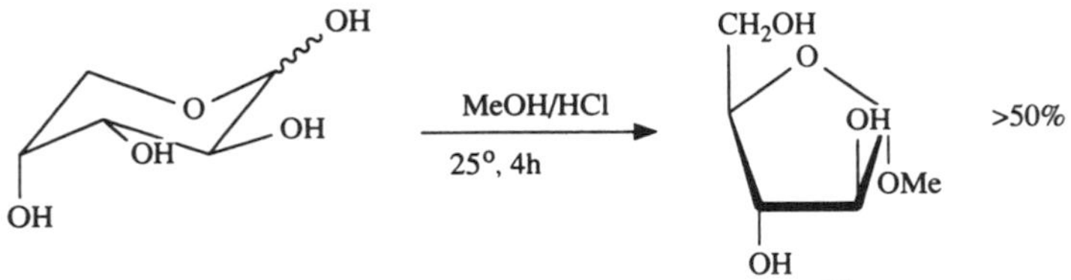

Scheme 2.18 *Kinetic control allows synthesis of methyl* α-D*-arabinofuranoside*

tests indicated no remaining reducing sugar. Studies of the changing compositions of such reaction mixtures have confirmed that furanosides are formed more rapidly than the pyranosides and that they are the major products in early stages being converted into the more thermodynamically stable pyranosides as the reaction proceeds. Since free sugars are mainly pyranoses one might infer that glycosides are formed via an acyclic intermediate, such as shown in Scheme 2.14 for the mutarotation of glucose, but actually the balance of evidence favours the intermediacy of cyclic carbocation intermediates (Scheme 2.16) formed by elimination of the 1-OH group from furanoses or pyranoses.

Glycosidation of sugars by acid-catalysed reaction with alcohols is restricted to use of volatile alcohols (to allow use of a large excess) and by the difficulty of isolating the required glycoside from the mixture produced. Other more selective methods of glycoside synthesis are available but since these use protected sugars their discussion is left to later sections (2.4.3 and 2.5.4).

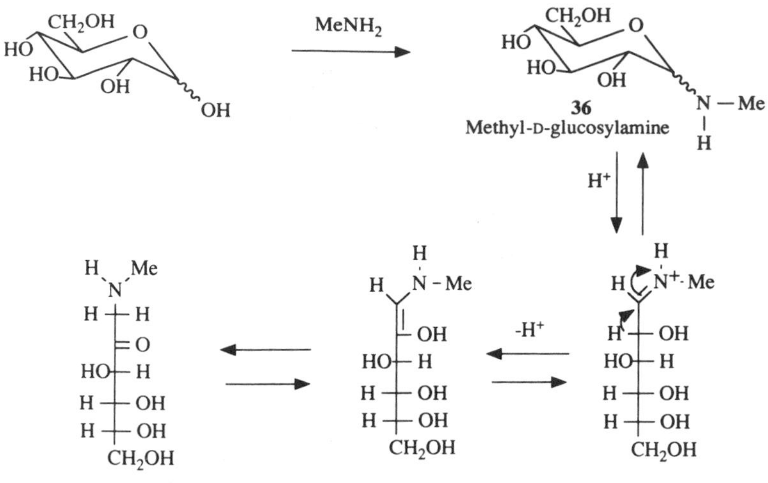

$MeNH_2$

36
Methyl-D-glucosylamine

H^+

$-H^+$

37

Scheme 2.19 *Reaction of glucose with methylamine followed by an Amadori rearrangement*

2.3.5 Glycosylamines and Aminoketoses from Amines

Aldoses react with ammonia and primary or secondary amines to give glycosylamines. Glycosylamines readily interconvert between their various cyclic and α- and β-forms in solution and although often crystalline most are mixtures of unknown compositions. Thus D-glucose gives methyl-D-glucosylamine (**36**) on heating for one hour under reflux with methylamine in methanol. The ring-opened form may tautomerize to an enamine and then to a 1-amino-1-deoxy-2-ulose (**37**). This process, known as the Amadori rearrangement (Scheme 2.19), frequently accompanies attempts to prepare glycosylamines and is catalysed by acids.

Reaction of sugars with the amino groups of amino acids and peptides occurs in foods and subsequent complex reactions are responsible for the production of much of the colour, flavour, and, since CO_2 is released, texture of baked foods.

2.3.6 Hydrazones and Osazones from Phenylhydrazine

Hydrazines react with sugars in a similar fashion to amines giving acyclic hydrazones or pyranoid ring structures depending on the sugar (Scheme 2.20). Phenylhydrazine was widely used by Emil Fischer in early investigations of sugar structures and it was found that using three moles of the hydrazine per mole of sugar the reaction went further to produce phenylosazones with aniline and ammonia as by-products. Since the osazone has an extra double bond, C2 is no longer a chiral centre and the same osazone is produced from both the aldoses and the 2-ulose having identical configurations at the remaining carbons. Thus D-glucose, D-mannose, and D-fructose all produce the same osazone (**38**). The structural similarities thus established between sugars were crucial in proving their structures.

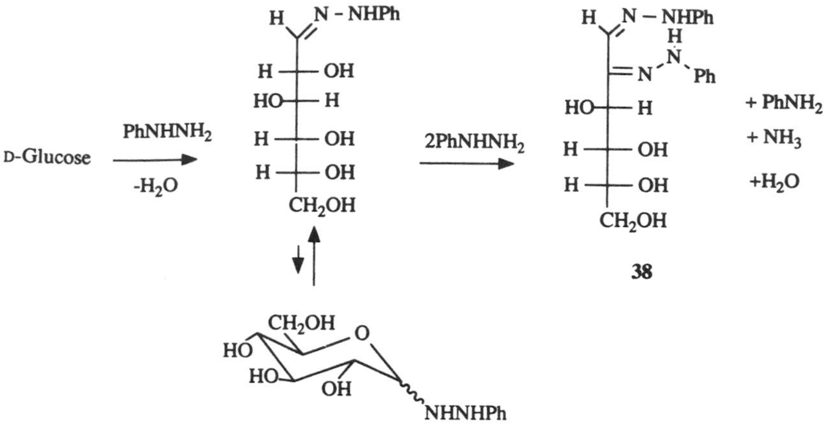

Scheme 2.20 *Formation of a hydrazone and an osazone from glucose*

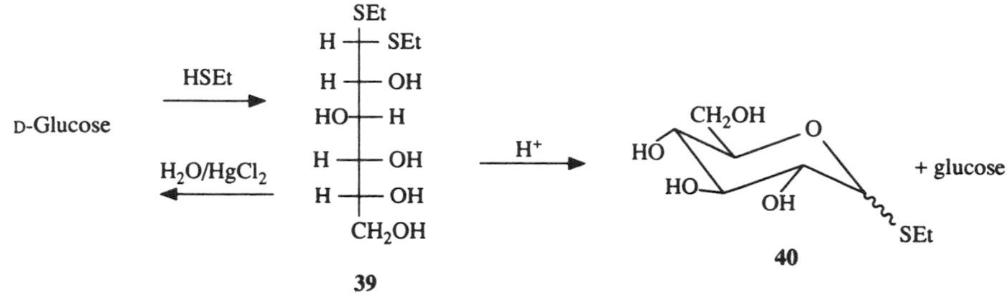

Scheme 2.21 *Preparation of glucose diethyl thioacetal and the thioglucoside*

2.3.7 Dithioacetals and Thioglycosides from Thiols

Alkane thiols may also act as nucleophiles in reaction with aldoses and produce acyclic dialkyl dithioacetals as the thermodynamically controlled products without the need of acid catalysts. D-Glucose thus gives (Scheme 2.21) D-glucose diethyl dithioacetal (**39**) on reaction with ethane thiol. Dialkyl dithioacetals are used in synthesis when the ring-open form of a sugar is required. The aldehyde group may be regenerated by treatment with aqueous mercury(II) chloride.

Dithioacetals of sugars are hydrolysed by aqueous acid to give thioglycosides such as 1-*S*-ethyl-1-thio-D-glucopyranoside (**40**). Many 1-thioglycosides occur in small amounts in certain plants and bacterial cells but seem not to have vital functions.

2.3.8 Alditols by Reduction

Aldoses are easily reduced to alditols in aqueous solutions using sodium amalgam or sodium borohydride. Alditols have CH_2OH groups at both ends of the chain giving additional symmetry possibilities. D-Allose, for example, is reduced to

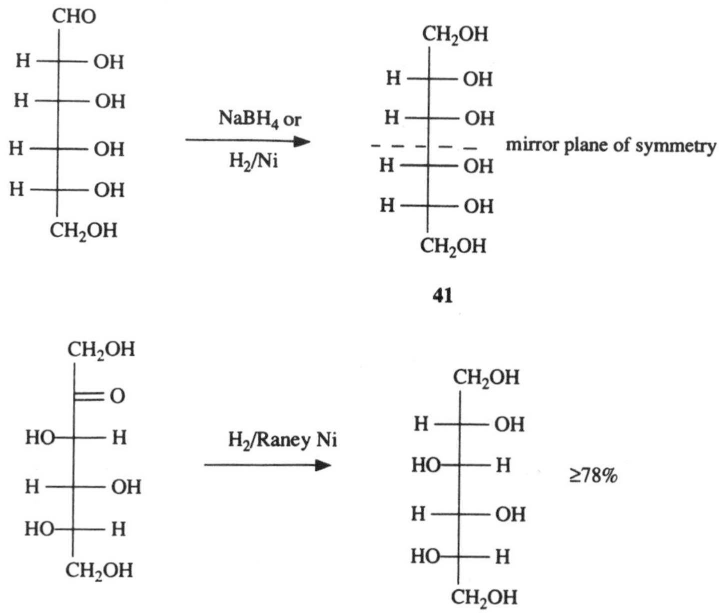

$NaBH_4$ or H_2/Ni

mirror plane of symmetry

41

H_2/Raney Ni

≥78%

L-sorbose
(L-*xylo*-hexulose)

L-iditol **(42)**

Scheme 2.22 *Reduction of allose and sorbose*

allitol (**41**, Scheme 2.22) whose lack of optical activity demonstrated that it is a *meso*-compound and helped establish the configuration of allose. Ketoses may also be reduced but this produces an extra chiral centre and so generally gives a mixture of two alditols. High-pressure hydrogenation over Raney nickel usually reduces aldoses and ketoses to alditols in high yield and may also be quite selective in the reduction of ketoses, enabling a single alditol such as **42** to be isolated in favourable cases.

2.3.9 Oxidation and Oxidative Cleavage

Although the aldehyde group in aldoses is masked as a hemiacetal its presence is indicated by reduction of the copper(II) in Fehling's solution to red copper(I) oxide. Ammoniacal silver(I) nitrate is also reduced by aldoses to an elemental silver(0) mirror and these two reactions are standard tests for a reducing sugar, *i.e.* one with a free anomeric OH group. In synthetic work bromine water is used to oxidize the anomeric centre alone, more strongly oxidizing conditions are employed for primary or secondary alcohols, and using certain difunctional oxidizing agents, such as periodate, 1,2-diols may be cleaved to pairs of carbonyl compounds.

2.3.9.1 Bromine Water. Oxidation of an aldose at the anomeric position produces an aldonic acid and is conveniently performed with bromine. D-Glucose gives D-gluconic acid (**43**) with bromine water and this cyclizes on warming to the 1,4-

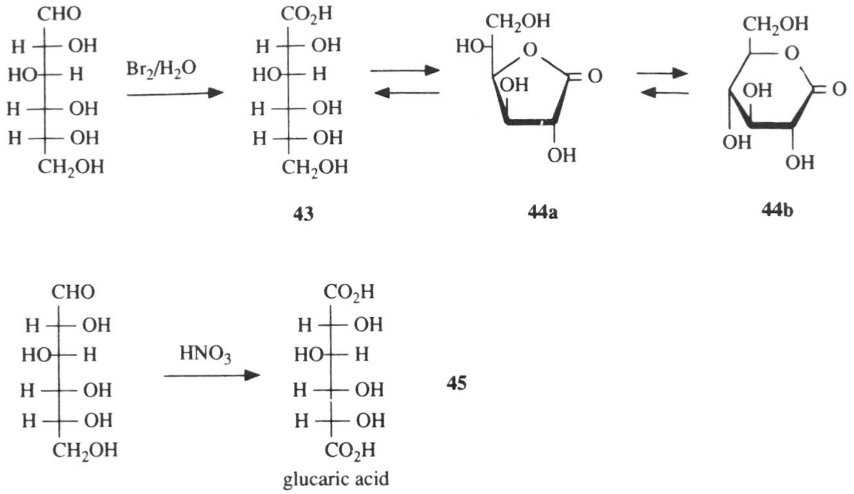

Scheme 2.23 *Oxidation of* D*-glucose*

(or γ-) and 1,5- (or δ-) lactones (**44a** and **44b**). The temperature at which the oxidation reaction mixture is evaporated chiefly determines which product is isolated: below 25 °C the free acid (**43**) is obtained, concentration at 35 °C gives the 1,5-lactone (**44b**), while higher temperatures produce mainly D-glucono-1,4-lactone (**44a**).

2.3.9.2 Nitric Acid. Oxidizing conditions stronger than bromine water convert primary alcohols into carboxylic acids and so give dicarboxylic aldaric acids from aldoses. D-Glucaric acid (**45**, Scheme 2.23) is thus produced by oxidation of D-glucose with strong nitric acid.

2.3.9.3 Strong Oxidizing Agents. Secondary hydroxyl groups on sugars are oxidized to ketones with agents such as ruthenium tetroxide which are more strongly oxidizing than nitric acid. This is of little value with free sugars but is useful in the preparation of uloses from sugars having most hydroxyls protected (example in Section 2.5.1).

2.3.9.4 Periodate and Lead Tetra-acetate Oxidations. Several reagents, including lead tetra-acetate in acetic acid and aqueous periodate salts, cleave the carbon–carbon bonds of vicinal diols. Of these periodic acid and its salts are the most widely used in carbohydrate chemistry, giving highly selective, quantitative reactions, useful both in the synthesis of carbohydrate derivatives and in their analysis. The one-carbon fragments produced in the oxidative cleavages are easily determined and may differentiate between different types of sugar: terminal diols produce formaldehyde, vicinal triols give formic acid and mid-chain diols give neither.

(i) *Diols.* Periodate oxidation (Scheme 2.24) of simple vicinal diols breaks the carbon–carbon bond and produces two carbonyl (normally aldehyde) groups. Methyl 4,6-*O*-benzylidene-α-D-glucopyranoside (**46**) for example is oxidized to the dialdehyde (**47**).

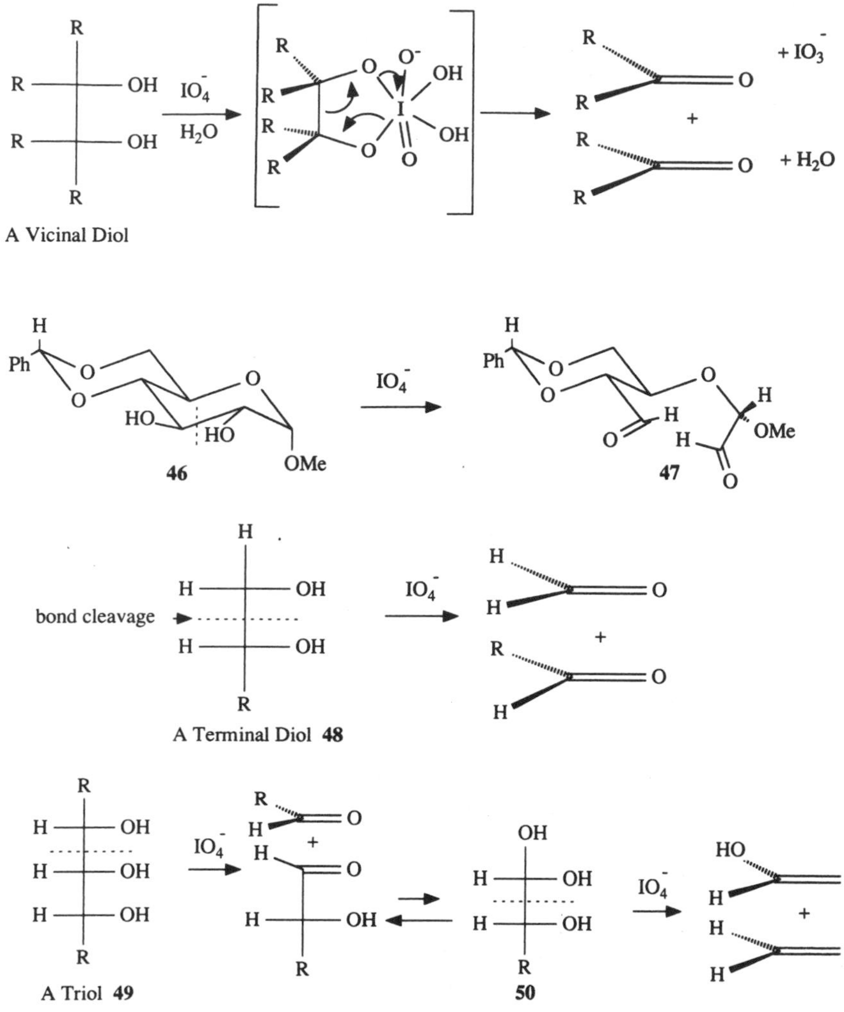

Scheme 2.24 *Periodate cleavage reactions*

(ii) *Terminal diols.* Periodate oxidation of a CH(OH)CH_2OH group (**48**) is identical to that of a simple vicinal diol except that the cleavage produces formaldehyde which may be assayed by various colorimetric reactions.
(iii) *Vicinal triols.* Periodate oxidation of triols (**49**) cleaves out the centre carbon as formic acid. This is easy to understand since preliminary cleavage of either diol would produce an α-hydroxyaldehyde whose hydrated form (**50**) is a vicinal diol and will therefore be cleaved again by periodate to give the observed products. The proposed intermediates, α-hydroxyaldehydes, also produce formic acid on periodate oxidation so providing support for this mechanism.

Periodate oxidation of free monosaccharides is of value to confirm their ring sizes. Glucose rapidly consumes three moles per mole of periodate to give the formate ester (**51**, Scheme 2.25) and two molar equivalents of formic acid. The

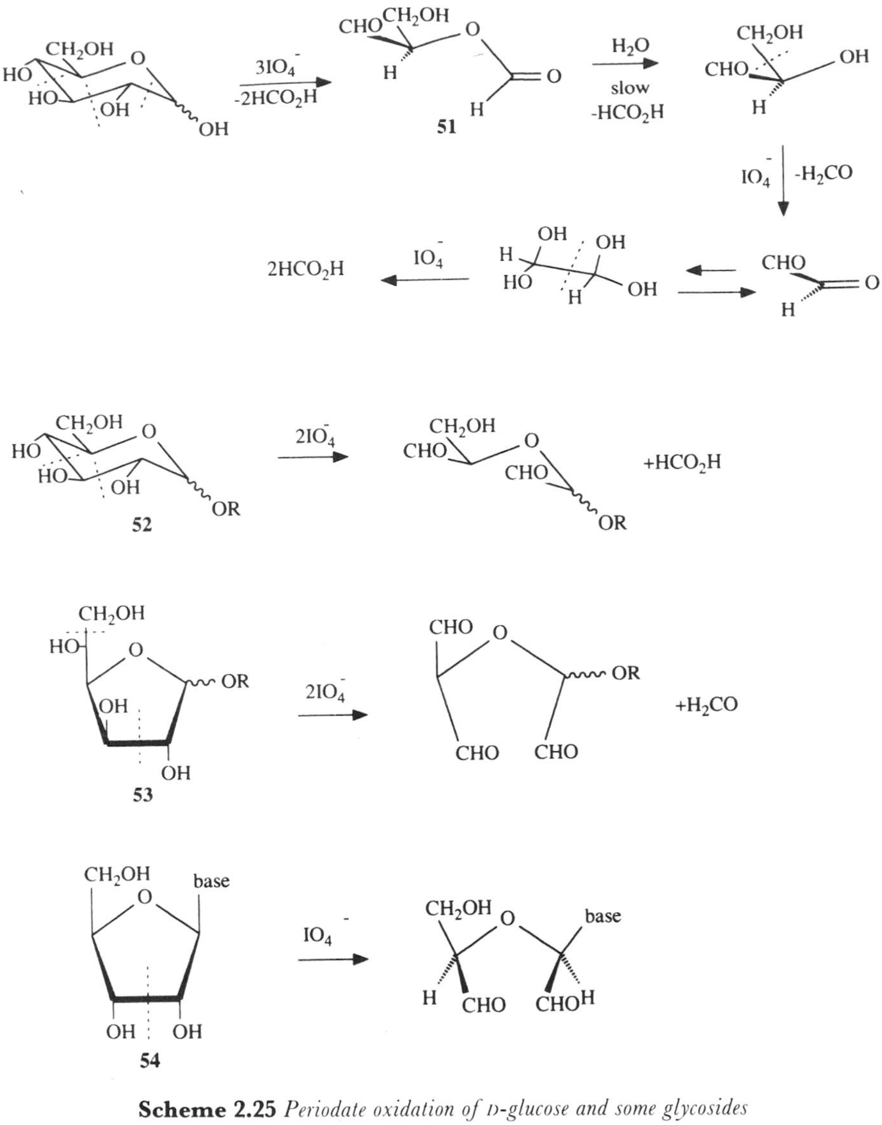

Scheme 2.25 *Periodate oxidation of D-glucose and some glycosides*

reaction stops here using a diol cleavage reagent like lead tetra-acetate in acetic acid which is used anhydrous, but in water the formate ester is hydrolysed to glyceraldehyde which is oxidized to a further two molar equivalents of formic acid and one of formaldehyde. If glucose were a furanose one molar equivalent of both formaldehyde and formic acid would be released in the initial oxidation and then four molar equivalents of formic acid would be produced after hydrolysis. Periodate oxidation is particularly good for determining the ring size of glycosides; alkyl hexopyranosides (**52**), for example, produce formic acid as the only one-carbon fragment while alkyl hexofuranosides (**53**) give only formaldehyde.

An additional source of information in periodate oxidations, particularly of

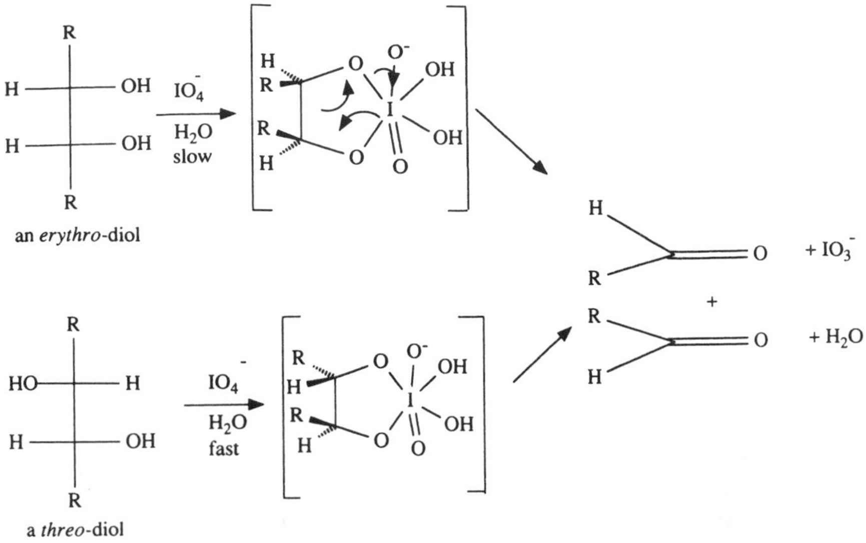

Scheme 2.26 *Periodate cleavage of* threo- *and* erythro-*diols*

substituted sugars, is the configuration of the asymmetric carbons in chiral degradation products. The configurations of such chiral centres are often easier to establish than those of the original sugars and this has been crucial in many structural investigations of sugar derivatives including the assignment of the anomeric configuration of nucleosides (**54**), components of nucleic acids.

Periodate oxidation of diols is quickest when the dihedral angle between the hydroxyl groups is small since the reaction proceeds via a five-membered ring intermediate which is least strained if approximately planar. Cyclic *cis* vicinal diols (*e.g.* **54**) generally react faster than their *trans* isomers which, especially in rigid situations, are oxidized relatively slowly or not at all by periodate. Of the two types of acyclic vicinal diol, an *erythro*-diol reacts significantly more slowly than a *threo*-diol because closing the five-membered ring of its reaction intermediate brings the two ends of its carbon chain into eclipse (Scheme 2.26). Under the oxidation conditions normally employed almost all vicinal diols in carbohydrates are cleaved but useful configurational information may be obtained by comparison of relative rates of periodate oxidation.

2.4 PROTECTIVE GROUPS FOR MONOSACCHARIDES

All saccharides have many reactive functional groups but the synthesis of specific oligosaccharide or monosaccharide derivatives generally demands reaction at selected hydroxyl groups with conservation of the rest of the molecule. Occasionally one may achieve selective reaction at a chosen site simply by suitable choice of reaction conditions but more often protective groups must be attached to reduce the reactivity of the other functional groups. Such protective groups must be preserved during the appropriate structural modifications, but easily removed when necessary.

2.4.1 Esterification

The hydroxyl groups of monosaccharides are readily acylated using acid anhydrides or acyl chlorides under acidic or basic catalysis. Acetates and benzoates are the most frequently used for hydroxyl protection and full acetylation or benzoylation is also useful to prepare crystalline, easily characterized, derivatives of syrupy saccharides. Sulphonate esters are the best choice when nucleophilic displacement of a (substituted) hydroxyl group is required and phosphate esters are of interest in the synthesis of nucleic acids. Cyclic esters tend to be more troublesome to prepare and their use is generally confined to applications which require their special properties.

2.4.1.1 Acetates. Acetylation of free sugars using acetic anhydride always gives an anomeric mixture but the desired anomer can usually be prepared by suitable choice of conditions. α-D-Glucose, for example, is acetylated (Scheme 2.27) rapidly by acetic anhydride in pyridine at 0 °C to give penta-*O*-acetyl-α-D-glucopyranoside (**55**) but heating α-D-glucose under reflux with acetic anhydride containing sodium acetate gives penta-*O*-acetyl-β-D-glucopyranoside (**56**). Using pyridine as base the sugar acetylates faster than it can anomerize and so gives the product without anomeric change. In contrast, using the weakly basic sodium acetate, acetylation is slow and allows anomerization of free glucose to an

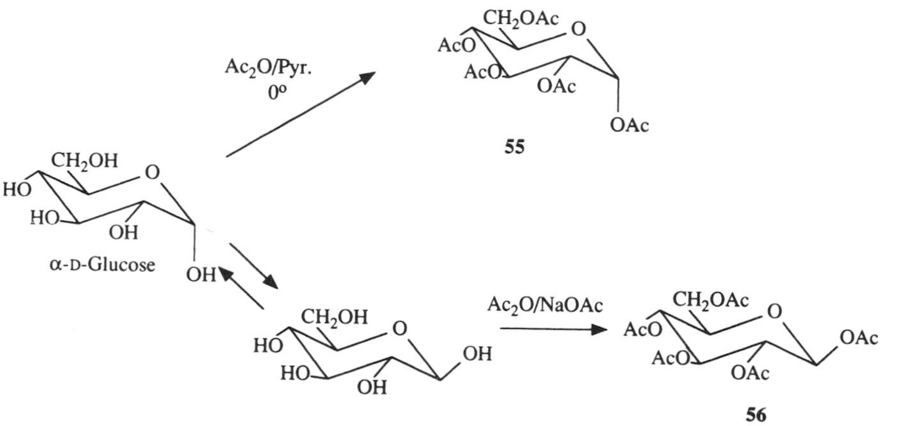

Scheme 2.27 *Acetylation of glucose*

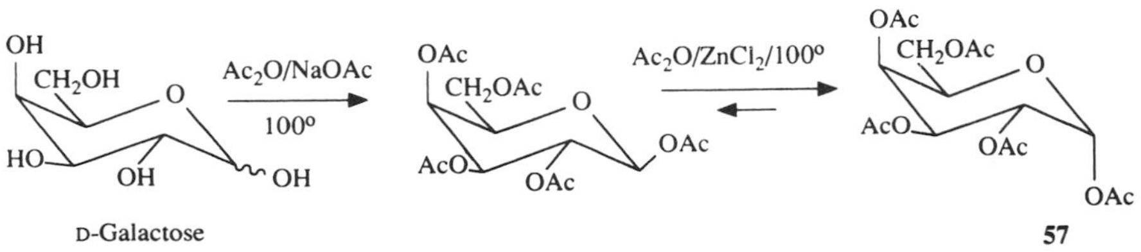

Scheme 2.28 *Acetylation of galactose*

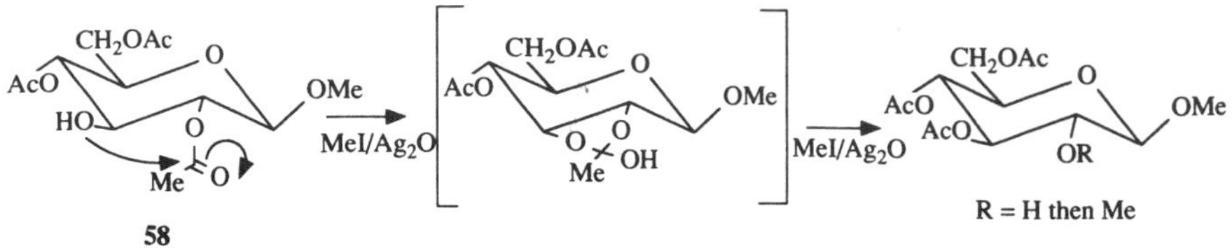

Scheme 2.29 *Migration of an acetate group during methylation*

α/β mixture from which the major product arises from the β-form since its equatorial 1-OH acetylates faster than the more hindered axial 1-OH of the α-anomer.

Some free sugars are not available in their α-forms but the fully acetylated α-anomers are generally more stable than their β-anomers owing to the anomeric effect and may be prepared by acid-catalysed equilibration of the acetylated sugar. α-D-Galactopyranoside penta-acetate (**57**, Scheme 2.28) is thus prepared by anomerization of the β-anomer with zinc chloride in acetic anhydride.

One complication arising in the reactions of partially acetylated saccharides, especially under basic conditions, is that acetate groups will migrate onto adjacent free hydroxyl groups. This can make them unreliable protecting groups as with methyl 2,4,6-tri-*O*-acetyl-β-D-glucopyranoside (**58**) which methylates (Scheme 2.28) at the 2-position instead of the initially free 3-OH.

Deacetylation may be effected by transesterification or by acidic or basic hydrolysis. Deacetylation is rapid using aqueous base but care is required since aqueous free sugars are base-labile; cold, saturated aqueous barium hydroxide solutions are generally successful. The most convenient method is usually transesterification (Scheme 2.29) catalysed by methoxide. Thus addition of a small piece of sodium, or a little sodium methoxide, to a methanolic solution of an acetylated sugar results in rapid production of the deacetylated sugar which is stable at room temperature under the anhydrous basic conditions. The product generally separates as crystals in high yield—the deacetylation of methyl tetra-*O*-acetyl-α-D-mannopyranoside (**59**) is typical. Most *N*-acetyl groups (*i.e.* CH_3CO—N) are unaffected by such mild deacetylation procedures and their removal usually requires treatment of the acetamido derivative with boiling aqueous acid or base. Tetra-*O*-acetyl-2-deoxy-2-(*N*-methylacetamido)-α-L-glucose (**60**), for example, may be deacetylated by heating with 4M hydrochloric acid for 45 minutes.

2.4.1.2 Benzoates. Benzoates are used in much the same ways as acetates but are more stable and tend to reduce the reactivity of neighbouring hydroxyl or halide groups. This additional stability can be useful when saccharide derivatives containing reactive groups are required. Acetylated glycofuranosyl halides, for

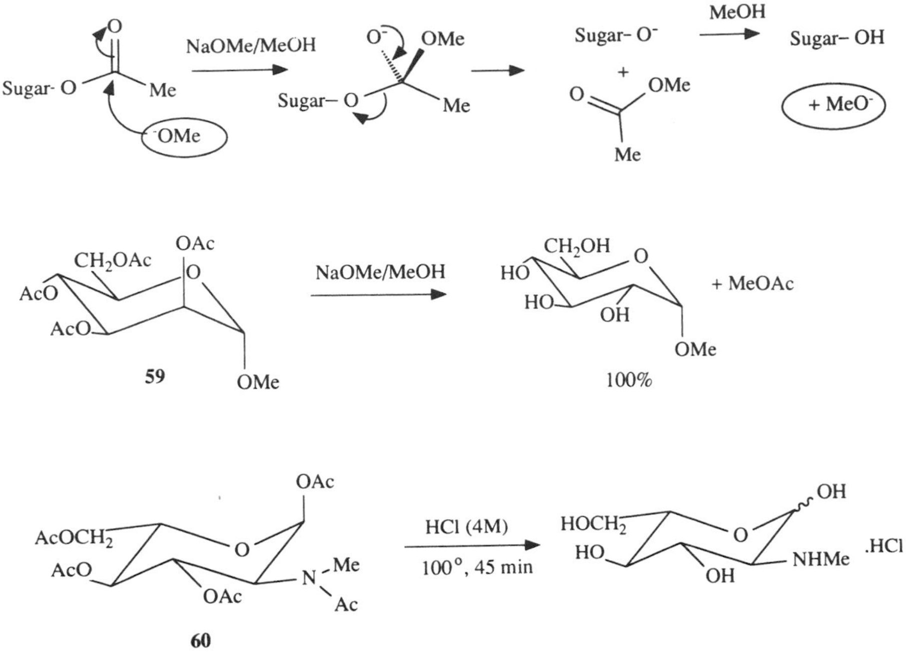

Scheme 2.30 *Deacetylation using methoxide ion or HCl*

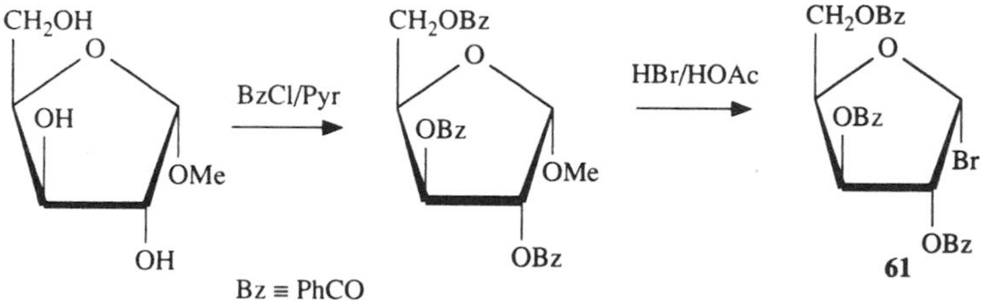

Scheme 2.31 *Preparation of an arabinofuranosyl bromide*

example, are unstable but benzoylated derivatives such as tri-*O*-benzoyl-α-D-arabinofuranosyl bromide (**61**, Scheme 2.31) are readily prepared.

The usual means of preparing sugar benzoates is addition of benzoyl chloride to a pyridine solution of the sugar which gives a rapid esterification resulting in mainly the same anomer as the starting material. Anomerization prior to addition of the benzoyl chloride can be used to prepare the alternative anomeric configuration as shown for arabinose (Scheme 2.32). If the reagent is not in excess preferential reaction at primary hydroxyl groups is observed.

Another strategy (Scheme 2.33) for selective esterification is first to produce a stannyl derivative by reaction of the most reactive hydroxyl group(s) with an alkyl tin compound and then to acylate these activated positions as in the preparation of methyl 3,6-di-*O*-benzoyl-β-D-galactopyranoside (**62**).

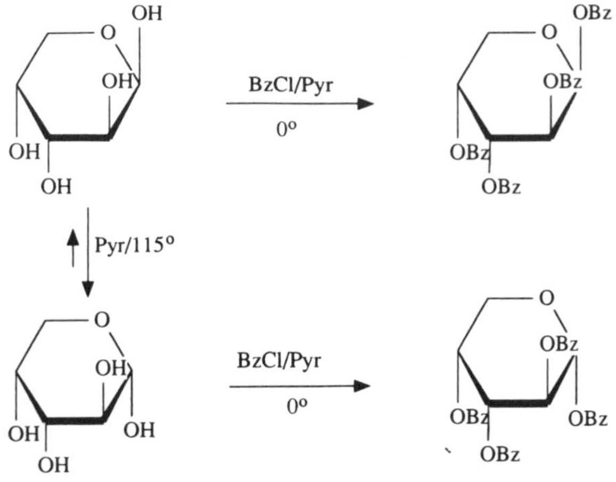

Scheme 2.32 *Preparation of* D*-arabinopyranose tetrabenzoates*

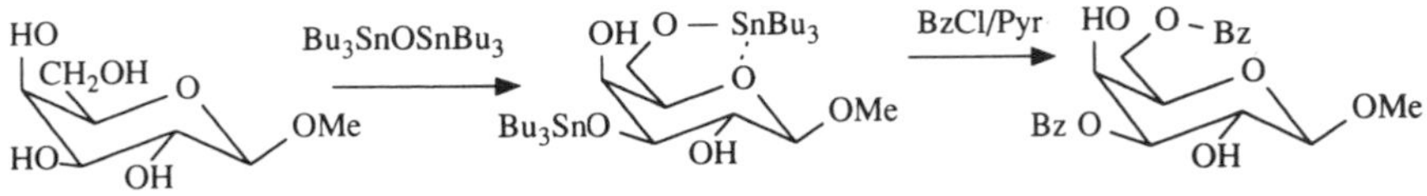

Scheme 2.33 *Stannylation permits selective benzoylation of methyl* β-D*-galactoside*

2.4.1.3 Cyclic Esters. Di-acid derivatives may react with two saccharide hydroxyl groups to give cyclic diesters. The two important ones for use as protective groups are carbonates and phenylboronates while cyclic phosphates have biological importance.

Carbonates are produced on reaction of sugars with the highly toxic gas phosgene ($COCl_2$) or by reaction with methyl chloroformate. Apart from being less toxic, the latter reagent has the advantage of producing methoxycarbonyl derivatives from hydroxyl groups which are not *cis* vicinal diols, and so gives 1,5-di-*O*-methoxycarbonyl-D-ribofuranose 2,3-carbonate (**63**, Scheme 2.34) from D-ribose whereas phosgene gives polymeric products. Reaction usually occurs at the same diol sites as with cyclic acetals (Section 2.4.2) but carbonates are removed by base rather than acid-catalysed hydrolysis.

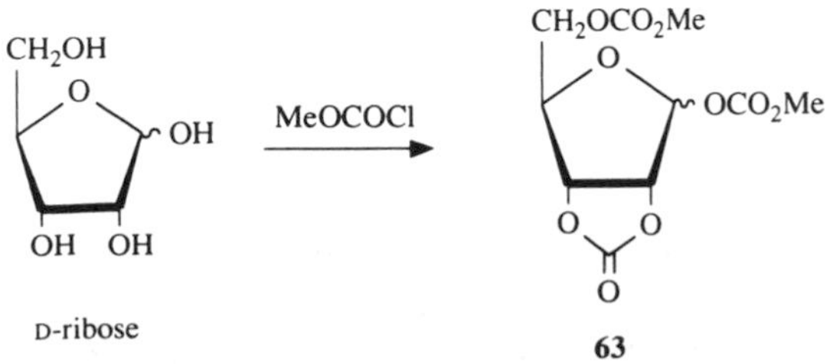

Scheme 2.34 *Preparation of a cyclic carbonate using methyl chloroformate*

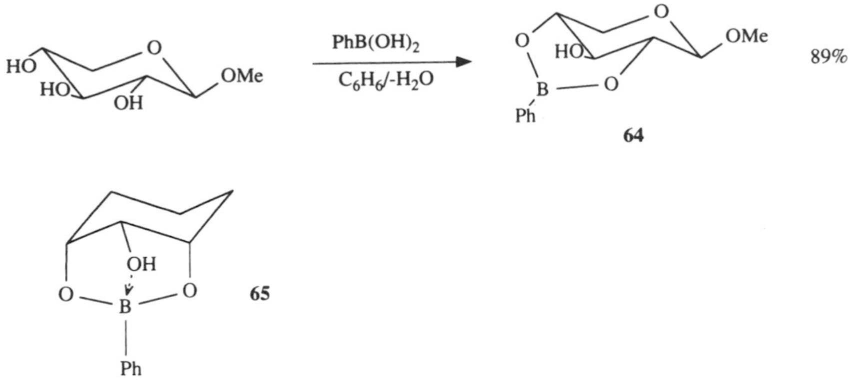

Scheme 2.35 *Phenyl boronates*

The difunctional phenylboronic acid [$PhB(OH)_2$] and its cyclic trimeric anhydride, triphenylboroxole, have the unusual property of condensing preferentially with 1,3-*cis*-related diols to give six-membered cyclic diesters such as the methyl β-D-xylopyranoside 2,4-phenylboronate (**64**, Scheme 2.35). Phenylboronates are stable enough to permit selective oxidation or esterification of remaining hydroxyls, or the selective addition of sugar or phosphate residues (for disaccharide and nucleotide syntheses) but are easily removed under mild neutral conditions by hydrolysis or alcoholysis. Phenylboronic acid reacts with compounds containing a *cis*-1,2,3-triol group to give cyclic esters (**65**) in which the third hydroxyl is co-ordinated to the boron. Sugars complexed in this way differ markedly in solubility and chromatographic properties from close analogues without a *cis*-1,2,3-triol group, permitting facile separations in favourable cases.

Monosubstituted phosphates may be obtained by reaction of phosphoric acid or phosphorus acid chlorides with sugars or partially protected sugars. 1-Phosphoryl glycosides such as the glucose β-phosphate (**66**, Scheme 2.36) are easiest to prepare, but syntheses of the biologically crucial ribose 5-phosphates are also well developed and will be described in Section 5.1.2.4. Cyclization of furanoside monophosphates to 2,3- or 3,5-diesters does not usually occur spontaneously but may be achieved (Scheme 2.37) using condensing agents such as DCC (its use in condensation reactions is also considered in Section 4.5.4.1). Such diesters are important synthetic targets because some, like the 3′,5′-cyclic adenosine monophosphate **67** are vital biochemical messengers.

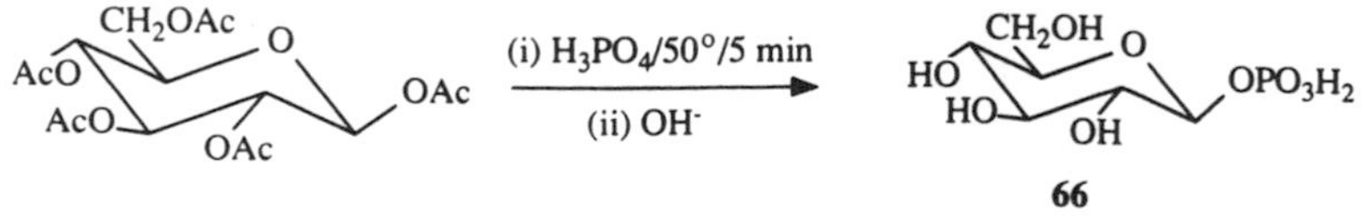

Scheme 2.36 *Synthesis of glucose β-phosphate*

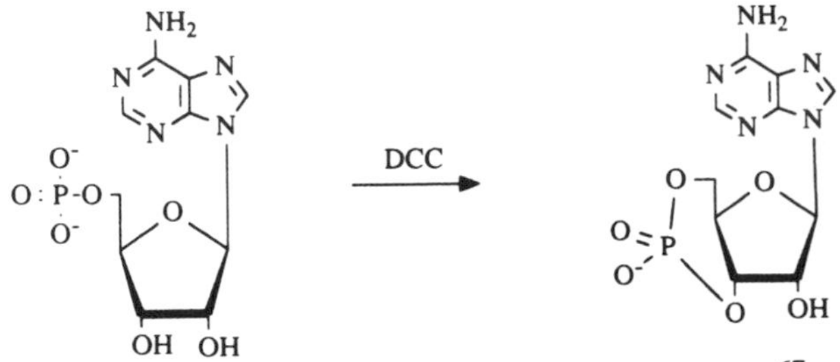

Scheme 2.37 *Synthesis of a cyclic phosphate*

2.4.1.4 Sulphonate Esters. The preparation of sulphonate derivatives of monosaccharides is similar to the preparation of benzoates; the sugar is dissolved in pyridine and a sulphonyl chloride added. Selectivity for primary hydroxyl groups is observed for reaction at room temperature especially with more bulky reagents such as *p*-toluenesulphonyl chloride (p-CH_3PhSO_2Cl, tosyl chloride) used in the preparation of tosylates such as **68** (Scheme 2.38).

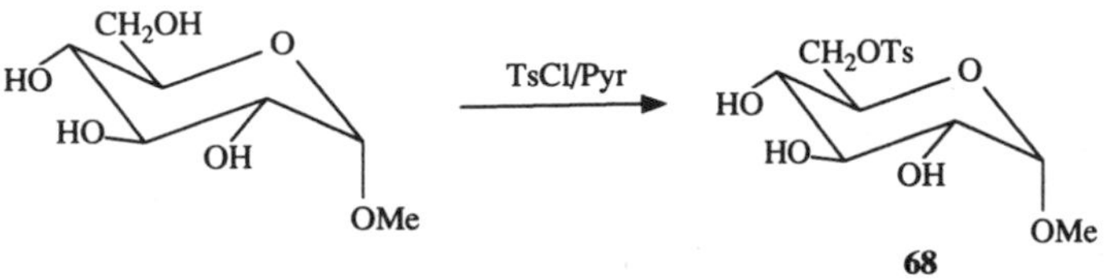

Scheme 2.38 *Synthesis of methyl 6-*O*-*p*-toluenesulphonyl-α-*D*-glucopyranoside*

The importance of sulphonate esters in carbohydrate chemistry lies in their being very good leaving groups, easily displaced by various nucleophiles (Scheme 2.39). One of the best leaving groups is trifluoromethanesulphonate (triflate, $CF_3SO_3^-$) since the electron-withdrawing fluorines provide additional stabilization of the ion. However, the methanesulphonate (mesylate) and *p*-toluenesulphonate (tosylate) esters are more commonly used even though the latter is displaced slowly from hindered secondary positions. Displacements by sulphur or halide nucleophiles, as in the preparation of **69**, are used frequently and displacement by nitrogen nucleophiles is important for amino sugar synthesis (Section 2.5.1). Intramolecular displacement, particularly by unblocked OH groups to give anhydro sugars, is also common; a simple example is base treatment of methyl β-D-galactopyranoside 6-mesylate (**70**) to give the 3,6-anhydro derivative (**71**).

Removal of sulphonate esters generally occurs with retention of configuration and may be accomplished using aqueous alkali. However, complications, such as anhydro sugar formation, are less frequently encountered using Raney nickel or sodium amalgam which give the sulphinic acid, RSO_2H, and release the free parent sugar. Lithium aluminium hydride is also used for reductive removal of sulphonate esters but may give deoxy sugars. At secondary positions the hydride

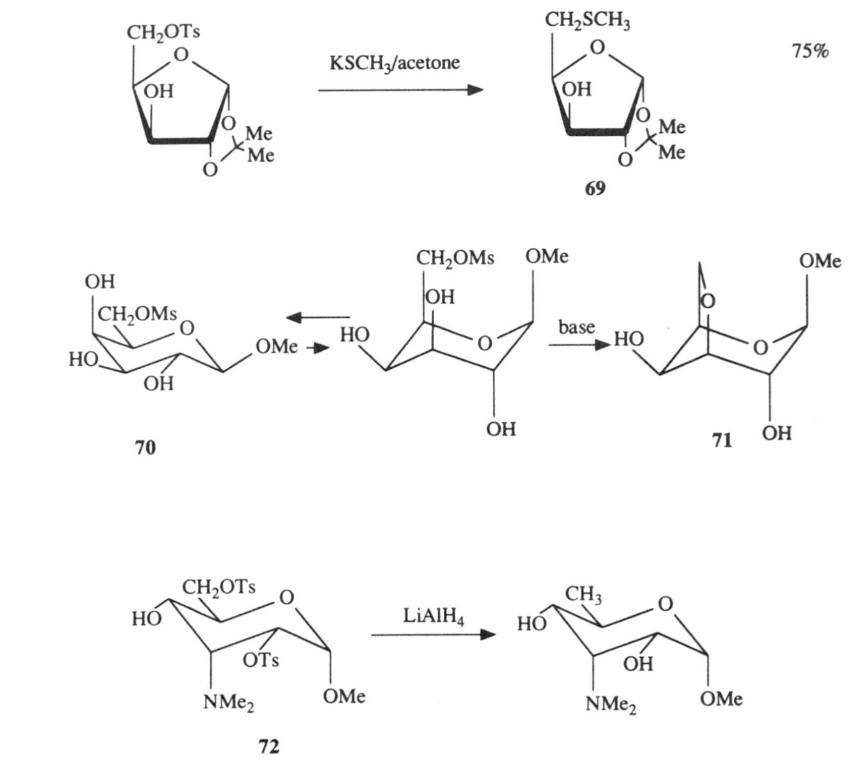

Scheme 2.39 *Reactions of sulphonate esters*

ion usually attacks at sulphur to give the free sugar hydroxyl group in the normal way but at primary positions attack generally occurs at carbon, displacing the sulphonate to give the deoxy sugar. Reaction of the ditosylate (**72**) with lithium aluminium hydride shows both types of desulphonation to give a deoxy sugar which is an antibiotic component.

2.4.2 Cyclic Acetals and Ketals

One of the most common ways of protecting sugar hydroxyl groups is to react the sugar with an excess of a ketone or aldehyde in the presence of an acidic catalyst. Initially the aldehyde or ketone reacts with a sugar hydroxyl group to give a hemiacetal (or hemiketal) which then reacts with a second sugar hydroxyl group giving a cyclic acetal or ketal (consider the two OR groups of Scheme 2.2 to be the sugar). Two sugar hydroxyl groups are thus blocked on forming a cyclic acetal and the process may be repeated to protect another two hydroxyl groups or even a total of six by formation of three acetal rings.

The reaction conditions used for acetalation (not to be confused with acetylation) of free sugars normally allow acetal isomerization and so produce the thermodynamically most favoured acetals. The major products from reactions of hexoses with aldehydes are six-membered cyclic 4,6-*O*-acetals. Further reaction to add a five-membered ring acetal may occur and where there is a choice the thermodynamically preferred *cis*-fused cyclic acetals will be produced. An important example is the reaction of D-glucose with benzaldehyde, catalysed by anhydrous zinc chloride, to produce 4,6-*O*-benzylidene-D-glucopyranose (**73**) in 42% yield with 1,2:4,6-di-*O*-benzylidene-α-D-glucopyranose (**74** which may be removed from an aqueous solution of the products by extraction with ether).

Acetalation using ketones, R_2CO, normally gives five-membered ring ketals since the alternative six-membered cyclic ketals must have one of the alkyl groups, R, in a hindered axial position of chair conformations. *cis*-Fusion is again preferred and always found with furanosides since a *trans*-fused pair of five membered rings is highly strained. The most stable products from hexoses and pentoses generally have two cyclic ketals per molecule and are the furanose forms if this permits more *cis*-ring fusions than the pyranose forms. D-Glucose reacts with acetone (propanone) in the presence of sulphuric acid or zinc chloride to give (Scheme 2.40) 1,2:5,6-di-*O*-isopropylidene-α-D-glucofuranose (**75**). 1,2-*O*-Isopropylidene-α-D-glucofuranose may be extracted from a dichloromethane solution of the products using water or produced separately by selective hydrolysis of the diketal (**75**). Other readily produced, energetically favoured, isopropylidene derivatives are 1,2:3,4-di-*O*-isopropylidene-α-D-galactopyranose (**76**), 2,3:5,6-di-*O*-isopropylidene-α-D-mannofuranose (**77**), 1,2:3,5-di-*O*-isopropylidene-α-D-xylofuranose (**78**), and 1,2:3,4:5,6-tri-*O*-D-glucitol (**79**).

Strained acetals may be synthesized using reagents which do not produce water, so allowing isolation of the kinetically controlled product. Methyl-α-D-glucopyranoside thus gives methyl 4,6-*O*-benzylidene-α-D-gluco-

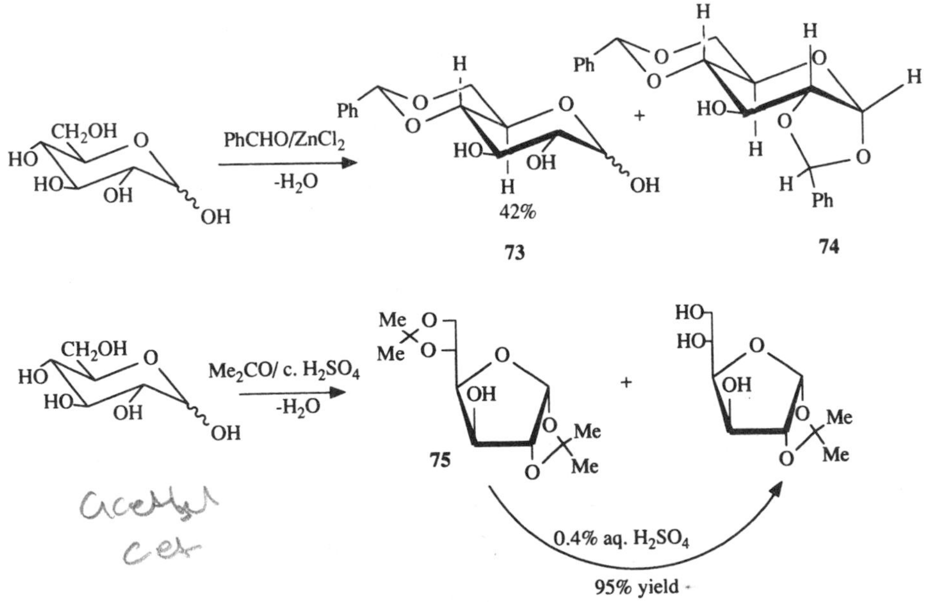

Scheme 2.40 *Cyclic acetals from glucose*

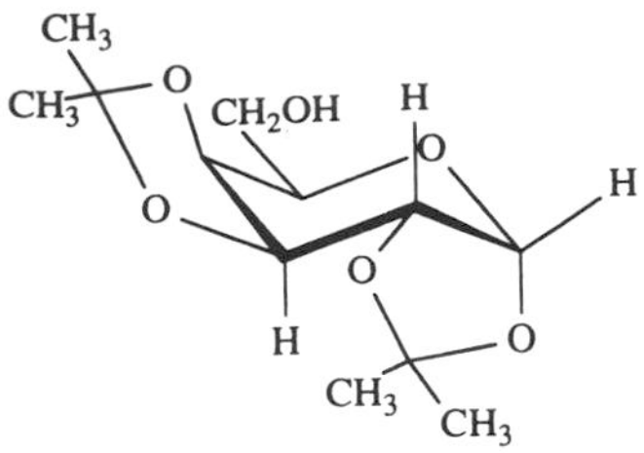

Scheme 2.41 *Acetals from free sugars and methyl* α-D-*glucopyranoside*

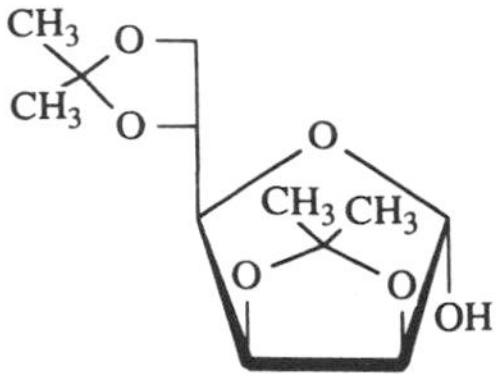

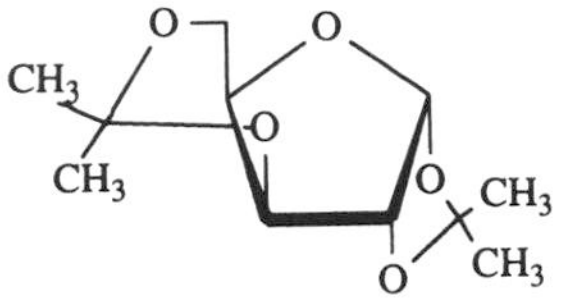

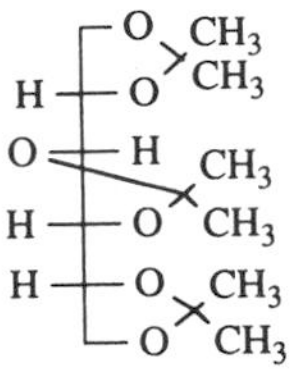

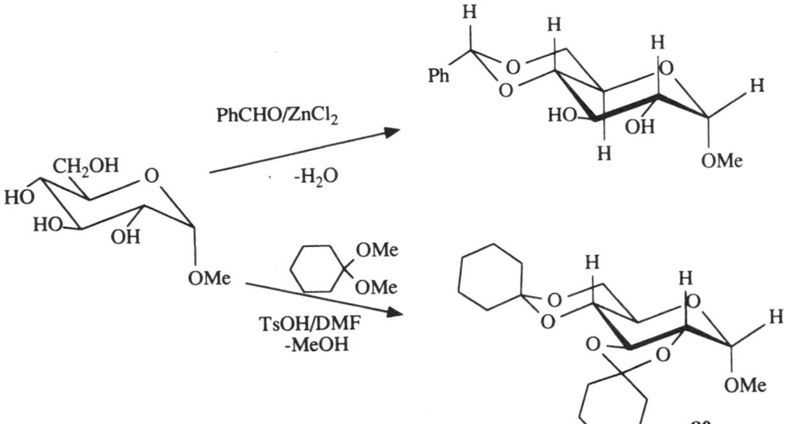

pyranoside using benzaldehyde–zinc chloride but the 2,3:4,6-diacetal (**80**, Scheme 2.41) on reaction with cyclohexanone dimethyl acetal in dimethylformamide.

Cyclic acetals are stable to basic conditions and most nucleophiles but are readily removed with retention of configuration by hydrolysis with aqueous acid (*e.g.* Scheme 2.40). Benzylidene acetals may also be removed by palladium-on-carbon catalysed reaction with hydrogen (hydrogenolysis).

Other acetals encountered in carbohydrate chemistry are the glycosides, covered in Sections 2.3.4 and 2.4.4, and 'tetrahydropyranyl ethers' whose introduction and removal is described during prostaglandin partial synthesis (Section 7.3.1.2).

2.4.3 Ether Preparation

2.4.3.1 Methylation and Silylation. Methyl ethers are stable to most reagents, including acids and bases; this makes them poor protecting groups but has suited them for use in the elucidation of the structure of carbohydrates. Many mixtures of reagents including CH_3I/Na, CH_3I/BaO/dimethylformamide (DMF), dimethyl sulphate/NaOH, and diazomethane/BF_3–Et_2O, have been used but the two most commonly used nowadays are CH_3I/Ag_2O (Purdie methylation) and CH_3I/NaH/dimethyl sulphoxide (DMSO) (Hakomori methylation).

The Purdie methylation, which reacts the carbohydrate dissolved in methanol with methyl iodide catalysed by silver oxide, is restricted to non-reducing sugars such as glycosides since the silver oxide will oxidize reducing sugars to carboxylic acid esters. Apart from its oxidizing nature it is a very mild procedure and often several treatments are required to obtain complete methylation. Methyl tetra-*O*-methyl-α-D-glucopyranoside (**81**, Scheme 2.42) is obtained in 97% yield from methyl α-D-glucopyranoside after four methylation cycles. A modern variation giving more rapid reactions is to substitute the dipolar aprotic solvent dimethylformamide for the methanol and to use a strong base such as sodium hydride. 1,2:5,6-Di-*O*-isopropylidene-α-D-gulofuranose (**82**) is methylated in 94% after one treatment.

The Hakomori procedure uses the methylsulphinyl anion produced from sodium hydride and dimethyl sulphoxide:

$$NaH + MeSO_2Me \rightarrow Na^+ \ {}^-CH_2SO_2Me + H_2$$

as base. It is added to a DMSO solution of the carbohydrate, thus converting the hydroxyls to alkoxides and subsequent addition of methyl iodide gives complete methylation of all OH, NH, and COOH groups within a few minutes. This methylation procedure is widely used to give rapid, complete methylation of polysaccharides.

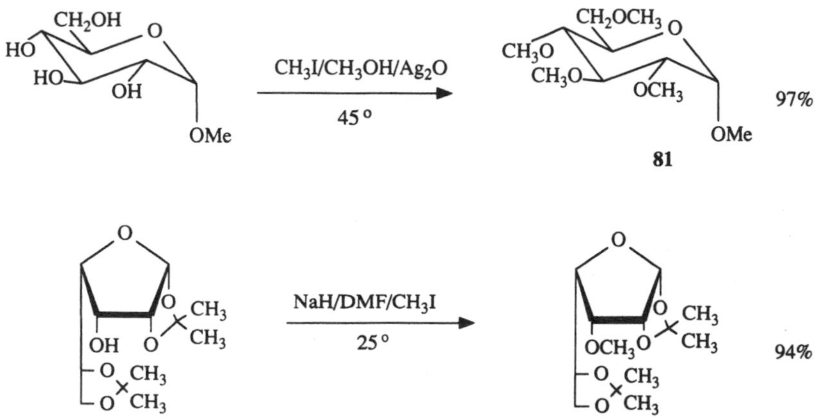

$CH_3I/CH_3OH/Ag_2O$, 45°: **81** 97%

NaH/DMF/CH_3I, 25°: **82** 94%

Scheme 2.42 *Methylation of sugar derivatives*

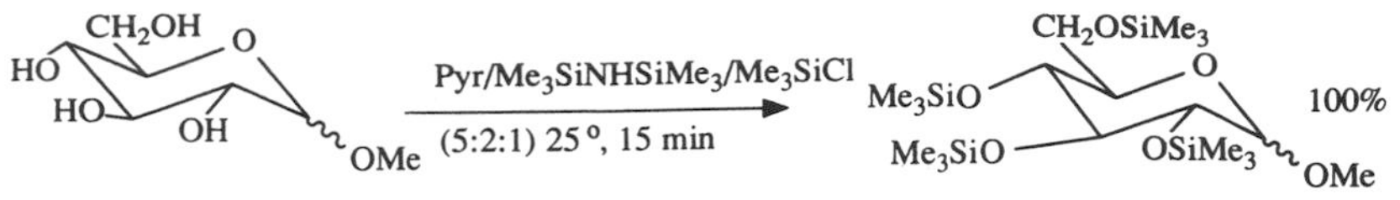

CH_2OH O HO HO OH OMe — $Pyr/Me_3SiNHSiMe_3/Me_3SiCl$ (5:2:1) 25°, 15 min → CH_2OSiMe_3 O Me_3SiO Me_3SiO $OSiMe_3$ OMe 100%

Scheme 2.43 *Trimethylsilylation of methyl glucosides*

Removal of methyl ethers may be accomplished using boron trichloride or by oxidation to formate esters using chromium trioxide in acetic acid followed by hydrolysis to the demethylated sugar. These methods are far less destructive than the older hydrogen iodide treatment but methyl protective groups are not normally removable in high yield.

An important use of methylation is to render sugars sufficiently volatile for analysis by gas–liquid chromatography (GLC) or mass spectrometry; trimethylsilylation is an alternative which gives even more volatile derivatives, easily hydrolysed by dilute aqueous acid to the original sugars. Trimethylsilyl chloride (Me_3SiCl) and/or hexamethyldisilazane ($Me_3SiNHSiMe_3$) in pyridine or triethylamine are the usual silylating mixtures and every free hydroxyl group reacts rapidly (Scheme 2.43).

2.4.3.2 Allyl Ethers. Allyl ethers may be prepared by substituting allyl chloride for methyl iodide in the Hakomori procedure. They are stable to a wide variety of reagents but may be removed when necessary in a two-step process: first isomerization to a vinyl ether (enol ether) using potassium t-butoxide or a ruthenium or palladium catalyst and then mild acid hydrolysis using aqueous mercury(II) salts (Scheme 2.44). Variation in ease of removal may be achieved by use of methylated allyl groups ($OCH_2CHR{=}CR_2$, R = H or Me) and allows flexible strategies for synthesis in their presence.

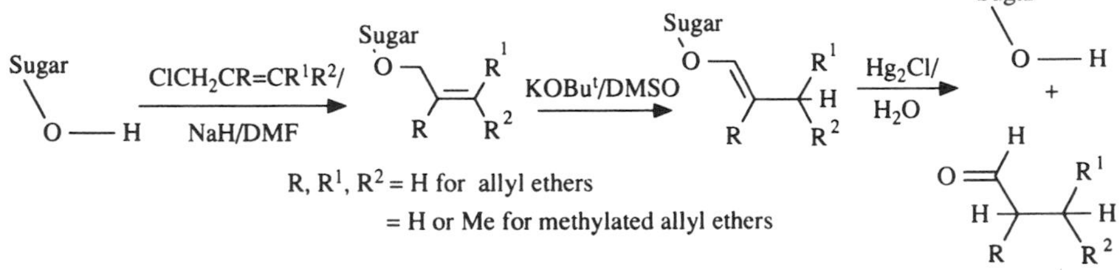

Sugar–O–H — $ClCH_2CR{=}CR^1R^2$/ NaH/DMF → Sugar–O–$CH_2C(R){=}CR^1R^2$ — $KOBu^t$/DMSO → Sugar–O–$CH{=}C(R)CHR^1R^2$ — Hg_2Cl/ H_2O → Sugar–O–H + O=CH–CH(R)–CHR^1R^2

R, R^1, R^2 = H for allyl ethers
= H or Me for methylated allyl ethers

Scheme 2.44 *Attachment and removal of allyl ethers*

2.4.3.3 Benzyl Ethers. Benzyl ethers ($ROCH_2Ph$) are widely used hydroxyl-protecting groups since they are readily prepared and removed. The usual preparative routes involve treatment of a DMF solution of the carbohydrate with a benzyl halide and a base such as sodium hydroxide, sodium hydride, barium oxide or silver oxide, but base-sensitive carbohydrates may also be benzylated using benzyl triflate in the hindered base 2,6-di-t-butylpyridine. Benzyl groups

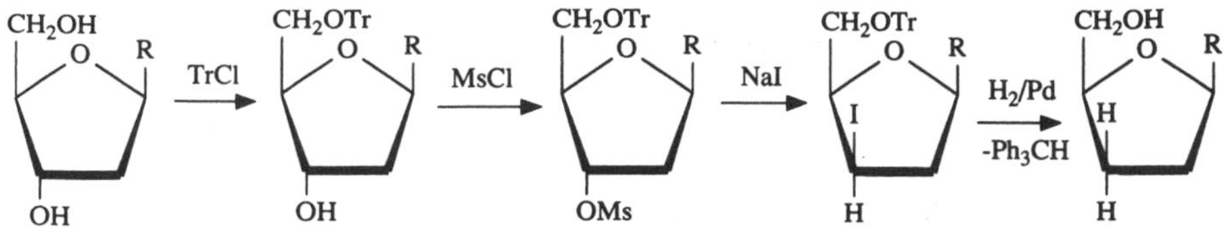

Scheme 2.45 *Use of a trityl protective group in synthesis*

are usually removed by catalytic hydrogenation. An example of their use during oligosaccharide synthesis is given in Section 2.5.4.3.

2.4.3.4 Trityl Ethers. Triphenylmethyl (trityl) ethers are bulky groups easily placed selectively on primary hydroxyl groups by use of trityl chloride in pyridine. Catalytic hydrogenation or mild acid hydrolysis may be used for their removal and even passing through a column of silica gel may result in detritylation. An example of temporary protection of a primary hydroxyl by a trityl group is shown in the synthesis of the dideoxy sugar shown in Scheme 2.45.

2.4.4 Glycosidation

An obvious way to protect the anomeric position of carbohydrates is formation of a glycoside. The Fischer glycoside synthesis (Section 2.3.4) is fine for alkyl glycosides but uses excess alcohol and is not convenient for the synthesis of aryl glycosides since phenols are insufficiently volatile to be removed from the product by distillation. Aryl glycosides are usually prepared instead by the Koenigs–Knorr method (Scheme 2.46) in which the phenol is dissolved in sodium hydroxide solution and then added to the fully protected glycosyl halide. An ester protective group at C2 of the glycosyl halide participates in the reaction and ensures that the product has the 1,2-*trans*-configuration. *p*-Nitrophenylglycosides are used in the clinical estimation of glycosidase enzymes such as β-galactosidase, essential in lactose metabolism, since the extent of hydrolysis may be followed by spectrophotometric determination of the bright yellow *p*-nitrophenoxide ion released (Scheme 2.46).

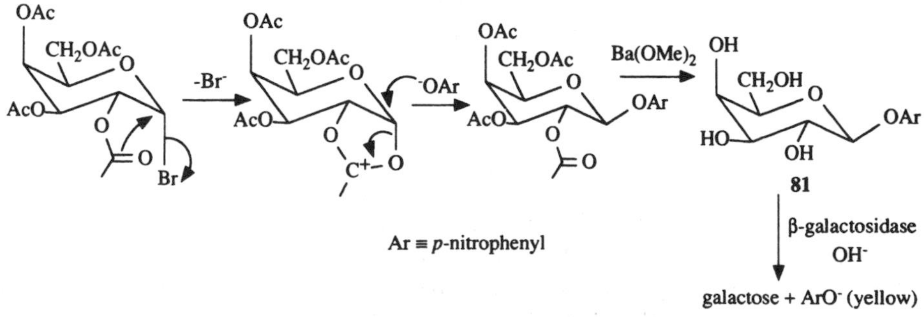

Scheme 2.46 *Aryl glycoside synthesis using the Koenigs–Knorr reaction*

The removal of an aglycone (the non-sugar part of a glycoside) is achieved by acidic aqueous hydrolysis and is often used to release monosaccharides from natural products. The hydrolysis mechanism is the reverse of glycoside formation (Scheme 2.16)—protonation of the aglycone oxygen is followed by elimination of the alcohol to give a cyclic carbocation, addition of water, and loss of a proton. If selective hydrolysis is required glycosidase enzymes may be employed and are often highly selective for the type of sugar and its anomeric configuration (α or β).

2.5 SYNTHESIS OF MONOSACCHARIDE DERIVATIVES

Ambitious biological chemists should consider specializing in carbohydrate synthesis since this skill is likely to be crucial in solving the last great mystery of biological polymers—the biological functions of complex carbohydrates. Naturally occurring saccharides frequently bear functional groups other than the hydroxyls present in simple monosaccharides and it is thus common for synthetic carbohydrate targets to include amino, carboxylate, sulphate, phosphate or other functional groups. Previous sections have considered the synthesis of many such monosaccharide derivatives and the synthesis of amino sugars, branched-chain monosaccharides, and glycosyl halides is covered in the following sections, 2.5.1 to 2.5.3. Such syntheses are relatively simple but most biologically important sugars, such as those responsible for intercellular recognition, are oligosaccharides and their synthesis remains a great synthetic challenge.

2.5.1 Amino Sugar Synthesis

The wide biological importance of amino sugars and the potency of aminoglycoside antibiotics such as kanamycin A (**83**) has stimulated a multitude of synthetic approaches. Nucleophilic displacement of a sugar sulphonyloxy group by ammonia is a simple route to amino sugars which usually works well with primary sulphonate groups as in Scheme 2.47. Sometimes the initial amino sugar product, being nucleophilic, reacts with further starting material to give a secondary amine and in such cases it is better to displace the sulphonate with azide and subsequently reduce the azido sugar to an amine. Azide ion and hydrazine are more nucleophilic than amines and so displace secondary sulphonates under mild reaction conditions. Subsequent hydrogenation of the azide or

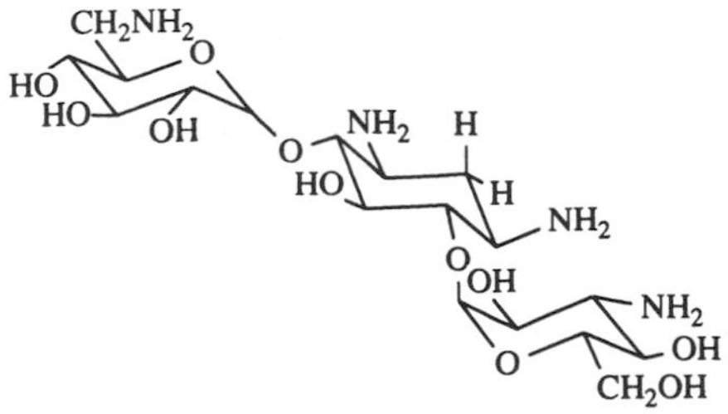

Kanamycin A **83**

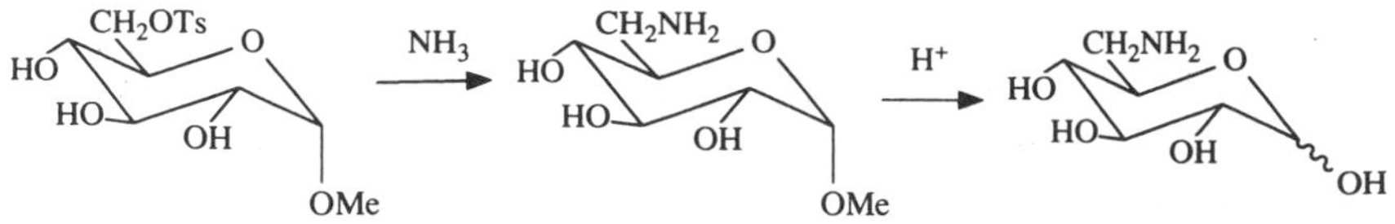

Scheme 2.47 *Amino sugar synthesis* via *sulphonate ester displacement*

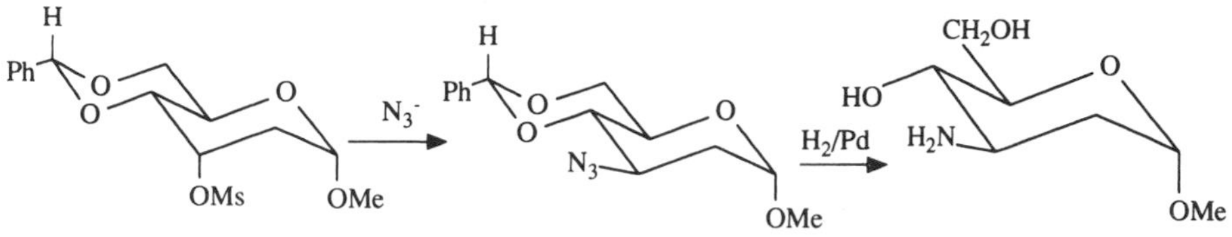

Scheme 2.48 *Amino sugar synthesis* via *azide displacement of a sulphonate ester*

hydrazine intermediate to give an amino sugar is easy and usually proceeds in high yield (*e.g.* Scheme 2.48).

A variation (Scheme 2.49) on the sulphonate displacement reaction is to allow an internal displacement of the sulphonate by a neighbouring *trans*-related OH group. The resulting epoxide is then ring-opened with ammonia which attacks the less hindered and/or more electronegative carbon and gives the *trans*-amino-alcohol. The 6-aminoglucose derivative (**84**) is made in this way.

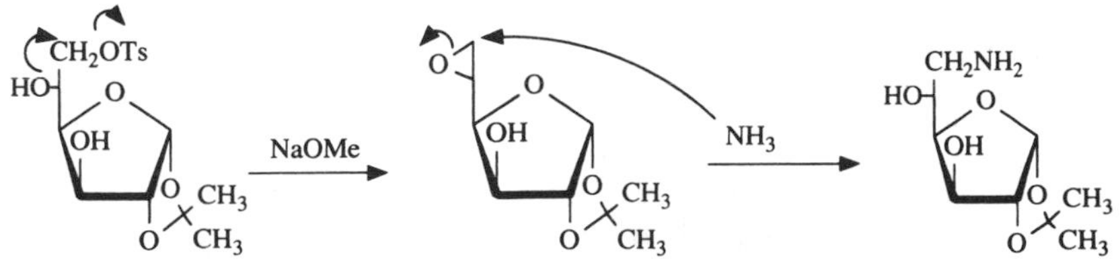

Scheme 2.49 *Synthesis of a 6-amino-D-glucose derivative*

An alternative approach (Scheme 2.50) is to oxidize a single OH group to a ketone, react with hydroxylamine to give an oxime, and then reduce to the amino sugar. Thus diacetone glucose (**85**) may be oxidized to the ulose, converted into the oxime (**86**), and then reduced stereospecifically (the reducing agent approaches from the less hindered upper face) to the 3-amino-allose derivative (**87**). Direct reduction of the ulose gives access to diacetone allose.

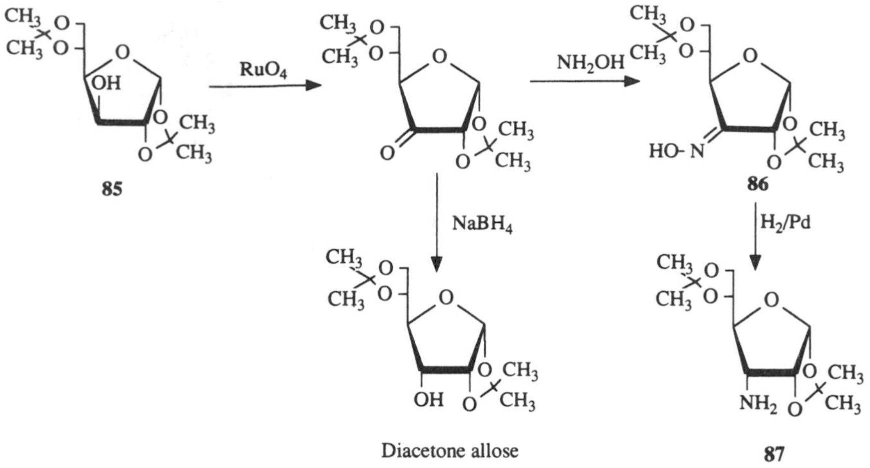

Scheme 2.50 *Synthesis of a 3-amino-D-allose derivative*

2.5.2 Glycosyl Halides

Glycosyl halides are important synthetic intermediates, particularly for oligosaccharide and nucleoside synthesis. Pyranosyl halides show a strong anomeric effect and so the anomer having an axial halide (usually the α-anomer) is favoured thermodynamically. The normal starting material is a fully blocked sugar having an acetate, benzoate, or alkoxy group at the anomeric position. Although blocked this position retains considerable reactivity and the ester or alkoxy group may be displaced by halide ion under acidic catalysis. Since both acetylation and bromination are acid-catalysed they often may be performed sequentially without isolation of an intermediate and the preparation of acetobromoglucose (**88**) has been developed to an enjoyable one-pot sequence (Scheme 2.51) giving a yield of better than 80% from glucose. There is a strong anomeric effect in glycosyl halides and only the α-form of acetobromoglucose is obtained in this way.

To obtain the thermodynamically unfavoured anomer with an equatorial halide, kinetically controlled reaction conditions must be used. Tetra-*O*-acetyl-β-D-glucopyranosyl chloride may thus be prepared from β-D-glucopyranose penta-acetate by reaction with aluminium chloride and silicic acid at room temperature. Strictly anhydrous conditions are maintained because the equatorial chlorine of the β-glycosyl chloride is hydrolysed very rapidly. Acylated

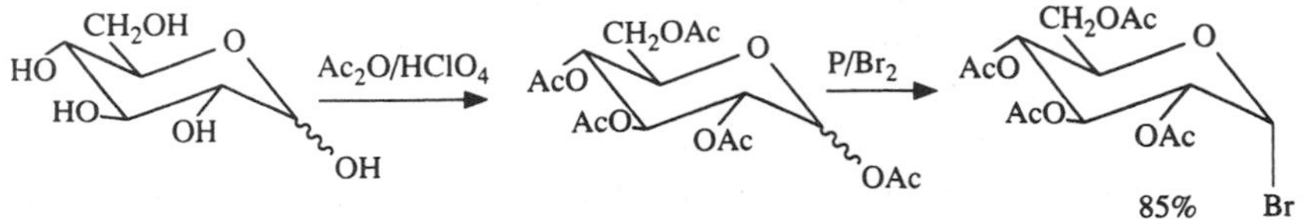

Scheme 2.51 *Synthesis of acetobromoglucose* (**88**)

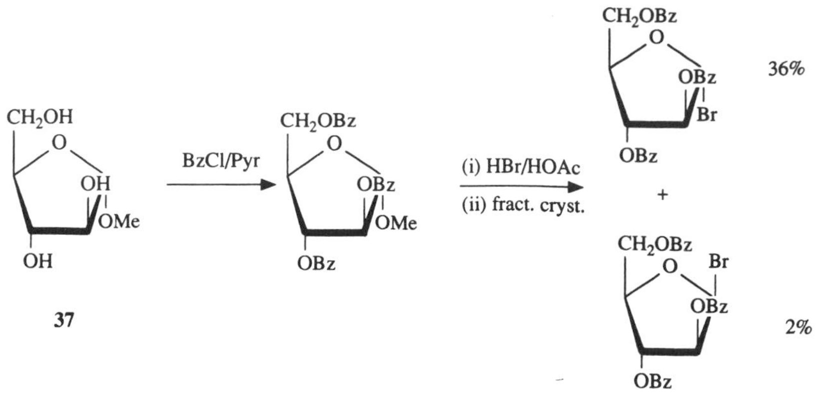

Scheme 2.52 *Preparation of Tri-O-benzoyl-α- and -β-D-arabinofuranosyl bromides*

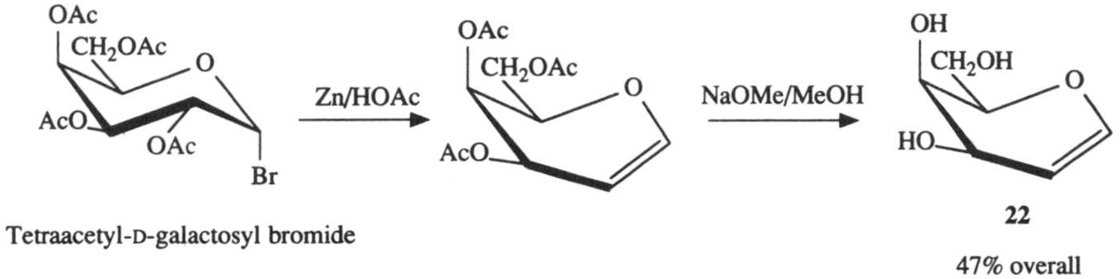

Scheme 2.53 *Synthesis of D-galactal* (**22**)

furanosyl halides are much less stable than their pyranose analogues but may be prepared by displacing the alkoxy groups from alkyl glycofuranosides as in the synthesis of tri-*O*-benzoyl-α-D-arabinofuranosyl bromide (Scheme 2.52).

Apart from their uses in joining sugars, alcohols or bases to the anomeric position, glycosyl halides may also be reacted with zinc in acetic acid to give 1,2-unsaturated sugars, known as glycals. D-galactal (**22**), for example, is easily produced from tetra-*O*-acetyl-galactosyl bromide (Scheme 2.53) by treatment with zinc followed by removal of the acetates using a catalytic amount of sodium methoxide in methanol. Such glycals readily undergo regiospecific addition reactions and may be used as intermediates to enable epimerization at the 2-position during the preparation of unusual sugars (see Section 2.2.2).

2.5.3 Branched-chain Monosaccharides

Sugars bearing additional carbons—methyl, ethyl, aldehydo, or hydroxymethyl groups—attached mid-chain are found in many naturally occurring saccharides especially macrolide antibiotics (12-, 14-, or 16-membered ring lactones bearing one or more glycosides) such as erythromycin (**89**). The syntheses of branched-chain monosaccharides generally involve attack of a carbanion (organometallic reagent, Wittig reagent, *etc.*) on a sugar epoxide or ketone. Representative examples are given in Scheme 2.54. The sugar starting materials are attacked

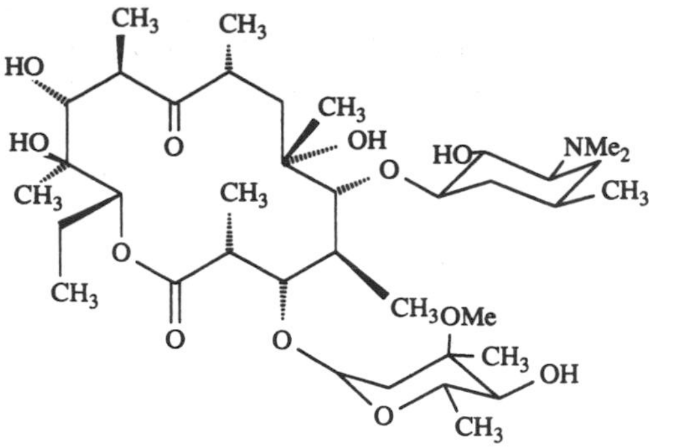

Erythromycin A **89**

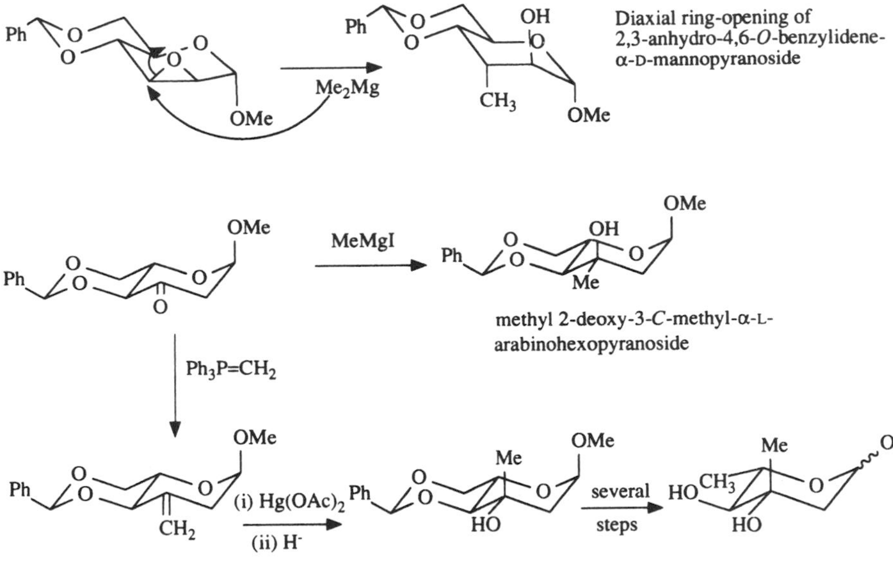
Diaxial ring-opening of
2,3-anhydro-4,6-*O*-benzylidene-
α-D-mannopyranoside
Me2Mg
MeMgI
methyl 2-deoxy-3-*C*-methyl-α-L-
arabinohexopyranoside
Ph3P=CH2
(i) Hg(OAc)2
(ii) H-
several
steps
L-olivomycose

Scheme 2.54 *Synthesis of branched-chain sugars*

preferentially on the less hindered side or atom and the selectivity is often good enough to obtain a single isomer.

2.5.4 Oligosaccharide Synthesis

The majority of natural oligosaccharides are linked entirely by glycosidic bonds between the anomeric position of one sugar (the glycosyl component) to an OH of the next (the hydroxyl component) and are composed of derivatives of pento- and hexo-pyranosides, furanosides being infrequently encountered. In principle oligosaccharide syntheses consist of a sequence of glycoside bond forming reactions similar to disaccharide synthesis and so the basic method comprises:

(i) Blocking all hydroxyl groups of the glycosyl component except the anomeric position which should be activated for reaction;
(ii) Blocking all hydroxyls apart from the chosen linkage position in the hydroxyl component;
(iii) Reacting the two components such that the correct configuration (α or β) of the new anomeric centre is obtained;
(iv) Selectively deblocking a chosen hydroxyl of the oligosaccharide ready for reaction with a new glycosyl component.

We have already covered most of the chemistry necessary for (i), (ii), and (iv) in Section 2.4 but the most difficult part—joining the sugars stereospecifically as well as regiospecifically—remains. The formation of 1,2-*trans*-glycopyranosides is comparatively simple since this geometry may be enforced on a glycosidation by a participating blocking group at position-2 of the glycosyl component. Synthesis of 1,2-*cis*-glycopyranosides is more difficult and generally relies on complete inversion at the anomeric centre of the glycosyl component during glycosidation.

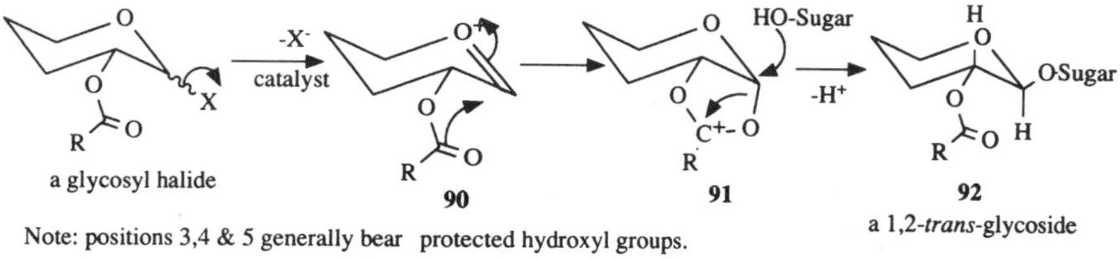

Note: positions 3,4 & 5 generally bear protected hydroxyl groups.

Scheme 2.55 *Mechanism of the Koenigs–Knorr synthesis of 1,2-*trans*-glycosides*

*2.5.4.1 Formation of 1,2-*trans*-Glycopyranoside Bonds.* The Koenigs–Knorr synthesis (Section 2.4.4) employing a glycosyl halide is the most commonly employed to synthesize 1,2-*trans*-glycosides (**92**, Scheme 2.55). The catalyst is usually a silver or mercury salt which encourages departure of the halide ion to give an oxocarbenium ion (**90**) which reacts with acetate or benzoate blocking groups to give a more stable dioxocarbenium ion (**91**). Nucleophilic attack by the hydroxylic component results in ring opening to give a glycosidic bond *trans* to the participating acetate or benzoate group at O2. Primary hydroxyl groups (6-OH) react most easily followed by 3-OH, 2-OH and least reactive 4-OH. The more biologically common glycosides synthesized in this way are β-*gluco*-, β-*galacto*-, α-D-*manno*-, and α-L-*rhamno*- (*i.e.* 6-deoxymanno-) pyranosides. The β-D-galactoside **93**, a precursor of determinants of blood-group substances, is made (Scheme 2.56)

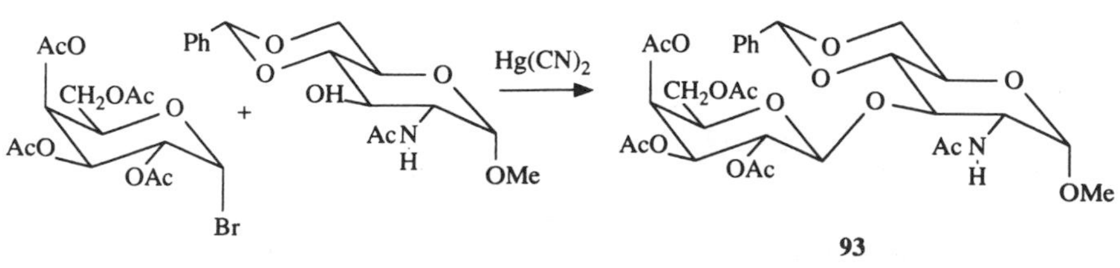

Scheme 2.56 *Synthesis of a 1,2-*trans*-galactoside*

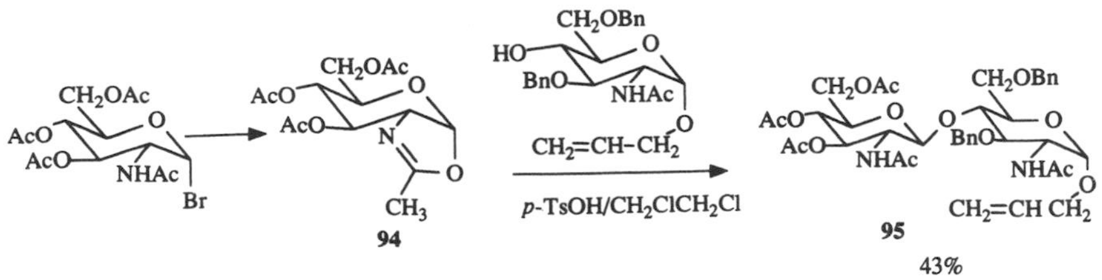

Scheme 2.57 *Synthesis of a 1,2-*trans*-glucoside* (**95**) via *an oxazoline* (**94**)

in 80% yield by $Hg(CN)_2$-catalysed reaction of the appropriate galactosyl bromide and glucosamine derivatives.

Glycosyl *chlorides* with an acetamido group at C2 react normally in Koenigs–Knorr reactions but 2-acetamido-2-deoxy-glycosyl *bromides* are very labile, giving stable 1,3-oxazolines (**94**, Scheme 2.57) analogous to the acetoxonium ion produced with acetylated sugars. However, these oxazolines may be isolated and reacted with hydroxyl components to link saccharides under acidic catalysis; synthesis of the chitobiose derivative **95** demonstrates that even weakly reactive 4-OH groups give reasonable yields. Koenigs–Knorr glycoside synthesis using 2-phthalimido-glycosyl halides is also used; the phthalimido group participates to give the 1,2-*trans*-glycoside and the free 2-amino sugars may be released from their phthalimides by treatment with hydrazine.

*2.5.4.2 Formation of 1,2-*cis*-Glycopyranoside Bonds.* Glycosyl halides with non-participating groups at C-2 must be used if 1,2-*cis*-glycosidic bonds are required on reaction with hydroxyl components. Although the main mechanistic route from glycosyl halides to glycosides is an S_N1 reaction involving an oxocarbenium ion (Scheme 2.58), this proceeds largely with inversion of configuration in low-polarity solvents. Thus *cis*-glycopyranosides, such as α-*gluco*- or α-*galacto*-pyranosides, may be obtained by silver- or mercury-catalysed reaction of β-glycosyl halides with appropriate hydroxyl components. This direct method is

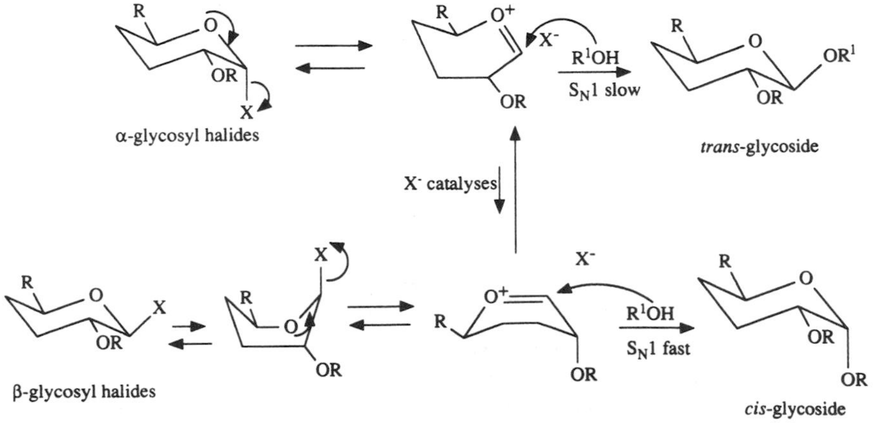

Scheme 2.58 *Mechanisms of glycoside formation from glycosyl halides*

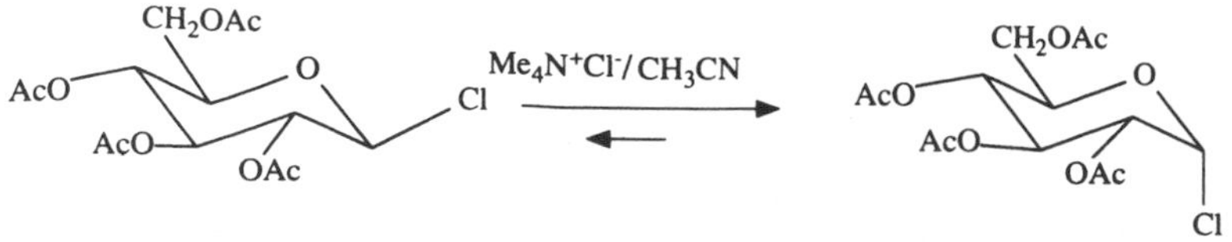

Scheme 2.59 *Equilibration of a glycosyl halide with halide ion*

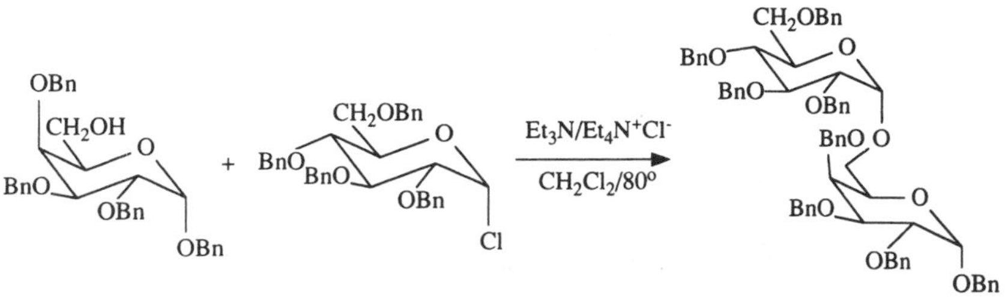

Scheme 2.60 *Synthesis of a 1, 2-*cis-α-D-*glucoside under kinetic control*

little used, however, since β-glycosyl halides are of limited availability, being destabilized relative to their α-anomers by the anomeric effect.

Alternatively it is possible to produce α-glycosides from anomeric mixtures of glycosyl halides using kinetically controlled reactions. Glycosyl halide anomers may be rapidly equilibrated (Scheme 2.59) by addition of halide ion—often as a tetraethylammonium halide which also serves as catalyst for glycosidations. Although the β-anomer is present only as a low equilibrium concentration it is the more reactive anomer and may give rise to the α-glycoside (Scheme 2.58) highly selectively provided the equilibration rapidly provides new β-anomer. Success in these halide-catalysed glycosidations is dependent on control of the relative energy barriers of competing reactions and often requires much preliminary work to establish the optimum conditions (catalyst, solvent, blocking groups, *etc.*). A simple example is the synthesis (Scheme 2.60) of the benzylated disaccharide α-D-Glu(1→6)-D-Gal (**96**) formed in 65% yield.

2.5.4.3 Imidates for Oligosaccharide and Glycoside Synthesis. Difficulties in the classical Koenigs–Knorr method, including the toxicity and expense of heavy metal salts, and the instability of glycosyl halides have led to the development of alternative synthetic methods for glycosides and oligosaccharides of which one of the most promising uses glycosyl imidates in place of glycosyl halides.

Glycosyl trichloroacetimidates may be prepared (Scheme 2.61) with either the α or β configuration by reaction of a 1-OH group of a protected sugar with trichloroacetonitrile and base. Using sodium hydride as base the reaction is reversible and so tetrabenzylglucose produces the α-D-acetimidate (**97**), favoured thermodynamically because of the anomeric effect. To obtain the β-anomer (**98**) a weak base such as potassium carbonate is used to produce a low equilibrium concentration of the deprotonated sugar and most of the product is then formed from the more reactive β-sugar alkoxide under kinetic control.

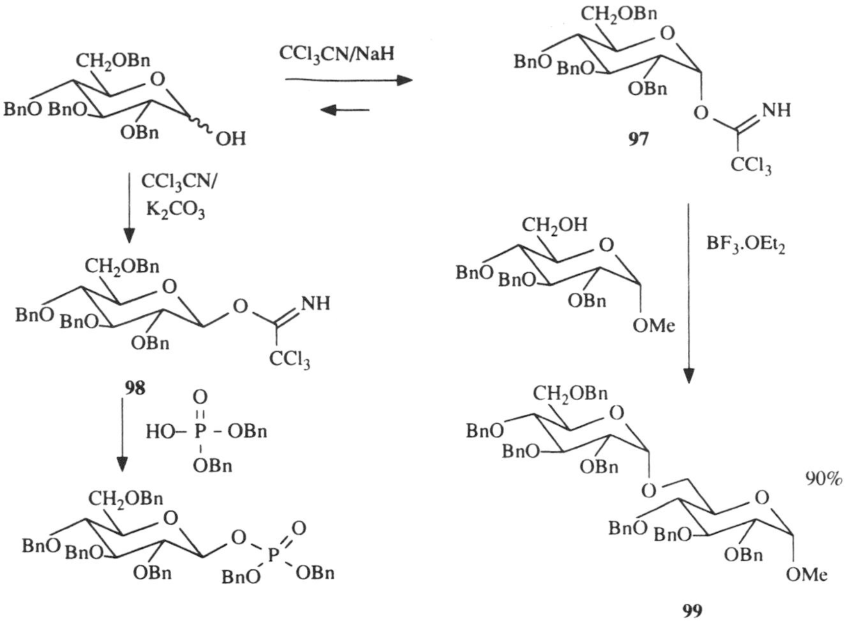

Scheme 2.61 *Synthesis and use of glycosyl imidates in glycoside synthesis*

The imidates so produced react with nucleophiles (Scheme 2.61) under acid catalysis with inversion of configuration to give the 1-*O*-derivatives. Using carboxylic or phosphoric acids no other catalyst is required and glycoside esters are produced, usually with inversion of configuration even when there is an acyl group at position-2. Reactions of glycosyl imidates with alcohols, including sugars, have produced several glycosides and oligosaccharides (*e.g.* **99**) in much improved yield over the Koenigs–Knorr route. The reactions are normally conducted at −40 °C, catalysed by an acid such as boron trifluoride, and go with inversion at position-1 except when there is a participating group at position-2 which dictates a 1,2-*trans* product.

2.5.4.4 Synthesis of a Blood Group Trisaccharide. Human cell surfaces have oligosaccharide sequences of one to six monosaccharide units responsible for blood type recognition and most of these have been synthesized and attached to proteins for use as artificial antigens in blood-typing. The specificity of their molecular recognition is remarkable: the only difference between the A and B human blood groups is the substitution of a 2-OH for a 2-NHAc on a single galactose residue of the hexasaccharide determinant, and yet this difference causes the sera of blood group type A to agglutinate the red cells of group B and *vice versa*, which can turn a life-saving blood transfusion into a life-threatening agglutination reaction.

The synthesis of such an oligosaccharide generally requires careful planning and many preliminary experiments. There are few generally applicable methods for the *selective* blocking and deblocking of hydroxyl groups crucial in oligo-

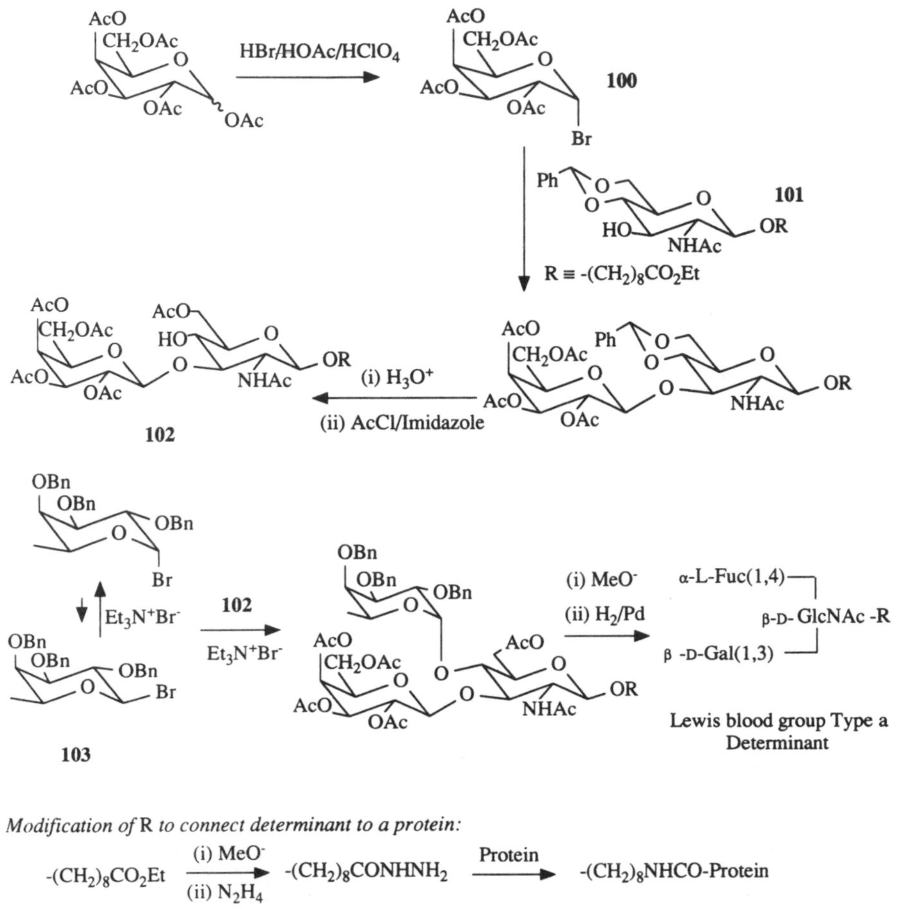

Modification of R *to connect determinant to a protein:*

$$-(CH_2)_8CO_2Et \xrightarrow[\text{(ii) } N_2H_4]{\text{(i) MeO}^-} -(CH_2)_8CONHNH_2 \xrightarrow{\text{Protein}} -(CH_2)_8NHCO\text{-Protein}$$

Scheme 2.62 *Synthesis of a trisaccharide for use as a synthetic antigen*

saccharide synthesis but typical tactics employed are illustrated in the synthesis of the trisaccharide responsible for the Lewis blood group type a (Scheme 2.62). The galactosyl bromide (**100**) has an *O*-acetate at position-2 which, being a participating group, ensures that the reaction with the glucosamine derivative (**101**) gives the 1,2-*trans*-galactoside. The benzylidene group is then removed by acid catalysis and selective acetylation on the primary OH is achieved using acetyl chloride with imidazole catalyst to give the disaccharide (**102**) having a single free OH. A 1,2-*cis*-fucoside link to this OH is now required and is made by equilibrating the L-fucosyl bromide (**103**) to an α/β mixture with halide ion followed by preferential reaction of the more reactive β-anomer with the disaccharide (**102**), under kinetically controlled conditions, producing the required α-L-fucoside. The acetyl groups are removed by treatment with methoxide and the benzyl ethers by hydrogenation and then the long-chain ester aglycone (R) is converted into the reactive acid hydrazide for coupling with the protein bovine serum albumin. The synthetic antigen so produced gives about 65% of the antibody agglutination reaction of natural Le^a antigen.

CHAPTER 3

Polysaccharides

The main biological roles of monosaccharides are as building blocks for construction of other biomolecules and to provide energy for organisms by breakdown, but once several saccharide units are joined together they usually have important biological activities of their own. Oligosaccharides are generally defined as having two to nine monosaccharide units while polysaccharides have ten or more units but this distinction is largely artificial and we will be considering both classes in this chapter.

3.1 NOMENCLATURE

Fully systematic names of oligosaccharides are very cumbersome and so shortened forms are employed for all but the smallest oligosaccharides. The full names (exemplified in Figure 3.1) treat the saccharide as a derivative of the reducing end sugar, so giving 4-*O*-(α-D-glucopyranosyl)-D-glucose for maltose (**1**). Short-form names use symbols for each monosaccharide (including rhamnose, fucose, neuraminic acid, *etc.*) consisting of their first three letters except glucose which has the abbreviation Glc or often just G. It is assumed that these have the pyranose form but *p* may be added to confirm this or an additional *f* used to indicate a furanose. Any substituents are indicated by additional letters; D-GlcN is 2-amino-2-deoxy-D-glucose (glucosamine) and D-GlcA6Et is ethyl D-glucuronate, for example. These symbols for the monosaccharides and their derivatives are prefixed by D-, L-, α-, or β- as appropriate and are linked with numbers indicating linkages between the monosaccharide units as shown in Figure 3.1 for the commonly occurring disaccharides. Minor variations of these conventions are frequently encountered but the intended structures usually remain unambiguous.

Polysaccharides may be fully described by similar use of the short-form notation, but regular polysaccharides are conveniently named by adding -an to the constituent monosaccharide to indicate a polysaccharide. Cellulose, for example, is a (1→4)-α-D-linear glucan, glycogen is a (1→4)-α-D-,(1→6)-α-D-branched glucan, and pectic acid is a a (1→4)-α-D-linear galacturonan (*i.e.* derived from galacturonic acid). The name dextran is derived from dextrose (an alternative name for glucose), and refers to a family of (1→6)-glucans common in bacterial secretions. Linear dextran polymers are rare in nature—various degrees of branching are seen, usually from the 3-position.

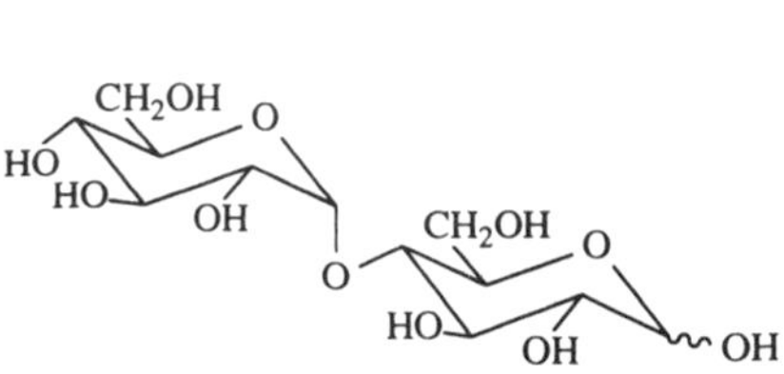

Maltose **1**

4-*O*-(α-D-glucopyranosyl)-D-glucose

α-D-Gal*p*-(1→4)-D-Glc

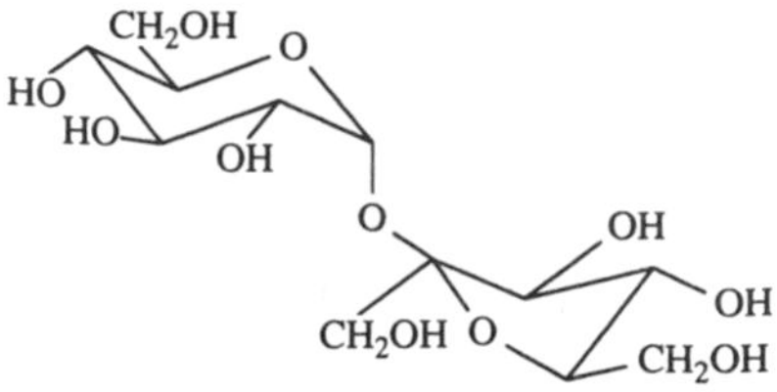

Sucrose **2**

α-D-glucopyranosyl-(1↔2)-α-D-fructofuranoside

α-*D*-Glc*p*(1↔2)-*D*-Fru*f*

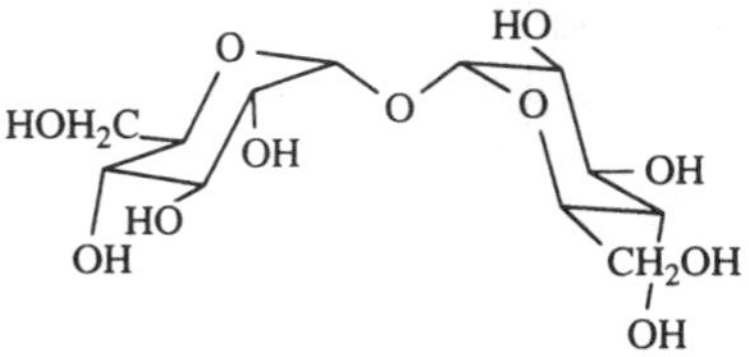

Trehalose **3**

α-D-glucopyranosyl-(1↔1)-α-D-glucopyranoside

α-*D*-Glc*p*(1↔1)-α-*D*-gluc*p*

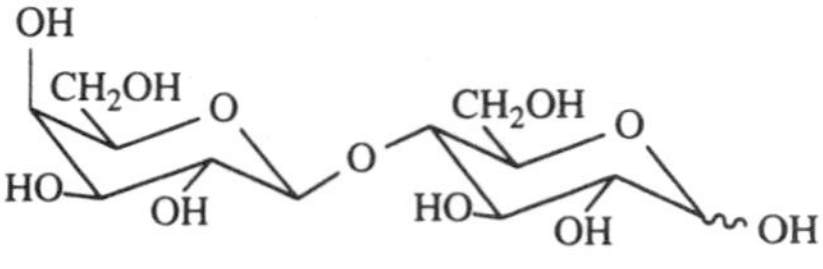

Lactose **4**

4-*O*-β-D-Galactopyranosyl-D-Glucopyranose

β-D-Gal*p*-(1→4)-D-Glc

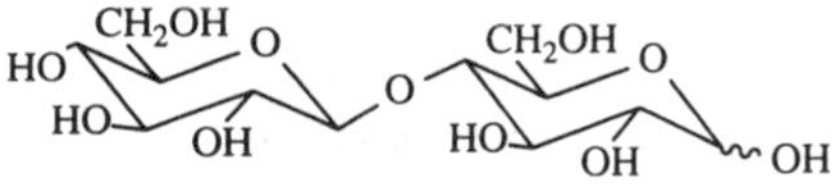

Cellobiose **5**

4-*O*-β-D-Glucopyranose-D-Glucopyranose

β-D-Glc*p*-(1→4)-D-Glc

Figure 3.1. *Common disaccharides with full and short-form names*

3.2 OLIGOSACCHARIDES

Eleven different glycosides are possible from an identical pair of hexopyranoses and the total number of conceivable disaccharides is immense. However, the only free disaccharides which occur naturally in significant quantities are sucrose, trehalose and lactose and each is important in nutrition.

Sucrose (**2**) is the most familiar as 'sugar' in foods and it is the main water-soluble energy reserve in plants. Although one of the commonest carbohydrates its structure contains three uncommon features: a ketose (fructose); a furanose ring (drawn here in its favoured envelope conformation); and linkage of the two anomeric centres to give a non-reducing sugar. The non-reducing nature of sucrose makes it unreactive to amino groups in food constituents and thus particularly versatile for food use. Treatment of sucrose with mild aqueous acid hydrolyses it to a mixture of glucose and fructose known as invert sugar (since the optical rotation inverts to a negative value). Fructose is roughly as sweet as sucrose, but yields far fewer calories and is used as a cheaper, less-fattening sweetener. It may be prepared from glucose (in invert sugar or from other sources) by the action of the enzyme glucose isomerase attached to a solid support. Another cheap sweetener consisting of glucose and fructose is high fructose corn syrup (HFCS) which is produced enzymatically from starch (using potatoes, wheat, or rice) and accounts for around one third of the sugar in Europeans' food and drink. Neither fructose nor HFCS may completely replace sucrose since they contain reducing sugars and so may undergo condensation reactions with proteins or amino acids.

Trehalose (**3**), which serves as the energy reserve disaccharide in insects and fungi (up to 15% of the dry weight of mushrooms), has the same structure as sucrose except that the fructofuranoside residue is replaced by glucopyranoside. The linkage may be either α or β at each glucose, giving three different trehaloses but the α,α-trehalose is most common. It is much less easily hydrolysed than sucrose and, since it is also a non-reducing disaccharide, it is relatively inert.

The principal free oligosaccharide in mammals is lactose (**4**) which can comprise up to 8.5% of mammalian milk and is prepared as a by-product of cheese manufacture. Interestingly the galactose portion of lactose is rapidly converted into glucose by infants and where this ability is absent or lost, as in most non-Caucasian adults, lactose causes ill-effects.

Maltose (**1**) and cellobiose (**5**) are reducing disaccharides which do not occur naturally but are released on hydrolysis of the polysaccharides starch and cellulose respectively.

Higher oligosaccharides (trisaccharides, tetrasaccharides, *etc.*) occur naturally in lower concentrations than disaccharides. The two most common are the non-reducing trisaccharides raffinose, α-D-Gal*p*-(1→6)-α-D-Glc*p*-(1↔2)-β-D-Fru*f*, and melezitose, α-D-Glc*p*-(1→3)-β-D-Fru*f*-(2↔1)-α-D-Glc*p*. Raffinose occurs widely in plants and although present in low concentrations it is available commercially as a by-product of sucrose production from beet molasses. Melezitose is present in honey (from which it may be isolated) and in the exudate from trees damaged by insects.

Cyclic oligosaccharides such as cyclomaltoheptaose (**6**, also named cyclohepta-amylose or β-cyclodextrin) have the shape of a squat cylinder with a non-polar cavity and a polar external surface bearing all the OH groups. Cyclomalto-oligosaccharides containing six, seven, or eight α-D-glucopyranoside units (previously known as α-, β-, or γ-cyclodextrins respectively) may be prepared from starch by the action of an enzyme, cyclodextrin transglycosylase, and in contrast to most oligosaccharides they are readily crystallized. They have attracted much recent attention because of their abilities to complex small organic molecules (such as benzaldehyde) forming 'inclusion compounds' which are usually crystalline even when the complexed molecule is a liquid. Inclusion compounds are generally less volatile, less reactive, and more soluble than their uncomplexed counterparts and are being increasingly used in the food, cosmetic

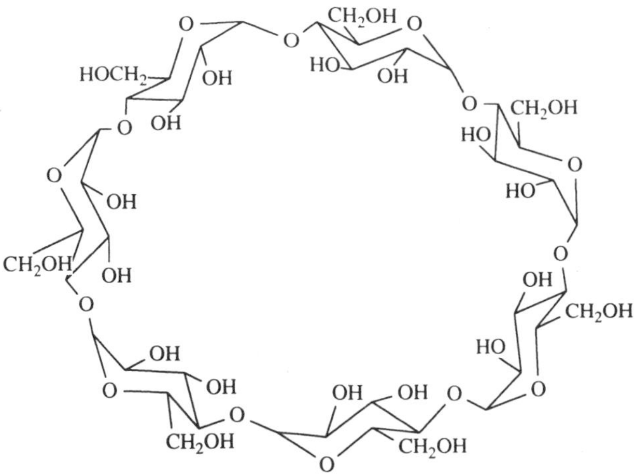

β-Cyclodextrin **6**

and drug industries. Sometimes the included compound displays selectivity not found in the free molecule. Chlorination of anisole (methoxybenzene) with HOCl (10^{-2} M), for example, is catalysed at the *para*-position but blocked at all other positions by complexation with α-cyclodextrin enabling a 96% yield. Such reactions are used to model hydrophobic interactions which are important in many biological processes, particularly enzyme mechanisms.

3.3 PROPERTIES AND THREE-DIMENSIONAL STRUCTURES OF POLYSACCHARIDES

Although many polysaccharides are known which have complex repeating groups of monosaccharides, or even no repeating units, those based on a small regular repeating unit are most common and many of the most abundant polysaccharides are homopolysaccharides—comprised of a single sugar always linked in the same way. The properties of polysaccharides are largely governed by the three-dimensional shapes adopted by the polymer chains. The regular structures seen in homopolysaccharides are described as extended ribbons, egg-boxes, crumpled ribbons, hollow helixes, or flexible coils and more complex polsaccharides are composites of these basic structures. The favoured shape for a particular homopolysaccharide is mainly dependent on the distance and angle between the two linkage bonds and little dependent on which sugar is present.

3.3.1 Ribbon Polysaccharides

Polysaccharides having the two linkage bonds to each monosaccharide unit roughly parallel to each other and only slightly offset give a ribbon structure (Figure 3.2) typically extending for several thousand monosaccharide units. The most obvious and important examples are 1,4-diequatorially linked polysaccharides such as cellulose (**7**, 1,4-β-glucan), 1,4-β-xylans (**8**), and 1,4-β-mannans (**9**)

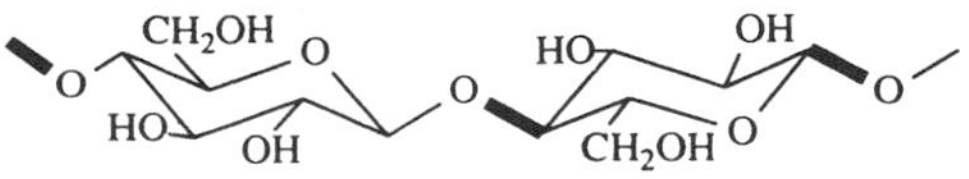

7 Cellulose, (1→4)-β-D-glucan
Adjacent glucoside units are
twisted through 180°

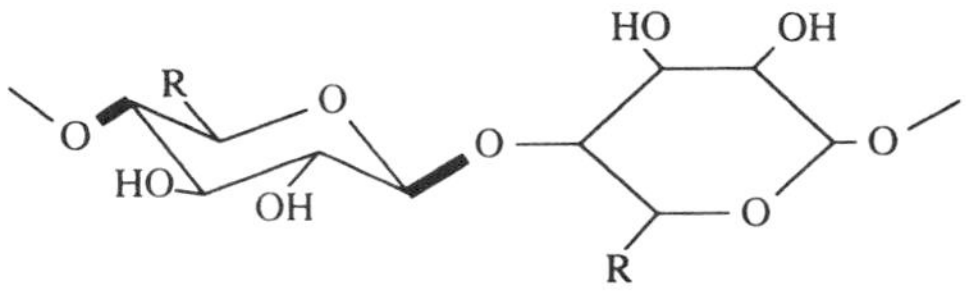

8 R = H (1→4)-β-D-xylan
Adjacent xyloside units are
twisted through ≈120°

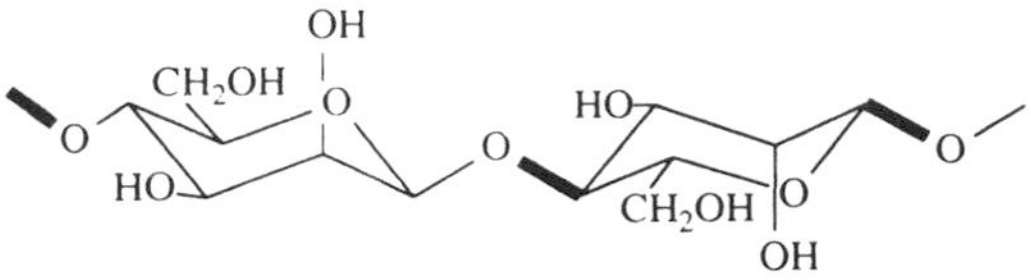

9 (1→4)-β-D-mannan

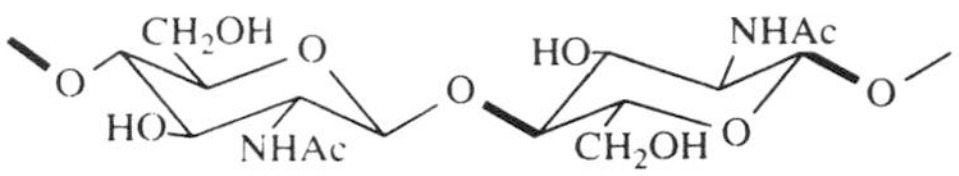

10 Chitin
(1→4)-2-acetamido-2-deoxy-
β-D-glucan

Figure 3.2. *Ribbon polysaccharides with parallel linking bonds highlighted*

but 1,3 axial–equatorial linkages generally lead to ribbon structures as well. A consequence of the ribbon structure is that individual polysaccharide molecules may align closely with others allowing hydrogen bonding between the chains. In cellulose, bundles of aligned molecules pack together giving microfibrils which also associate to give strong water-insoluble microcrystalline fibres. If, however, regular packing is prevented, as in oat glucan in which about 30% of the linkages are 1→3 (diequatorial) instead of 1→4 (diequatorial), the glycan is water soluble. Cellulose is the main constituent of plant cell walls (virtually 100% in cotton fibres) and is used with the other, more flexible, ribbon polysaccharides xylan (**8**) and mannan (**9**) to provide rigidity and strength. Mature plant cell walls, in tree trunks for example, are composed of densely packed microfibrils of cellulose (about 50%) and xylans (about 20%) cemented together by polymers of coniferyl alcohol (**11**) known as lignin (**12**). In young shoots the polysaccharide microfibrils are much less ordered and have less lignin and a high water content, so giving them greater flexibility.

In chitin (**10**), the major constituent of crustacean (crab, lobster) shells and

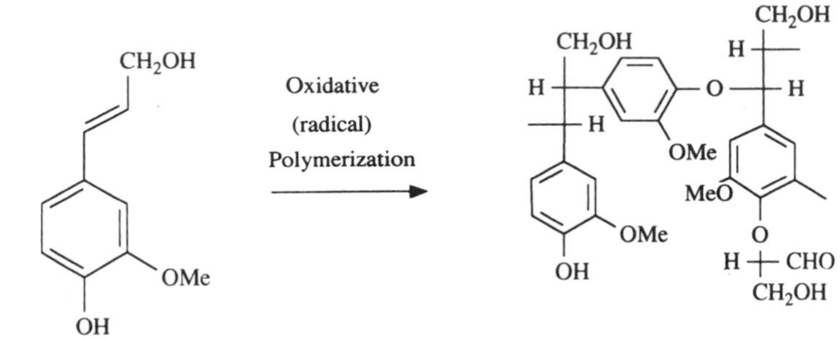

Coniferyl Alcohol **11** Lignin (representative portion) **12**

insect skeletons, the molecules pack side by side in crystalline, strongly hydrogen-bonded, water insoluble rigid sheets. Bacterial cell walls (particularly of the Gram-positive type) contain over 40% of a peptidoglycan similar to chitin. This peptidoglycan (Figure 3.3) is a 1→4 linked polymer of 2-acetamido-2-deoxy-D-glucose alternating with its 3-*O*-lactic acid ether, muramic acid. The polysaccharide chains are shorter than in chitin but the muramic acid moieties are cross-linked by short peptide bridges to give a very strong continuous network.

3.3.2 'Egg box' Polysaccharides

Polysaccharides having the two linkage bonds to each monosaccharide unit roughly parallel to each other but offset by the length of the monosaccharide units (1,4-diaxial linkages typify the group) have buckled ribbon structures which leave cavities between adjacent monosaccharide pairs (Figure 3.4). The close interaction between atoms in these pairs is unfavourable unless, as fre-

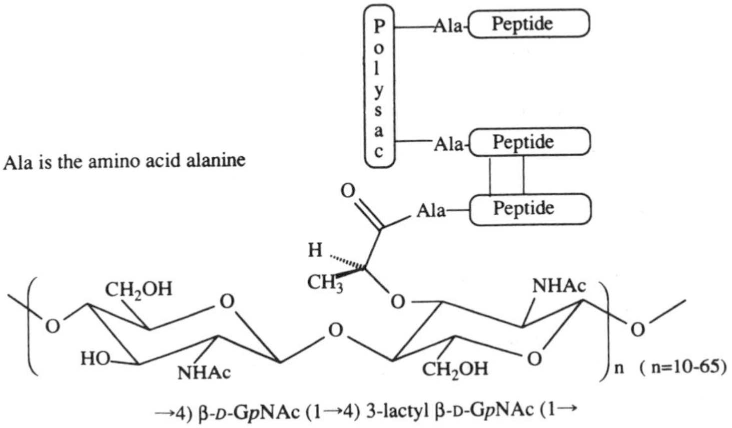

Figure 3.3 *Repeating disaccharide unit in the glycopeptide of bacterial cell walls. A diagrammatic representation of the cross-linking between chains is also shown*

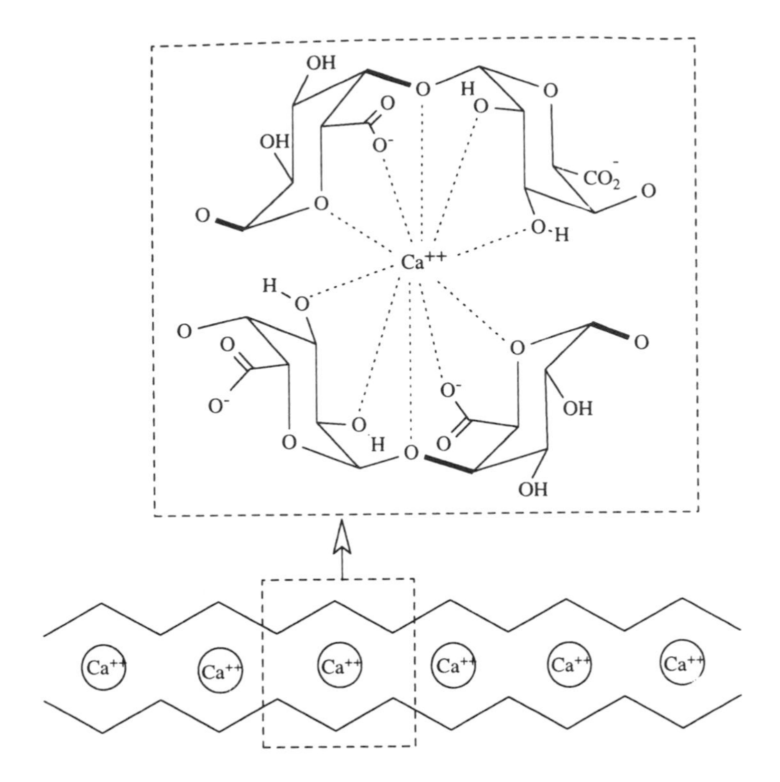

Figure 3.4 *'Egg-box' binding in polygalacturonate (pectin). Dashed lines indicate possible chelation, highlighted bonds are parallel linking bonds*

quently happens, the cavities can be filled with small species such as water or ions. Pectins are a group of substances found in fruit and many other plant cell walls which exhibit such a structure with calcium ions encased in an 'egg box' of polygalacturonate chains (Figure 3.4) thus conferring an ability for strong cohesion. Below pH 3.6 the negatively charged galacturonate residues are protonated to galacturonic acid residues which are much less effective at complexing ions, so providing a possibility for regulation of interchain bonding.

3.3.3 Hollow Helix Polysaccharides

Polysaccharides in which the two linkage bonds to each monosaccharide unit define a V shape tend to form a helix provided that hydrogen bonding or some other favourable interaction can be established down the axis of the helix. In the absence of such favourable interactions random coils—that is continually fluctuating conformations—are observed. Helices and random coils are difficult to pack regularly and so inter-chain bonding is weak and these polysaccharides are usually partially water soluble and unsuitable as structural materials.

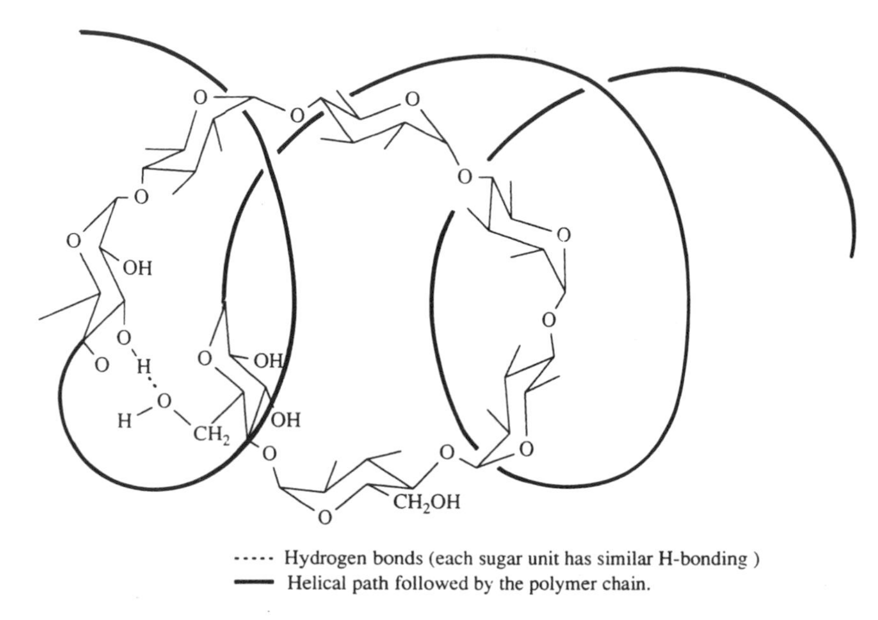

Figure 3.5 *Helical structure of amylose V (most OH groups omitted)*

Polysaccharides having hollow helices are found in starch, which occurs in large amounts in plants. In common with its animal counterpart glycogen, starch is used as an energy reserve and so must be able to pack together in concentrated deposits but be accessible enough to be released when required. It consists of two polysaccharides, amylose and amylopectin, both of which consist of (1→4)-α-D-glucan (as does glycogen).

Amylose consists largely of long (1000–2000 units) linear chains each of which, depending on the source, may coil into helices having between four and eight glucose units per turn held together by hydrogen bonds such as those between O2 and O6 of vertically adjacent glucose units in Figure 3.5. The coiling leaves a hollow hydrophobic tube and is encouraged by inclusion of other molecules within the cavity; with iodine a chain of iodine atoms line up along the axis to give the blue–black 'starch–iodide complex'. A wide variety of low-polarity organic molecules will also form complexes and fat is often found included in natural amylose. In neutral aqueous solution pure amylose tends to form random coils rather than helices but intertwined double helices (each chain filling the cavity of the other) and even aligned linear chains are found in crystalline particles of starch.

Amylopectin consists of very large molecules having about 10^6 (1→4)-α-D-glucoside units but about every 20 units it is branched through position-6. This frequent branching makes extended helical structures impossible and amylopectin gives a red coloration with iodine rather than the blue–black complex from amylose. Glycogen is very similar to amylopectin but is even more highly branched—about every eleven units.

3.3.4 Crumpled Ribbons and Loosely Jointed Chains

Homopolysaccharides linked 1→2 are rare in nature. They have adjacent monosaccharide units twisted both horizontally and vertically, and non-consecutive glycoside units suffer steric clashes along the chains. These crowded structures are described as crumpled ribbons but ordered conformations are difficult to achieve.

Another linkage type non-conducive to ordered structures is 1→6. Adjacent monosaccharide units are linked by —CH_2O— instead of just —O— and the increased distance between the sugar residues gives greater chain flexibility so encouraging disorder. Such linkages are widespread in nature, although mainly as branch points such as in amylopectin (see Section 3.3.3), and their flexibility may be important in allowing access to the interior of bulky polysaccharides.

3.3.5 Chemically Modified Celluloses

Chemical modification of cellulose allows production of many materials widely used in the clothing, paper, and plastics industries.

One of the purest forms of natural cellulose is the seed hairs of the cotton plant, cotton. Cotton yarns are manufactured by separating cotton fibres from their seeds, followed by cleaning, forming into loose ropes, stretching to align the fibres, and spinning to twist the fibres together. The native form of cellulose, still existing in the yarn, has all the chains parallel to each other with the reducing ends of each chain at the same end of the bundle. A finishing process known as mercerization is often applied to the yarn to increase lustre, improve strength, and enhance its affinity for dyes. Mercerization is achieved by immersion under tension in caustic soda solution which produces a permanent swelling of the fibres and causes the cellulose chains to become anti-parallel and more flexible. The cotton yarn is finally neutralized in acid.

Rayons are fabrics manufactured by dissolving cellulose as some complex or derivative and then forcing the liquid through tiny holes in a spinneret to regenerate cellulose fibres. The long fibres or filaments that emerge are hardened by air-drying or chemical means. The most common types are viscose and cuprammonium rayon and they are similar to silk, being lustrous, able to absorb moisture readily, and almost as strong as nylon.

Viscose rayon is produced by treatment of soft wood or cotton with 5 M hydroxide (mercerization) followed by dissolution in carbon disulphide. This converts (Scheme 3.1) some cellulose OH groups into xanthates (otherwise called dithiocarbonates) and long cellulose fibres are regenerated by running spinneret filaments into dilute sulphuric acid. The carbon disulphide is also regenerated and recycled. Extrusion of viscose through a narrow slit instead of a spinneret leads to thin, flat, transparent sheets known as cellophane. Cuprammonium rayon is produced by dissolving cellulose in cuprammonium hydroxide solution and regenerating cellulose by treatment with water. It is of slightly better quality than viscose but more expensive.

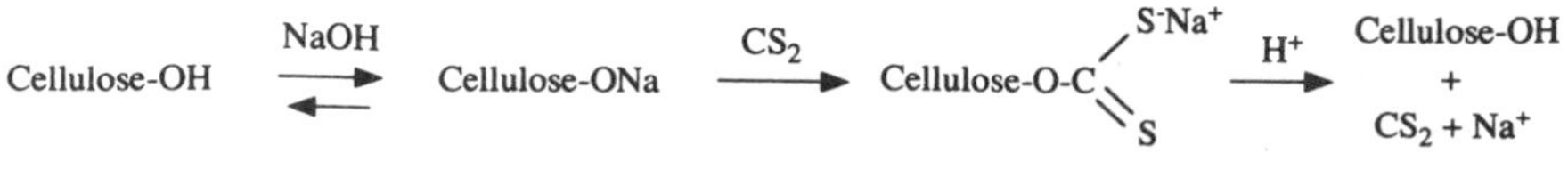

$$\text{Cellulose-OH} \underset{}{\overset{\text{NaOH}}{\rightleftharpoons}} \text{Cellulose-ONa} \xrightarrow{CS_2} \text{Cellulose-O-C}(=S)S^-Na^+ \xrightarrow{H^+} \text{Cellulose-OH} + CS_2 + Na^+$$

Scheme 3.1 *Reactions involved in the generation of viscose rayon*

Paper is essentially a thin matt of fibres, usually cellulose, formed on a wire screen. Starting from debarked rolled wood chips, treatment with sodium hydroxide solution at 100 °C and 7 atm dissolves the lignin (**12**, which contains phenolic OH groups) and separates and wets the cellulose fibres. An alternative to this soda process for dissolving lignin is the sulphite process which treats the bruised wood chips with sulphite salt solutions containing excess sulphur dioxide (addition of SO_3^{2-} to benzyl alcohol or aldehyde groups gives sulphonic acid salts). The wood pulp is then bleached, beaten to swell the fibres, and filtered under suction. Additions of sizing (to reduce penetration of water), pigments and fillers (often China clay) are used to produce paper of the required quality.

Acetates and nitrates of cellulose are also important materials. Cellulose acetate is produced by acetylation of cellulose fibres with acetic acid and acetic anhydride to give cellulose triacetate on full acetylation or 'secondary acetate' if the acetylated material is partially hydrolysed. Both are used in textiles. Cellulose nitrate (nitrocellulose) is prepared by nitration of cellulose with concentrated nitric acid. It retains some of the fibrous properties of cellulose but is explosive and is used as gun cotton and in gelatin dynamites. A colloidal dispersion of cellulose nitrate (or other low esters of cellulose) and camphor gives celluloid, the first plastic material ever made. Although highly flammable it has been used extensively for imitation ivory, coral, and amber.

Several ethers of cellulose are commercial water-soluble materials used as thickeners, binders, and stabilizers in many products including foods, detergents, and pharmaceuticals. Carboxymethylcellulose, prepared by treatment of cellulose with sodium hydroxide and sodium chloroacetate, is one of the most important and is cellulose with OCH_2OCH_2COONa in place of most 6-OH groups. Hydroxyethylcellulose (OCH_2CH_2OH in place of 6-OH) has similar properties and is prepared from treatment of cellulose with sodium hydroxide and ethylene oxide (oxirane, $\overline{CH_2CH_2O}$). It has not been approved for food use and so its main uses are in textiles, paper, adhesives and paints. Hydroxypropylcellulose is prepared by treatment of cellulose with sodium hydroxide then propylene oxide (methyloxirane). The reaction is run at elevated temperature and pressure to give hydroxypropyl groups [$OCH_2CH(OH)CH_3$] at most 6-positions and some 2-positions. Treatment of cellulose fibres with sodium hydroxide then chloromethane gives methylcellulose having OMe substitution at 2- and 6-positions. A variety of methylcellulose products is produced by incorporation of propylene oxide and 1,2-butylene oxide (ethyloxirane) in the reaction mixture to give partial substitution with hydroxypropyl and hydroxybutyl groups.

3.3.6 Glycoconjugates

Specific biological activity due to carbohydrates is commonly found in glycoconjugates, that is carbohydrates having an attached lipid or protein, known as glycolipids and glycoproteins. This was first recognized with blood group types and assumed greater importance when it was discovered that glycoconjugates of cell surfaces are profoundly modified in cancer cells.

Certain classes of glycoconjugate are rather heterogeneous and it seems likely that they possess general structural characteristics to give them useful bulk properties. Typical of such glycoconjugates are mucus glycoproteins (mucins) which are typically composed of about 80% carbohydrate which, for humans, is made up from only L-fucose, *N*-acetylgalactosamine, galactose, *N*-acetylglucosamine, and *N*-acetylneuraminic acid. Mucins are *O*-glycosylproteins, having high viscosity and viscoelastic and gel characteristics, employed in mucus secretions which provide protection for underlying tissues such as the stomach, cervix, and bronchial passages. The protein (Chapter 4) forms the backbone to which hundreds of short, often branched, oligosaccharides are attached giving a loosely coiled structure of about 10^6 relative molecular weight. Although heterogeneous, there is considerable similarity in the assembly and conformation of mucins, even those from different sources.

Of the glycoproteins showing specific interactions with other biomolecules the blood group antigens, which occur in tissue fluids, in secretions, and at cell and tissue surfaces, are amongst the best characterized. The capacity of these *O*-glycosylproteins to elicit a response from antibodies is primarily determined by terminal tri- or tetra-saccharide sequences of the molecule. The three most common blood group types, O, A, and B, differ in just one monosaccharide unit (Figure 3.6) and enzymatic removal of the terminal Gal or GalNAc residues converts A or B antigens into O antigens. Similarly minor differences in cell surface oligosaccharides account for many other antigenicities. The same oligosaccharide determinants also occur attached to lipids in cell membrane glycosphingolipids (see Section 6.1.1).

Other highly specific biological recognition processes, shown to be dependent on the carbohydrate portions of glycoproteins, include removal of certain glycoproteins from the circulation system by human liver 'lectins', attachment of nitrogen-fixing bacteria to legume root cells, and adherence of bacteria to mammalian cells. Such recognition of the carbohydrate determinants may involve molecule–molecule, cell–molecule, or cell–cell interactions.

3.4 POLYSACCHARIDE STRUCTURE DETERMINATION

Until quite recently polysaccharide structures were determined largely by identification of the products of chemical degradations. Thus total hydrolysis of polysaccharides gives the component monosaccharides, while methylation or periodate cleavage followed by hydrolysis, gives linkage positions. The advent of high-field (>200 MHz) NMR spectrometry and improved methods of sample ionization in mass spectrometry (MS) led to physical methods for complete

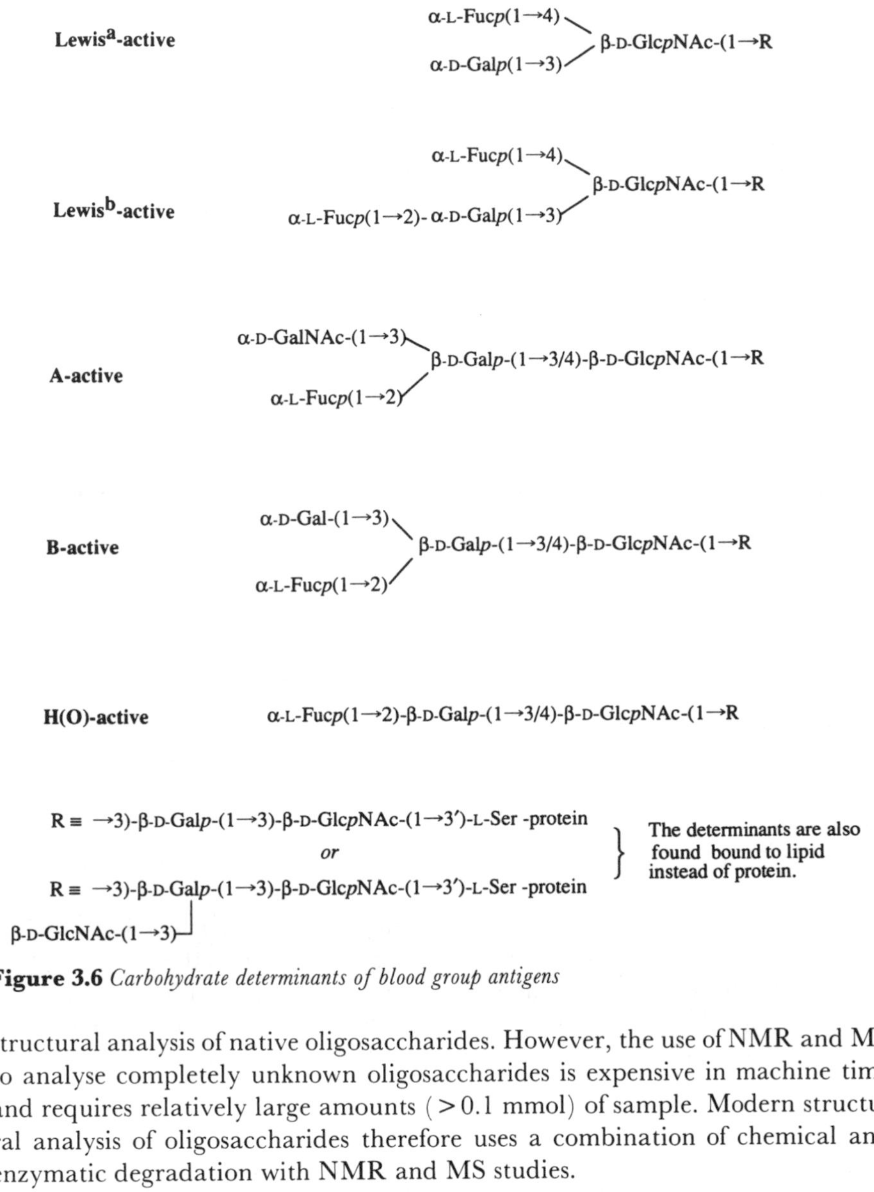

Lewis^a-active

α-L-Fuc*p*(1→4)
β-D-Glc*p*NAc-(1→R
α-D-Gal*p*(1→3)

Lewis^b-active

α-L-Fuc*p*(1→4)
β-D-Glc*p*NAc-(1→R
α-L-Fuc*p*(1→2)- α-D-Gal*p*(1→3)

A-active

α-D-GalNAc-(1→3)
β-D-Gal*p*-(1→3/4)-β-D-Glc*p*NAc-(1→R
α-L-Fuc*p*(1→2)

B-active

α-D-Gal-(1→3)
β-D-Gal*p*-(1→3/4)-β-D-Glc*p*NAc-(1→R
α-L-Fuc*p*(1→2)

H(O)-active

α-L-Fuc*p*(1→2)-β-D-Gal*p*-(1→3/4)-β-D-Glc*p*NAc-(1→R

R ≡ →3)-β-D-Gal*p*-(1→3)-β-D-Glc*p*NAc-(1→3′)-L-Ser -protein

or

R ≡ →3)-β-D-Gal*p*-(1→3)-β-D-Glc*p*NAc-(1→3′)-L-Ser -protein
β-D-GlcNAc-(1→3)

The determinants are also found bound to lipid instead of protein.

Figure 3.6 *Carbohydrate determinants of blood group antigens*

structural analysis of native oligosaccharides. However, the use of NMR and MS to analyse completely unknown oligosaccharides is expensive in machine time and requires relatively large amounts (>0.1 mmol) of sample. Modern structural analysis of oligosaccharides therefore uses a combination of chemical and enzymatic degradation with NMR and MS studies.

3.4.1 Degradations

3.4.1.1 Acidic Hydrolysis. Total hydrolysis of polysaccharides may usually be achieved by heating with 1 M sulphuric acid at 100 °C for four hours. This liberates the free monosaccharides which may be quantitatively estimated by high pressure liquid chromatography (HPLC). Gas chromatography (GC) is also used but then the monosaccharides must be converted into volatile derivatives, such as their trimethylsilyl ethers (Section 2.4.3.1), prior to analysis. Problems occasionally encountered in the acid hydrolysis include further degra-

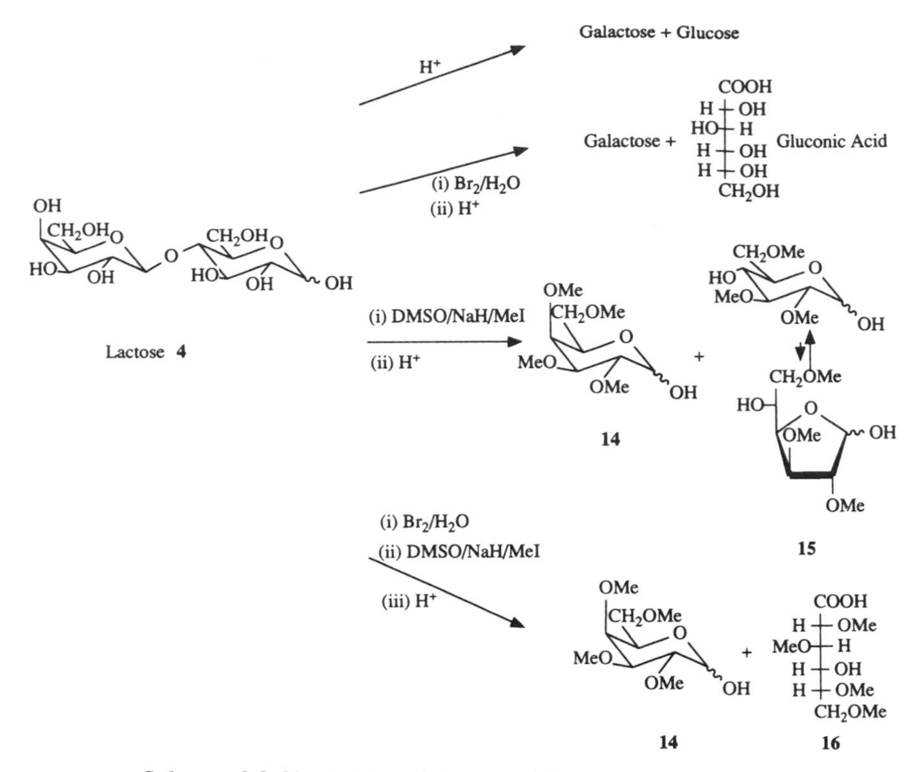

Scheme 3.2 *Chemical degradation to establish the structure of lactose*

dation of the more acid-sensitive monosaccharides and acid reversion (condensation of monosaccharides to oligosaccharides, Section 2.3.3).

Oxidation of the polysaccharide with bromine water prior to acid hydrolysis permits one to establish which sugar is at the reducing end of the chain. Lactose, for example (Scheme 3.2), is found to consist of equal amounts of galactose and glucose by total hydrolysis; but oxidation followed by hydrolysis yields equal amounts of galactose and gluconic acid thus establishing that galactose, but not glucose, is joined through position-1 in lactose.

Partial acid hydrolysis of a polysaccharide—stopping the reaction before completion—gives oligosaccharide fragments which may be purified by chromatography. Determination of the structures of these fragments is easier than that of the parent polysaccharide and often allows the structure of the polysaccharide to be deduced.

3.4.1.2 Methylation Analysis. Full methylation of all free OH and NH groups in a polysaccharide is achieved by reaction with dimethyl sulphoxide/sodium hydride/methyl iodide (the Hakomori reagent). Methylation is followed by total acid hydrolysis and yields monosaccharides which are fully methylated except at the anomeric and linkage positions. The mixture of partially methylated sugars is then analysed by HPLC or, after conversion into volatile derivatives, gas chromatography.

Lactose treated in this way (Scheme 3.2) would yield equimolar amounts of 2,3,4,6-tetra-*O*-methyl-D-galactose (**14**) and 2,3,6-tri-*O*-methyl-D-glucose (**15**), proving that galactose is linked only through position-1 and glucose is linked either through O4 as a pyranose or through O5 as a furanose. This uncertainty may be removed by oxidizing lactose prior to methylation, hydrolysis, and analysis. Once the reducing end is oxidized it is in a ring-open form and so O5 will be methylated yielding 2,3,5,6-tetra-*O*-methyl-D-gluconic acid (**16**) after hydrolysis, thus establishing lactose as D-Gal-(1→4)-D-Glc*p*.

3.4.1.3 Periodate and Lead Tetra-acetate Oxidation. The vicinal diol groups of polysaccharides are cleaved by periodate in much the same way as in monosaccharides (Section 2.3.9.4) and the linkage types may be deduced from the consumption of periodate and the products. Side-reactions are minimized by conducting the periodate oxidation with exclusion of light at pH 3–5 and the oxidation is normally followed by reduction (to stabilize the dialdehyde products), acidic hydrolysis, and derivatization of the products as their trimethylsilyl ethers. Periodate consumption is followed using the UV absorbance of IO_4^- at 223 nm, the relative molar amounts of the products are determined by GC, and formic acid (indicative of monosubstituted or 1,6-disubstituted residues) may be determined colorimetrically. These data are usually sufficient to give the relative incidence of 1,2-, 1,3-, 1,4-, and 1,6-disubstitution since each linkage type gives different products as indicated in Scheme 3.3. Determination of formaldehyde (produced from terminal diols), by colorimetric estimation of the yellow compound formed on reaction with pentane-2,4-dione and an ammonium salt, is also useful; it is produced from terminal diols and so indicates the presence of 2- or 3-monosubstituted pyranosides or of furanosides having both the 5- and the 6-OH groups free.

An indication of which sugars are present may be obtained from rates of reaction with periodate and from the configurations of products. Diaxial diols generally react very slowly under the normal conditions and so this (fairly rare) configurational feature is readily identified. Sugars linked 1→4 produce (*meso*)-erythritol or threitol according to the relative configurations of positions 4 and 5 of the original sugar. Thus sugars such as glucose having the *erythro*-configuration at positions-4 and -5 give erythritol whereas sugars such as D-galactose having the D-*threo*-configuration there give D-threitol. Glyceraldehyde is produced from 1→2-linked sugars and its configuration is dependent on the configuration of C2 in the sugar; D- or L-glucose gives D-glyceraldehyde, but D- or L-mannose gives L-glyceraldehyde.

3.4.1.4 Chain Shortening. The above degradative methods are sufficient to establish the structures of short oligosaccharides and regular polysaccharides. For irregular polysaccharides the chain is broken into smaller lengths by partial acid hydrolysis or enzymatic degradation. If this is done in a specific manner, the polysaccharide structure may be deduced from the structures of the shorter oligosaccharides. Partial acid hydrolysis can be quite selective—furanosides are hydrolysed about 100 times faster than pyranosides and mineral acids hydrolyse (1→4)-linkages faster than (1→6)-linkages—but enzymatic hydrolysis allows a wide choice of selectivities and is normally preferred.

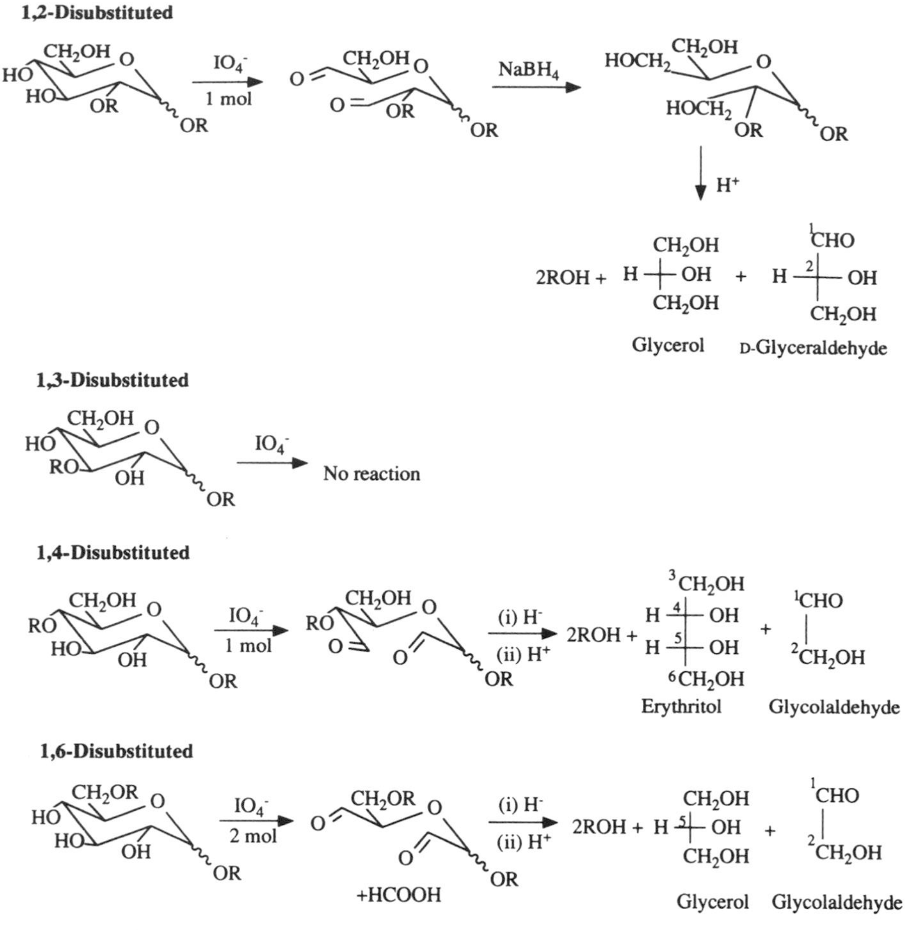

1,2-Disubstituted

1,3-Disubstituted

1,4-Disubstituted

1,6-Disubstituted

Scheme 3.3 *Periodate oxidation analysis of polysaccharides having disubstituted-glucose units. (Products bear numbers indicating the origin of their C-atoms)*

There is a wide variety of enzymes available for selective cleavage of polysaccharides. They are divided into *endo*-polysaccharide hydrolases—which split in the middle of a chain—and *exo*-polysaccharide hydrolases which remove a residue from one end, usually the non-reducing end, of the chain. The degree of branching in glycogen, for example, has been found using the *endo*-polysaccharide hydrolase pullulanase, which specifically cleaves the (1→6)-links of glucoside residues at 1,4,6-branch points to give linear (1→4)-α-D-glucose chains.

An advantage of enzyme cleavages is that, provided the enzyme preparation is pure, cleavage occurs specifically at certain linkage types; α-amylase cleaves only (1→4)-α-D-glucosidic linkages for example. Highly purified *exo*-polysaccharide hydrolases such as β-galactosidase are used to determine whether the monosaccharide unit for which they are specific is present (or absent) at the chain ends. After removal of the first monosaccharide from a chain the next unit may be identified and removed by treatment with its *exo*-polysaccharide hydrolase and so on along the chain. Unfortunately the *exo*-hydrolases are not specific for linkage type to the next sugar and so only the sequence and anomeric configurations are thus determined.

3.4.2 Physical Methods for Structural Analysis of Saccharides

3.4.2.1 Nuclear Magnetic Resonance. ^{1}H NMR spectrometry is of great value in the elucidation of the structures and conformations of carbohydrates especially since the dihedral angle between a pair of coupled hydrogens may be determined from their coupling constant using the Karplus relationship (Figure 3.7). In pyranoses (and other six-membered rings) protons may be either *trans*-diaxial with a dihedral angle of *ca.* 180°, which generally leads to a large coupling constant of 5–12 Hz, or vicinal *gauche* (equatorial/axial or equatorial/equatorial) with dihedral angles near 60° and associated J values of 1–4 Hz. The most common application is to establish the configuration of anomeric linkages since the anomeric proton is easily assigned, appearing well away from most other signals—between 4.4 and 5.2 δ in unsubstituted sugars (up to 7.2 in 2-*O*-acyl derivatives). Thus α-D-glucopyranose, β-D-mannopyranose and α-D-galactopyranose all have $J_{1,2}$ below 4 Hz while their anomers have higher J values.

Assignment of other signals is simplified by treating the sample with D_2O, which exchanges OH protons with deuterium and causes their signals to disappear along with any coupling of the OH groups to adjacent protons. At 200

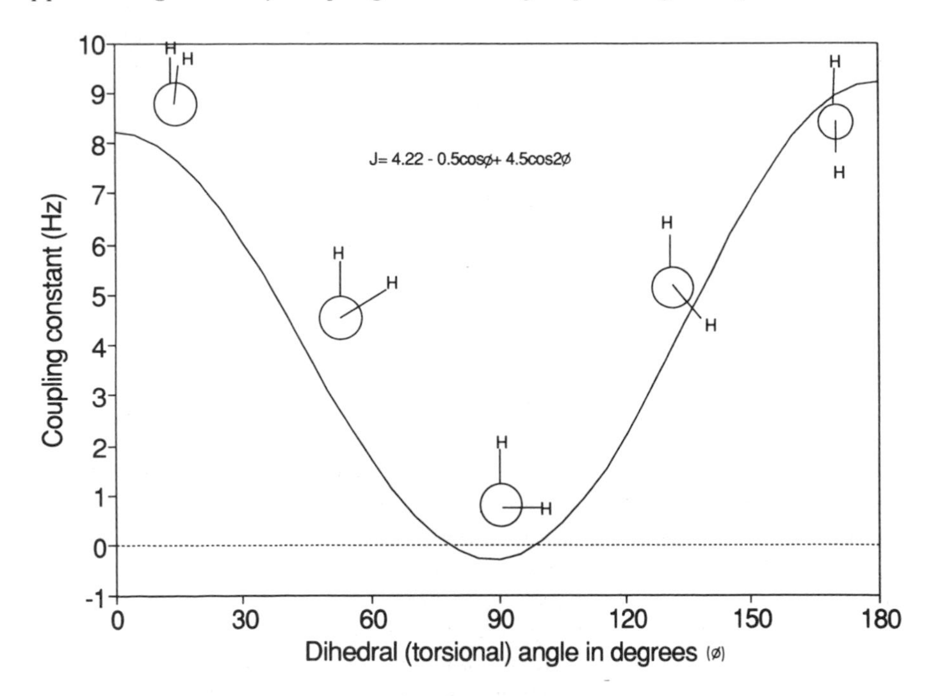

Figure 3.7 *The Karplus relationship*

MHz and higher fields most of the remaining signals from small oligosaccharides are separated and their identification is usually possible using double resonance techniques and/or comparison of chemical shifts with ^{1}H NMR databases. Finally measurement of the coupling constants for each pair of protons and application of the Karplus relationship gives the dihedral angles and hence conformations may be calculated. Such techniques often allow complete structural analysis of oligosaccharides.

Furanosides present difficulties since the J values of *cis*- and *trans*-related protons are quite similar in five-membered rings, but detailed decoupling studies (including nuclear Overhauser effects and two-dimensional NMR spectroscopy) usually allow their structures to be ascertained. NMR spectra from rigid molecules are easier to interpret and the anomeric configuration of furanosides may be determined by comparison of ^{13}C chemical shift values in their inflexible 2,3-*O*-isopropylidene derivatives.

3.4.2.2 Mass Spectrometry. Mass spectra suitable for structural analysis may be obtained from only 0.5–50 μg of fully methylated carbohydrates making this a highly sensitive technique. The fragmentation proceeds mainly at glycosidic linkages (Scheme 3.4) with high selectivities and propensities characteristic of the monosaccharide and its neighbours thus allowing the sequence to be assigned if it is known which monosaccharides are present. In favourable cases the linkage positions may also be deduced from the relative intensities of peaks.

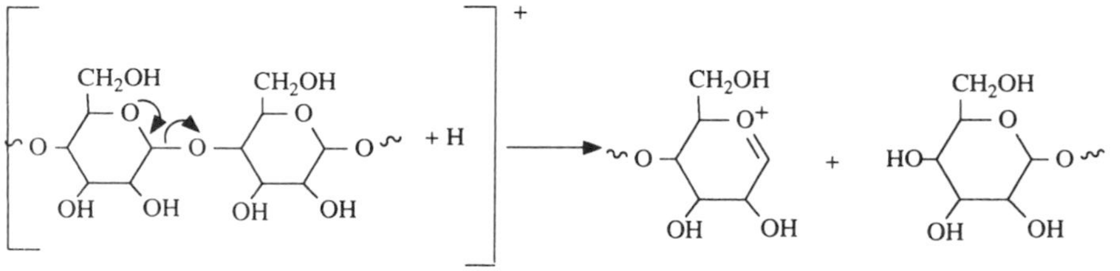

Scheme 3.4 *A common mechanism for glycosidic cleavage in mass spectrometry*

Using electron impact to ionize the sample gives relatively unstable ions and imposes an upper limit of about 1000 mass units (a pentasaccharide) but recently the development of fast atom bombardment (FAB) MS has raised this to over 4000 mass units. FAB-MS fires atoms at a solution (matrix) of the sample in a viscous low-volatility solvent such as glycerol. This causes sputtering of the sample, and the positive and/or the negative ions so vaporized may be analysed in the mass spectrometer. The original sample does not have to be volatile and so molecular masses of underivatized oligosaccharides are readily obtained by FAB-MS although their fragmentations are more complex to assign than for permethylated derivatives. Preliminary structural analysis of the components of carbohydrate mixtures from biological sources may be performed on a few nanomoles by HPLC separation followed by identification using FAB-MS. Such techniques have been applied to structural analysis of the oligosaccharides on the surface of HIV (AIDS) viruses and are valuable in studies of biological recognition processes which involve molecular interactions of carbohydrates with other biomolecules.

CHAPTER 4

Peptides and Proteins

Peptides and proteins are polyamides composed entirely of α-amino acids. Remarkably for such basically similar molecules they serve a multitude of purposes in living organisms and may display potent biological activity. They may, for example, function as enzymes (biological catalysts), hormones (biological messengers), pain-killers, structural molecules (to give tissues strength), or antibodies, or function in the interconversion of energy forms (chemical, mechanical, light). This chapter explains the properties of the amino acid building blocks, how they may be joined in chemical synthesis of peptides, and how some of the biological properties of peptides and proteins result from their three-dimensional structures. Their biosynthesis according to coded instructions contained in the DNA of genes is explained in the next chapter.

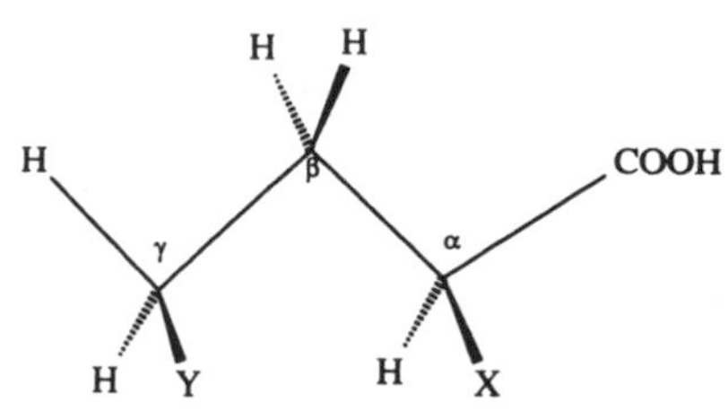

Amino acids (**1**)

α-L-amino acids (**1a**) have $X=NH_2$;
e.g. $X=NH_2$, $Y=SCH_3$ is L-methionine.

γ-amino acids have $Y=NH_2$;
e.g. X=H, $Y=NH_2$ is γ-aminobutyric acid (**1b**).

COOH

H_2N — H

R

α-L-Amino acids (Fischer projection)

2a

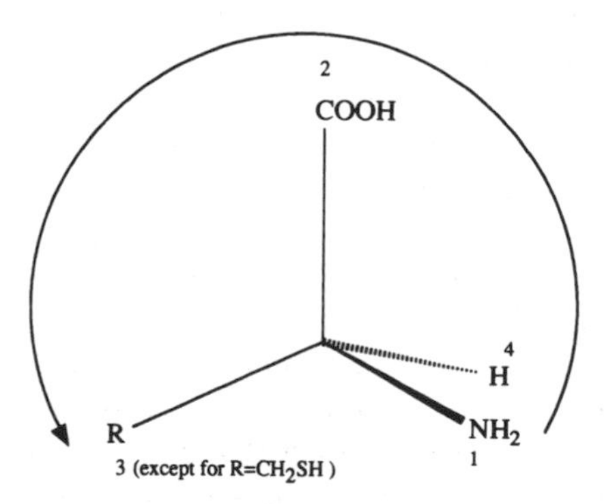

Most α-L-amino acids have the *S*-configuration

2b

4.1 AMINO ACIDS

4.1.1 Structures

Amino acids are molecules containing an amino (NH_2) group and a carboxylic acid (COOH) group. Peptides and proteins contain only α-amino acids (*e.g.* **1a**), in which both the amino and the carboxylic acid groups are attached to the same carbon. This is not the only type of biologically important amino acid—γ-aminobutyric acid (**1b**), for example, is involved in transmission of nerve impulses.

The α-carbon of α-amino acids bears a hydrogen and another group (R) and thus four different groups are usually attached and it is asymmetric. Of the two possible enantiomers, only that having the amino group to the left of the Fischer projection (**2a**), designated the L-enantiomer, is found in most peptides and proteins. Assigning priorities to the four groups according to the Cahn–Ingold rules (*R*/*S* nomenclature), **2b** shows that L-amino acids have the *S*-configuration. There is only one common exception: L-cysteine (**2**, R = CH_2SH) has the *R*-configuration since CH_2SH takes priority over COOH. Glycine (**2**, R = H) is also exceptional in being the only α-amino acid which is not chiral.

Twenty amino acids (all L-α-amino acids) are built into peptides and proteins according to codes in DNA and these are listed in Table 4.1. The three-letter and one-letter abbreviations shown for these amino acids are commonly used in specifying the sequences of peptides and proteins respectively. Some other amino acids are found in peptides and proteins (including D-amino acids) and usually arise by modification of these DNA encoded amino acids in precursor peptides or proteins. They are sufficiently uncommon not to merit an additional list.

4.1.2 Ionization and Polarity

The pK_a of the COOH group of amino acids is *ca.* 2, which is less than that of the protonated NH_2 group (pK_a *ca.* 9). This means that the carboxylic acid group will protonate the amino group and at pH 7 amino acids having neutral R groups will exist mainly in their neutral but doubly charged zwitterionic forms (**4**). Lesser amounts of the anions (**3**) and/or cations (**5**) will be present but the concentration of the uncharged $RCH(NH_2)COOH$ will be negligible at all pHs. If the side-chain (R) is acidic a protonated form will predominate at neutral pH. On the other hand basic side-chains will favour a negatively charged (anionic) form. This is illustrated by the titration curve for lysine (Figure 4.1). Also shown, midway between the positively and negatively charged species, is the isoelectric

R, CO_2^-, H, NH_2 (**3**) $\underset{OH^-}{\overset{H^+}{\rightleftharpoons}}$ R, CO_2^-, H, NH_3^+ (**4**) $\underset{OH^-}{\overset{H^+}{\rightleftharpoons}}$ R, CO_2H, H, NH_3^+ (**5**)

Table 4.1 *Structures and isoelectric points of amino acids coded for by DNA*

Name	*R in HCR (NH_2)COOH of full structure*	*Abbreviations*	*Isoelectric point*	*Side-chain pK_a, Type**
Alanine	CH_3	Ala, A	6.02	N L
Arginine	HN, NH_2, N, CO_2H, H NH_2	Arg, R	10.76	12.5 +H
Asparagine	CH_2CONH_2	Asn, N	5.41	N H
Aspartic Acid	CH_2COOH	Asp, D	2.98	3.9 -H
Cysteine	CH_2SH	Cys, C	5.02	9.3 -H
Glutamic Acid	CH_2CH_2COOH	Glu, E	3.22	4.3 -H
Glutamine	$CH_2CH_2CONH_2$	Gln, Q	5.70	N H
Glycine	H	Gly, G	5.97	N
Histidine	H, N, N, CO_2H, H NH_2	His, H	7.59	6.1 +H
Isoleucine	Me H, Me, CO_2H, H NH_2	Ile, I	6.02	N L
Leucine	CH_2CHMe_2	Leu, L	5.98	N L
Lysine	$(CH_2)_4NH_2$	Lys, K	9.74	10.5 +H
Methionine	CH_2CH_2SMe	Met, M	5.06	N L
Phenylalanine	CH_2Ph	Phe, F	5.48	N L
Proline	CO_2H, H, N, H	Pro, P	6.30	N L
Serine	CH_2OH	Ser, S	5.68	N
Threonine	Me H CO_2H, HO H NH_2	Thr, T	5.60	N
Tryptophan	CO_2H, N H, H NH_2	Trp, W	5.88	N L
Tyrosine	CO_2H, HO, H NH_2	Tyr, Y	5.67	9.1 -L
Valine	CH_2CHMe_2	Val, V	5.97	N L

* +,- or N (none) give charge if ionized; H = hydrophilic, L = lipophilic.

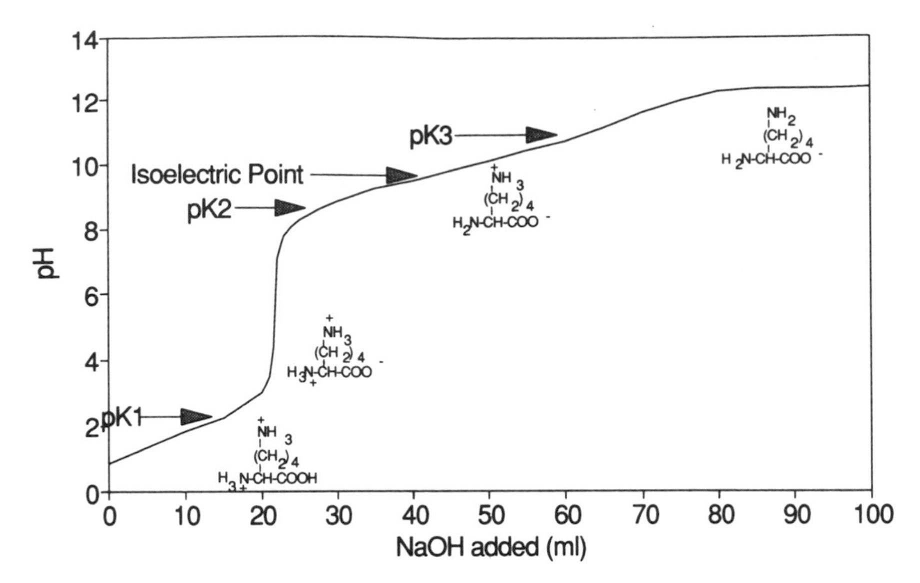

Figure 4.1 *Lysine dihydrochloride (1 M) titrated against NaOH (1 M)*

point—defined as the pH at which the molecule bears no net charge. This value is widely used to aid separation of amino acids from one another.

Since amino acids always carry at least one charged group they are very polar molecules which gives them high (>200 °C) melting points. The polarity also leads to moderate water-solubility for most amino acids although those with aromatic rings have low solubility owing to the lipophilic (non-polar or hydrophobic) nature of these rings. The nature of the side-chain, especially its pK_a, is important in determining the properties of both amino acids and the peptides and proteins in which they are found. From the list in Table 4.1 one can judge which side-chains will be charged at a given pH since the pK_a is the pH at which the group is half ionized. Also important is whether the side-chain readily associates with polar or non-polar structural fragments and an indication of this is given by H (hydrophilic) or L (lipophilic).

4.1.3 Separation of Amino Acids

That each amino acid has a different isoelectric point (Table 4.1) is exploited during separations on ion-exchange columns or by electrophoresis. Ion-exchange columns are packed with acidic or basic materials which are charged at normal pH values and will retain substances of the opposite charge. Thus a mixture of protonated amino acids will be retained on the top of a cation exchange resin, such as sulphonated polystyrene (**6**), while the eluant is acidic. If the pH is raised to the isoelectric point of one of the amino acids it loses its positive charge and is no longer retained with the rest; thus aspartic acid may be eluted at pH 3, glutamic acid at pH 3.2, cysteine at pH 5, and so on.

Electrophoresis is chromatography under the influence of an electric field. The

Polystyrene

(i) conc. H_2SO_4

(ii) NaOH

$SO_3^-Na^+$

6

mixture of amino acids is placed near the centre of a solid support such as paper, permeated with an electrolyte buffered to maintain a given pH. Under the influence of an applied potential, positively charged amino acids migrate towards the cathode while negatively charged ones move towards the anode. The direction and distance moved at that pH under the given conditions are characteristic for the particular amino acid, so permitting identification of each amino acid in the mixture. Ion-exchange chromatography and electrophoresis are also important in the identification and purification of peptides and proteins.

4.1.4 Detection of Amino Acids

A non-destructive method to detect amino acids in solution is to monitor the ultra violet (UV) absorption. This works best with aromatic amino acids since their $\pi \rightarrow \pi^*$ transitions give strong absorptions near 280 nm. Non-aromatic amino acids are detectable by the UV absorption of their carbonyl groups at lower wavelengths ($\pi \rightarrow \pi^*$ transition) but this absorbance overlaps with that of most solvents and so is much less convenient to observe.

Electrophoresis usually produces separated amino acids absorbed on paper or gels; the positions of the amino acids on such solid phases are conveniently made visible by spraying with a reagent such as ninhydrin or fluorescamine to give coloured and fluorescent derivatives respectively. The solutions eluted from ion exchange columns may also be assayed for amino acids by addition of these reagents.

Reaction of amino acids with ninhydrin (indanetrione hydrate, **7**) produces an intense blue–purple colour due to the highly conjugated imine product (**10**). First the amino acid reacts with ninhydrin to form an imine (**8**) which on decarboxylation and hydrolysis gives the amino-indanedione (**9**). This reacts with more ninhydrin to give the highly coloured product. Since dehydration to the imine (**8**) requires the original amino acid to have an NH_2 group, proline does not follow the same pathway and gives a yellow coloured product instead. Detection using ninhydrin is very sensitive but destroys the amino acids.

Detection of amino acids with fluorescamine is even more sensitive since fluorescent derivatives are produced. The reaction of amino acids with fluorescamine (**11**) is probably initiated by nucleophilic attack of the amino acid's NH_2 group on a masked carbonyl group in the reagent. Excess reagent is easily destroyed since it is unstable in water and as little as a picomole (10^{-12} mole) of an amino acid may be detected by observing the light emitted from the derivative at 475 nm when illuminated at 390 nm. Treatment of the fluorescent derivatives with acid hydrolyses them to the original amino acids.

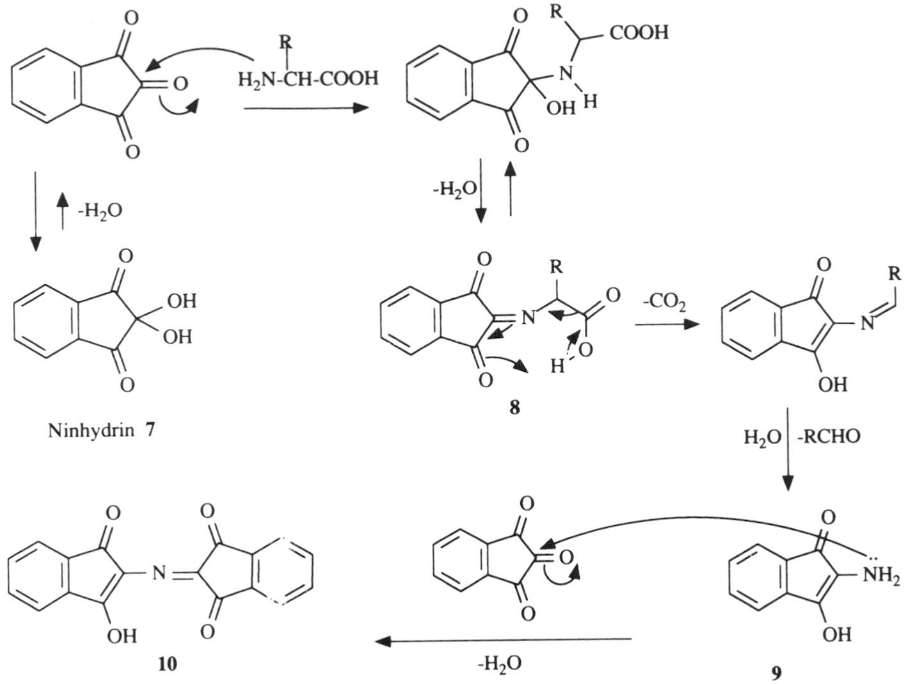

O
R
H_2N-CH-COOH
O
O
O R
COOH
N
OH H
O
$-H_2O$
$-H_2O$
O
OH
OH
O
Ninhydrin **7**
O R
N
O
$-CO_2$
H O
O
8
O R
N
OH
H_2O | -RCHO
O O
N
OH O
10
Purple (λ_{max} = 570 nm) 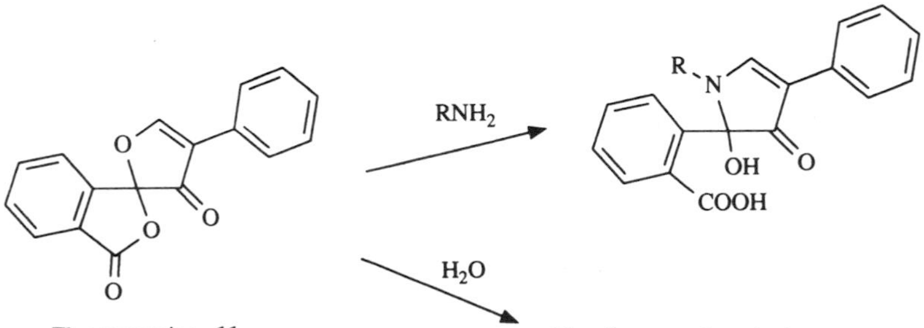
O
O
O
$-H_2O$
O
NH_2
OH
9

O
O
O
O
Fluorescamine **11**
RNH_2
R
N
OH O
COOH
H_2O
Non-fluorescent products

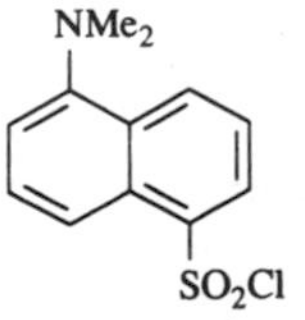

NMe_2
SO_2Cl
Dansyl chloride **12**

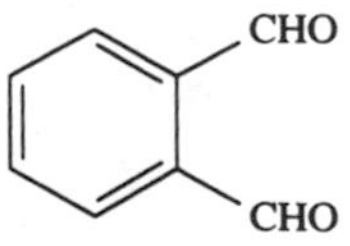

CHO
CHO
o-Phthalaldehyde **13**

Other reagents giving fluorescent derivatives include dansyl chloride (**12**), and *o*-phthalaldehyde (**13**) with mercaptoethanol ($HSCH_2CH_2OH$).

4.1.5 Chemical Synthesis of Amino Acids

All the naturally occurring DNA-encoded L-amino acids are readily available from natural sources but synthesis is useful for rare or isotopically labelled amino acids. Most methods for amino acid synthesis give the amino acid as a racemic mixture and so resolution is required to obtain pure enantiomers. However, direct preparation of either enantiomer by asymmetric synthesis is also possible.

4.1.5.1 Amination of α-Halogenated Acids. The α-position of carboxylic acids may be halogenated and then the halogen displaced by ammonia to give an amino acid. (±)-Alanine for example may be prepared by bromination of propionic acid using bromine with phosphorus followed by amination with ammonia:

$$CH_3CH_2COOH \xrightarrow{P/Br_2} CH_3CHBrCOOH \xrightarrow{NH_3} CH_3CH(NH_2)COOH$$

Provided that the bromopropionic acid is not in excess it reacts only with the ammonia and not the amino acid product since the nucleophilicity of the latter is reduced by the adjacent carboxylate group. At least two moles of ammonia are required per mole of bromopropionic acid since one is used to neutralize HBr liberated. This synthesis is particularly appropriate for amino acids with alkyl side-chains.

4.1.5.2 Alkylation of Malonic Acid Derivatives. The α-position of amino acids can be made sufficiently acidic for alkylation by attaching an additional carboxyl group. Provided the amino and carboxylate groups are protected, introduction of a wide range of alkyl groups is then possible. Starting from dimethyl aminomalonate (**14**), the amino group is protected by acylation, in this case with benzyl chloroformate to introduce the readily removable benzyloxy group (see Section 4.5.1.2), and then alkylated using an alkyl halide and base. The resulting aminomalonic acid derivative (**15**), which is usually crystalline and therefore easily purified, is then hydrolysed in acid to remove the methyl and benzyloxycarbonyl groups. Finally the extra carboxylate group is removed by heating to promote decarboxylation of the β-carbonylcarboxylic acid and yield the required amino acid. Thus using benzyl chloride, $PhCH_2Cl$, with sodium ethoxide in the alkylation step (±)-phenylalanine ($R = PhCH_2$) is produced in good yield.

COOMe
H_2N—C—H
COOMe
14

→ Z(H)N—C(COOMe)(COOMe)—H

RX/NaOEt →

Z(H)N—C(COOMe)(COOMe)—R
15

H^+/Δ →

H_3N—C(COOH)(H)—R
α-amino acid

Z ≡ $PhCH_2O$C(=O)—

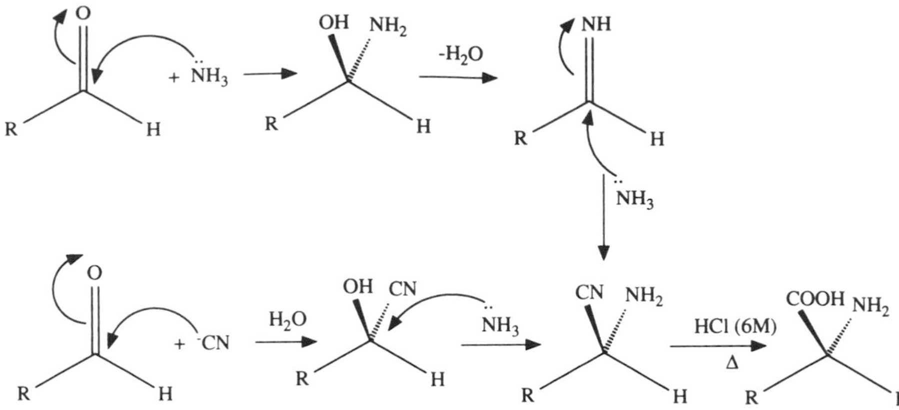

Scheme 4.1 *Strecker synthesis of amino acids*

4.1.5.3 The Strecker Synthesis. Another widely applicable route to α-amino acids, called the Strecker synthesis, reacts an aldehyde with a mixture of ammonia and cyanide. The aldehyde may react first (Scheme 4.1) with either cyanide or ammonia followed by the other reagent to give an aminonitrile (**16**). Some steps are catalysed by acid and others by base but good results are obtained using NH_4CN or NH_4Cl to buffer the reaction. Hydrolysis of the aminonitrile to an amino acid is accomplished by heating with aqueous acid.

Apart from its synthetic value the Strecker reaction is of interest as a possible route for the production of amino acids early in the Earth's history. Aldehydes, ammonia, water, and hydrogen cyanide were all present in the pre-biotic atmosphere and could thus have given the amino acids required for the emergence of simple life.

4.1.5.4 Resolution of Amino Acids. The above syntheses of amino acids give racemic products and separation (known as resolution) is necessary to obtain the two optically active enantiomers. The classical method of resolution is reaction of the mixture with a chiral acid or base to give a pair of diastereomeric salts. Enantiomers have identical physical and chemical properties (unless involving another chiral molecule or influence such as polarized light) but diastereoisomers do not. The salts may therefore be separated by physical means such as crystallization and the pure enantiomers regenerated by addition of mineral acid. Such chemical resolution allows the preparation of large amounts of optically pure amino acids but can be troublesome practically and so enzymic resolution is often employed instead.

A simple example is the enzymic resolution of (±)-leucine by acetylation to *N*-acetyl-(±)-leucine followed by treatment with the enzyme hog renal acylase which removes the acetyl group from the L-enantiomer only. Addition of hydrochloric acid to the mixture gives water-soluble L-leucine hydrochloride with water-insoluble *N*-acetyl-D-leucine making separation straightforward. The high

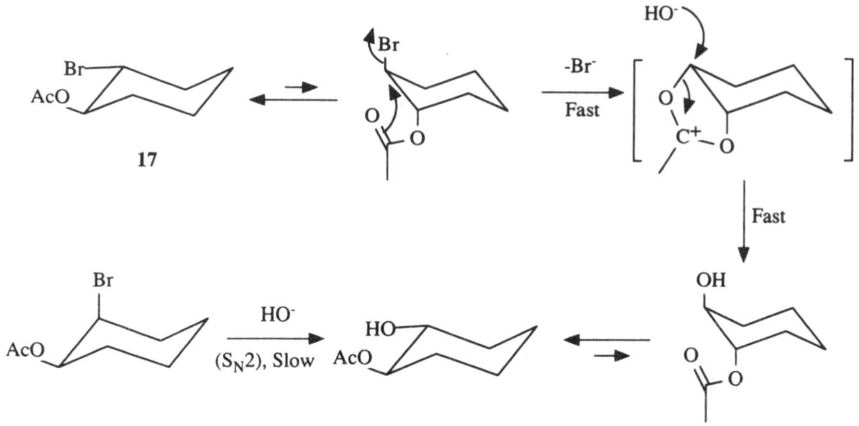

Scheme 4.2 *Neighbouring group participation in the hydrolysis of* trans-*2-acetoxycyclohexyl bromide* (**17**)

selectivity of the enzyme is readily understood. It forms two diastereomeric complexes with the mixture of *R*- and *S*-amino acids and one of the complexes is hydrolysed much faster than the other. Such behaviour is common in simple chemical systems such as the much more rapid hydrolysis of *trans*-2-acetoxycyclohexyl bromide (**17**, Scheme 4.2) than of its *cis*-diastereoisomer, due to neighbouring group participation.

4.1.5.5 Asymmetric Synthesis of Amino Acids. The maximum yield of a desired enantiomer achievable by resolution of a racemic mixture is 50% but if a synthetic route includes a chiral influence the desired enantiomer may be obtained in up to 100% yield; this is asymmetric synthesis. One of the best asymmetric syntheses of amino acids is outlined in Scheme 4.3. The asymmetric synthesis step is the aluminium amalgam reduction of the α-iminocarboxylic acid

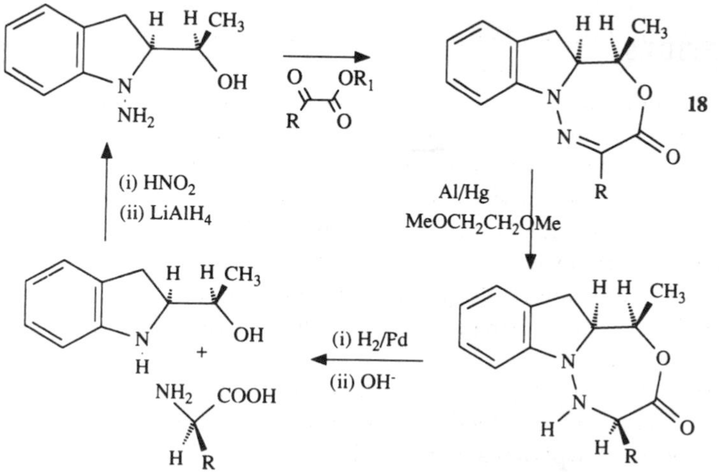

Scheme 4.3 *Asymmetric synthesis of* L-*amino acids*

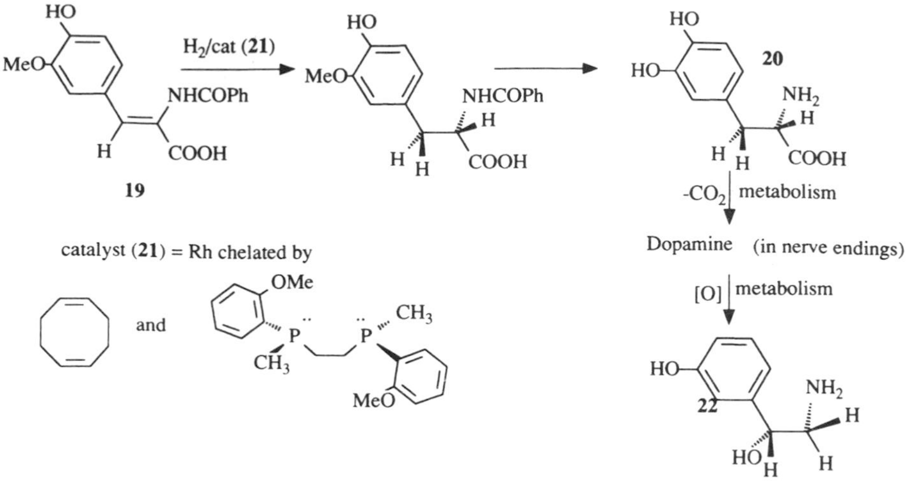

Scheme 4.4 *Asymmetric synthesis of L-dopa* (**20**) *and its conversion into noradrenalin* (**22**)

derivative (**18**) in which the hydrogen adds to the less hindered underside of the ring. This synthesis may be used for a variety of amino acids and the chiral hydrazine starting material may be regenerated for re-use as shown.

The L-α-amino acid L-dopa (**20**), which is formed *in vivo* by hydroxylation of tyrosine, is vital for transmission of nerve impulses and is converted into noradrenalin (**22**) in nerve terminals. It is used to treat Parkinson's disease and so considerable effort has been expended to develop an efficient asymmetric synthesis. One of the best routes is that shown in Scheme 4.4 in which the asymmetric induction occurs during hydrogenation of the alkene (**19**) under the influence of the chiral rhodium catalyst (**21**). The product is produced with less than 5% of D-enantiomer as contaminant.

4.2 PEPTIDES

Linking α-amino acids together, the amino group of one condensing with the carboxylate of another, gives an amide called a peptide. If many amino acids are linked the polyamide (or polypeptide) may be termed a protein. Both peptides and proteins therefore consist of amido groups —NH—CO— ('peptide bonds') joined by —CHR— (the R group being the side-chains of the α-amino acids). The biological properties of peptides and proteins vary greatly, not just with differences in their amino acid sequences, but also with the different three-dimensional shapes which may be adopted by a single polypeptide.

4.2.1 Geometry of the Peptide Bond

The backbone of peptides and proteins consists of a repeated sequence of a three-atom chain, (—CHR—CO—NH—)$_n$. The lone pair of the nitrogen is in conjugation with the carbonyl double bond as shown in Scheme 4.5 causing the

Scheme 4.5 *Conjugation in peptide bonds. Effective overlap between the nitrogen lone pair and the carbonyl group requires coplanarity of the* p*-orbitals of N, C, and O. Consequently the highlighted σ-bonds are all in the same plane*

C—N bond to have partial double bond character. Consequently there is an energetic advantage for each of the atoms attached to the C—N bond to be in the same plane and there is a moderately large barrier to rotation (*ca.* 80 kJ mol^{-1}); there is not free rotation in the peptide chain about NH—CO bonds. Although the atoms immediately attached to the peptide bond are effectively fixed coplanar, there are two, slowly interconverting, arrangements—*s-cis** (Scheme 4.6, **23**) (having the two —CHR— groups adjacent) and *s-trans* (**24**, having the two —CHR— groups on opposite sides). The *trans* form is energetically favoured owing to its lesser steric hindrance between non-bonded atoms, particularly between atoms in R and R_1. The dihedral angle about the peptide N—CO bond

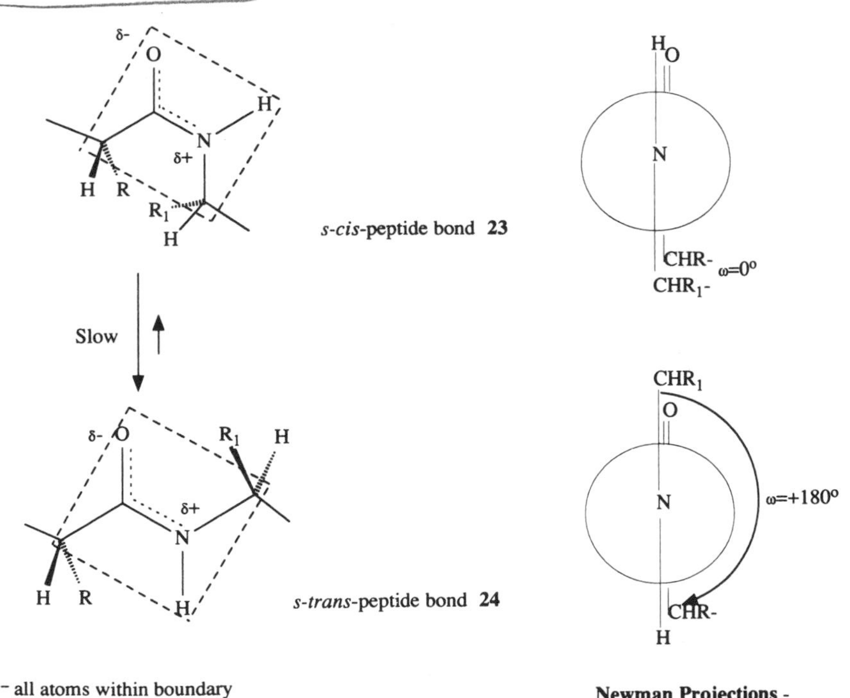

Scheme 4.6 cis- *and* trans-*Isomerism about peptide bonds*

* The *s*- stands for single bond, since *cis* and *trans* normally refer to configuration about a double bond or ring.

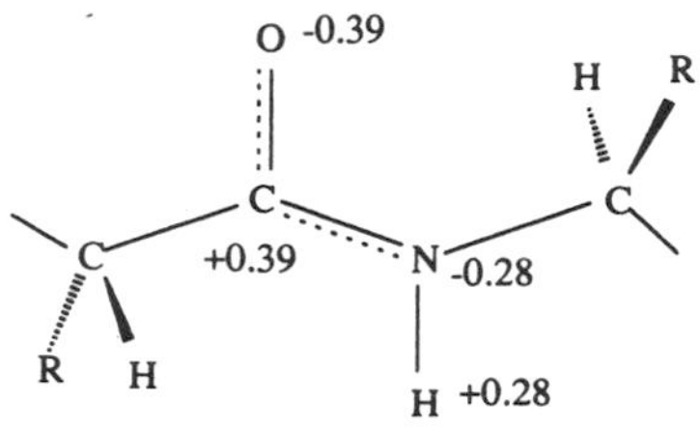

Figure 4.2 *Typical charges on atoms of the peptide bond*

(usually denoted ω) thus takes the values ± 180° (*trans*) or 0° (*cis*) but deformations of up to 20° can be achieved at the expense of only a little energy.

Proline, the only amino acid in which the side-chain is covalently linked to the peptide nitrogen, effectively has two similar alkyl groups attached to its nitrogen and so a peptide formed by attachment of amino acid to its nitrogen can adopt the *s-cis* form much more readily than normal. Proline also has no hydrogen on its nitrogen and so can not hydrogen-bond in the normal way. A proline residue in a protein is frequently the site of an abrupt change in direction of a peptide chain.

Typical charge separation in peptide bonds has been determined and is shown in Figure 4.2. It should be noted that the nitrogen has a negative rather than a positive charge and thus the prediction of properties from resonance structures, which is reliable for partial double bond character, is less dependable for charge densities. The charge separation gives each amino acid an associated dipole which interacts (favourably or unfavourably) with neighbouring dipoles. The peptide backbone is thus quite polar and particularly amenable to hydrogen bonding (Figure 4.3) since the positively charged hydrogen on nitrogen will readily associate with negatively charged carbonyl oxygens.

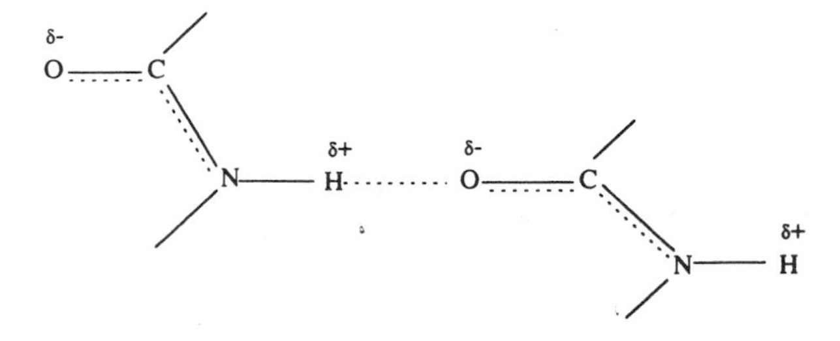

Figure 4.3 *Hydrogen bonding between two peptides*

4.2.2 Chemical Reactivity of Peptides

Amides are much less nucleophilic than amines because the lone pair on nitrogen is conjugated (Scheme 4.5) with the carbonyl and hence not readily available for donation to electrophiles. Peptides are therefore difficult to alkylate or acylate and are only weakly basic. Removal of H^+ from a peptide NH leaves an anion stabilized by the adjacent carbonyl and so amides are also weakly acidic. It requires a pH of *ca.* 16 to remove a proton from the nitrogen and − 1 to protonate a peptide (on oxygen rather than nitrogen) and so the weak acidity and basicity of peptides has little importance biologically. However, it does help explain why the hydrolysis of peptides requires fairly extreme conditions—6 M HCl, 36 hours, 100 °C is typical. Base-catalysed hydrolysis of peptides is also possible but the conditions (2M NaOH, 100 °C) are severe enough to destroy arginine, serine and threonine.

Such weak reactivity is typical of peptides and most other useful reactions of peptides depend on the special nature of neighbouring R groups to boost reactivity. Examples will be given when considering the analysis of peptides in Section 4.4.

4.2.3 Abbreviated Formulae for Peptides and Proteins

Peptides are always composed of α-amino acids linked by peptide bonds to give chains of amino acid residues which may be short (enkephalins have about five amino acids) or long (haemoglobin has four chains each containing between 140 and 150 amino acid residues). To specify the structures of even fairly small polypeptides using chemical formulae is impractical. Full chemical names, formed by concatenation of the names of constituent amino acids, *e.g.* alanylcysteinylphenylalanine for the tripeptide **25**, are little better and so shorthand

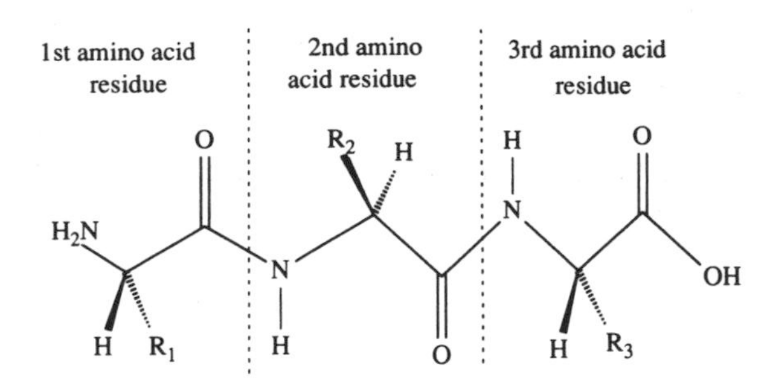

If R_1 = Me, R_2 = CH_2SH, & R_3 = CH_2Ph it is Ala-Cys-Phe (or H-Ala-Cys-Phe-OH).

forms, in which the amino acid abbreviations are linked by bonds (denoted by '-' or '.'), are universally used. To ensure that these abbreviated structures are unambiguous the terminal amino acid with a free amino group is normally placed on the left of the chain; thus the tripeptide **25** is Ala-Cys-Phe and not Phe-Cys-Ala. In situations where ambiguity is still possible an 'H' may be added to the amino acid abbreviation with a free NH_2 group (the *N*-terminal end) and an 'OH' to the carboxylate or *C*-terminal end. Other substituents are added at the appropriate place—thus GlyNH$_2$ means $H_2NCH_2CONH_2$ and $\overset{\text{Me}}{\overset{|}{\text{Glu}}}$ means $H_2NCH(CH_2CH_2CO_2Me)COOH$. Each abbreviated amino acid is assumed to have the L-configuration unless preceded by 'D-'.

4.2.4 Biological Activity of Peptides

Peptides with potent biological activities are found in most organisms. They provide much scope for development of pharmaceuticals, either by mimicking the action of the normal peptide, in which case the drug is known as an agonist, or by blocking the action of the normal peptide (for example by attaching irreversibly to the receptor sites) thus acting as an antagonist.

Sometimes natural peptides are protective in that they kill or deter competing organisms. The cyclic decapeptide gramicidin S (**26**) is a powerful antibiotic produced by certain bacteria, penicillins (see Chapter 8) are antibiotics produced by moulds, and phalloin (**27**) is a highly poisonous cyclic heptapeptide present in the death cap toadstool.

Other peptides are hormones—that is they are produced in one tissue but are transported to another tissue where they elicit their response. The nonapeptide hormone oxytocin (**28**), which is produced in the human pituitary gland in the brain, causes ejection of milk and uterine contraction and is used medically to induce labour. Other examples include gastrin (**29**), a 17-residue peptide which stimulates release of HCl in the stomach; angiotensin II (**30**), an octapeptide causing an increase in blood pressure; luteinizing hormone releasing hormone (LH-RH) (**31**), whose secretion from the hypothalamus gland in the brain controls the release of other hormones from the pituitary; and thyroid stimulating hormone.

Some peptides act as neurotransmitters—they are released at nerve endings and/or evoke a response when encountering a receptor on a nerve cell. It had been suspected that nerve cells possess receptors for pain-reducing compounds because the effect of morphine and its analogues is dependent on their three-dimensional arrangements of lipophilic and ionic sites. In 1975 two pentapeptides with opiate-like activity were isolated from pig brains. These peptides (**32**), called enkephalins, are produced from a larger peptide β-endorphin, and resemble (Figure 4.4) opiates such as morphine in their three-dimensional disposition of groups. Enkephalins are rapidly degraded in the body and so they make poor pain-killing drugs but elucidation of the relation between the solid geometry of opiates and their activity has culminated in the discovery of synthetic analgesics thousands of times more potent than morphine. Different types of nerve cell have different receptors and adjustment of the three-dimensional shape of opiate analgesics to fit receptors responsible for analgesia rather than those for respiratory depression has allowed development of safer pain-killers.

Many unusual amino acids, which can not be formed by modification of the normal 20 amino acids, are found in small peptides. The biosynthesis of such peptides is basically different from that of normal proteins; they are synthesized on an enzyme template rather than using nucleic acids.

L-Pro → L-Val → L-Orn → L-Leu → D-Phe
↑ ↓
D-Phe ← L-Leu ← L-Orn ← L-Val ← L-Pro

Gramicidin S[1,2] **26**

Ala → Tyr → HLeu → Ala
↖ ↙
HPro ← Cys ← D-Thr

Phalloin[1,2] **27**

H-Cys-Tyr-Ile-Glu-Asp-Cys-Pro-Leu-Gly-NH_2

Oxytocin **28**

Glp-Gly-Pro-Trp-Leu-$[Glu]_5$-Ala-
Tyr-Gly-Trp-Gly-Trp-Met-Asp-Phe-NH_2

Human Gastrin I[1] **29**

Asp-Arg-Val-Tyr-Ile-His-Pro-Phe

Angiotensin II **30**

Glp-His-Trp-Gly-Leu-Arg-Pro-Gly

LH-RH[1] **31**

Tyr-Gly-Gly-Phe-Met	Met-enkephalin **32a**
Tyr-Gly-Gly-Phe-Leu	Leu-Enkephalin **32b**

Notes

1. Names and structures of the unusual amino acids in these peptides:

Glp (pyroglutamic acid) H, CO_2H, N, O

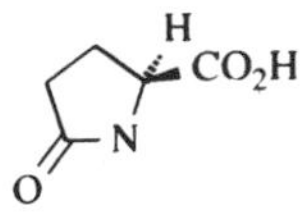

HPro (Hydroxyproline) HO, H, CO_2H, N

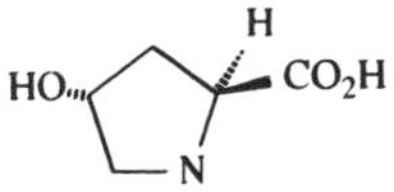

HLeu (Hydroxyleucine) $H_2NCH(CMe_2OH)COOH$ Orn (Ornithine) $H_2NCH(CH_2CH_2CH_2NH_2)COOH$

2. The arrows for peptide bonds indicate the bond direction: CO → NH

Examples of Biologically Active Peptides

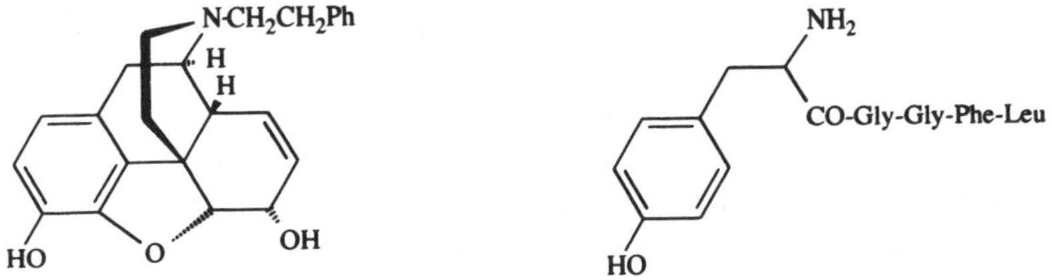

N-CH_2CH_2Ph, H, H, HO, O, OH

Morphine

NH_2, CO-Gly-Gly-Phe-Leu, HO

Leu-enkephalin

Figure 4.4 *Similarity of the Tyr residue of Leu-enkephalin to morphine*

4.3 THREE-DIMENSIONAL STRUCTURES OF PEPTIDES AND PROTEINS

4.3.1 Cross-linking

Many amino acids possess side-chains having reactive groups capable of covalently cross-linking polypeptide chains, but this capability is used with great discretion.

The most important cross-link between peptide chains is a disulphide bond between two cysteinyl (Cys) residues. Such links increase the stability of three-dimensional structures and are frequently found in proteins functioning outside the cells of their synthesis. These cross-links may either make loops in single protein chains or join separate ones together as in insulin which is comprised of two chains, of 30 and 21 residues, joined by two disulphide links. The disulphide bond is formed under mildly oxidizing conditions by reaction of the SH group in Cys with another Cys SH (Scheme 4.7). Oxidation may be by iodine or simply by

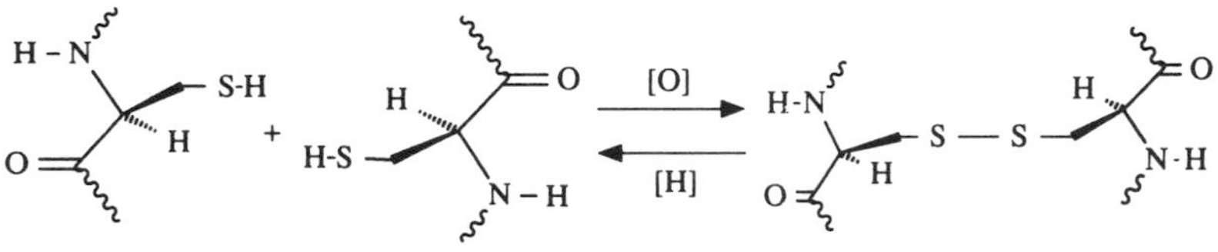

Scheme 4.7 *Formation of disulphide bonds between cysteinyl residues*

air and the latter process is accelerated by the presence of transition metal ions such as Cu^{2+} and Fe^{2+}. Thiols (RSH) are quite acidic compounds (pK_a *ca.* 9) and so formation of the non-acidic and non-polar disulphide gives a marked change in properties. Dimerization of Cys, for example, to the disulphide, cystine, reduces its water solubility by a factor of 10^3.

Cleavage of the disulphide link may be achieved in several ways. Interaction with other thiols, particularly under slightly basic conditions, gives an equilibrium:

$$RS{-}SR + {}^{-}SR^1 \rightleftharpoons RS^- + RS{-}SR^1$$

All the disulphide links of proteins may be broken by treatment with a large excess of a thiol such as mercaptoethanol ($HOCH_2CH_2SH$). A 'perm' or permanent wave in hair is induced by treating hair (a protein) with ammonium thioglycolate ($HSCH_2CO_2^-\ NH_4^+$), curling, and then reforming disulphide links in new positions by treatment with a mild oxidizing agent. Disulphide bonds may also be cleaved by other nucleophiles such as hydroxide:

$$RS{-}SR + OH^- \rightleftharpoons RSOH + RS^-$$

or by reduction to thiols using a chemical reducing agent such as sodium borohydride:

$$RS{-}SR \xrightarrow{[H]} RSH + HSR$$

or by oxidation to sulphonic acids:

$$RS{-}SR \xrightarrow{KMnO_4} RSO_3H + HO_3SR$$

In proteins which circulate in the blood-stream, the only other cross-link which may be found is cyclization of a side-chain onto either the *C*-terminal or *N*-terminal groups. Any other cross-linking so reduces water solubility that the protein is likely to precipitate; this is used to advantage by fibrin which forms amide cross-links by transamination between the $CONH_2$ of Gln and the NH_2 of Lys during blood clotting. Structural proteins such as in hair, wool, skin, and connective tissue (*e.g.* collagen and elastin), have a great variety of cross-links which greatly increase the strength and decrease the solubility of these polymers. Abnormal cross-links, especially those arising from the Amadori reaction (Section 2.3.5) between carbohydrates and proteins, are a feature of ageing, causing the opaque regions in eye lenses known as cataracts, for example.

4.3.2 Hydrogen Bonding and Protein Conformation

The sequence of amino acids together with the position of disulphide links is known as the primary structure of a protein but it does not specify the full structure. Protein chains are coiled or folded in an ordered way (known as secondary and tertiary structure) dependent largely on a maximization of hydrogen bonding between amino acid residues. Proteins having more than one polypeptide chain also pack these together in an ordered way (known as quaternary structure). Three types of conformation are commonly seen—an α-helix, a β-pleated sheet, and the irregular conformations of globular proteins.

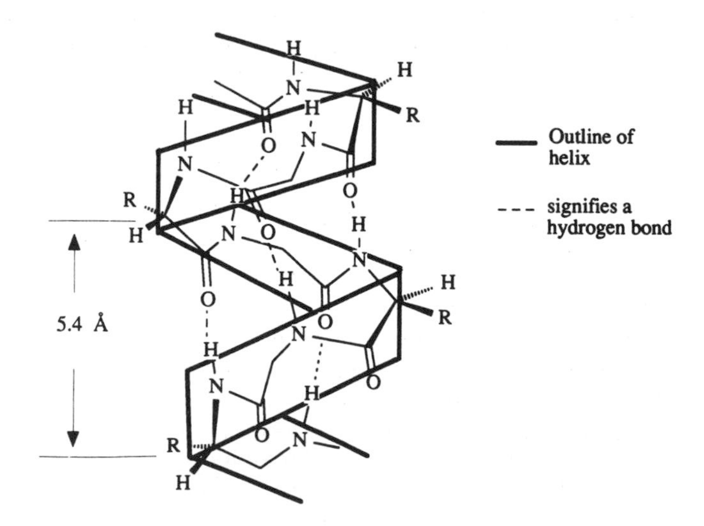

Figure 4.5 *The α-helix. Most R and H groups have been omitted to clarify the main features*

4.3.2.1 Helices. The most abundant conformation in proteins is an ordered coil known as the α-helix. The α-helix (Figure 4.5) is right-handed—if one slid down it one would be constantly turning to the right (no matter at which way up it was)—and the bulky side-chains project outwards from the helix thus experiencing little steric hindrance. The helix has 3.6 residues per turn and the backbone NH of each amino acid residue hydrogen bonds to the carbonyl of the fourth residue towards the *N*-terminal end. Since the first four residues can not hydrogen bond a fairly lengthy helical region is needed to produce a stable, virtually fixed conformation. Proline, having no NH, can not hydrogen bond and its side-chain, being attached to N, interferes with packing in the α-helix. All other L-α-amino acids suit the α-helix well (D-amino acids suit a left-handed helix better) and most proteins adopt this conformation over at least part of their length.

Structural proteins such as α-keratins (in skin, hair, and nail) and myosin (in muscle) have strong fibrous structures composed of coiled coils in which two or three α-helices are wound around each other to give a superhelix with a repeat distance of about 140 Å. The primary structures of these proteins show a regular occurrence of amino acids with non-polar side-chains. This allows favourable close contact between these side-chains at the centre of the superhelix. Side-chains on the outside are usually polar.

Other types of helix are often seen and may have either more or less residues per turn. Collagen forms one of the least coiled helices. It is a fibrous protein of high tensile strength but low elasticity which is the principal constituent of bone, tendons, and ligaments. There is a high proportion of proline residues which is incompatible with α-helix formation, and since proline has low conformational flexibility it also imposes rigidity on the molecule and hence on collagen as a material. The backbone is loosely twisted in a left-handed helix and three of these chains are wound round each other. Every third amino acid is glycine and its small side-chain (R = H) allows the close contact. Disruption of the helix by heating converts collagen into gelatin.

4.3.2.2 β-Pleated Sheets. The most common conformation of proteins after the α-helix is the β-pleated sheet in which protein chains align themselves side-by-side. The alignment may be either parallel (Figure 4.6) with all *N*-terminal groups at the same end or anti-parallel (Figure 4.7) with *N*-terminal ends close to the *C*-terminal ends of neighbouring chains. In either case the chains are held together by hydrogen bonding and the flat peptide regions alternate between two directions to give a pleated shape. In small peptides a parallel sheet is only feasible intermolecularly but an intramolecular antiparallel β-sheet structure becomes likely if a peptide chain undergoes a hairpin bend.

The commonest type of hairpin bend is called the β-turn (Figure 4.8). There is steric crowding in the region of the turn which restricts the type of amino acid which may be present. Glycine, proline, and D-amino acids are favoured in these positions and thus the likely three-dimensional structure of gramicidin S (**26**) can be easily (and correctly) guessed to involve turns at both the -L-Pro-D-Phe-positions with hydrogen bonding between the resulting anti-parallel chains.

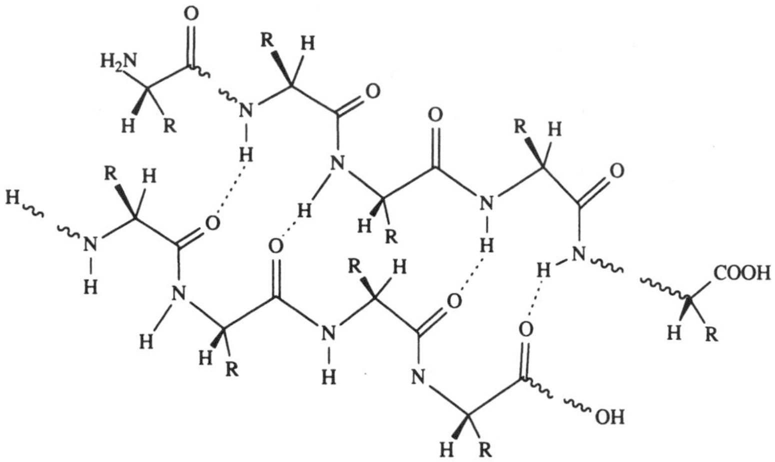

Figure 4.6 *Hydrogen bonding in a parallel* β*-pleated sheet*

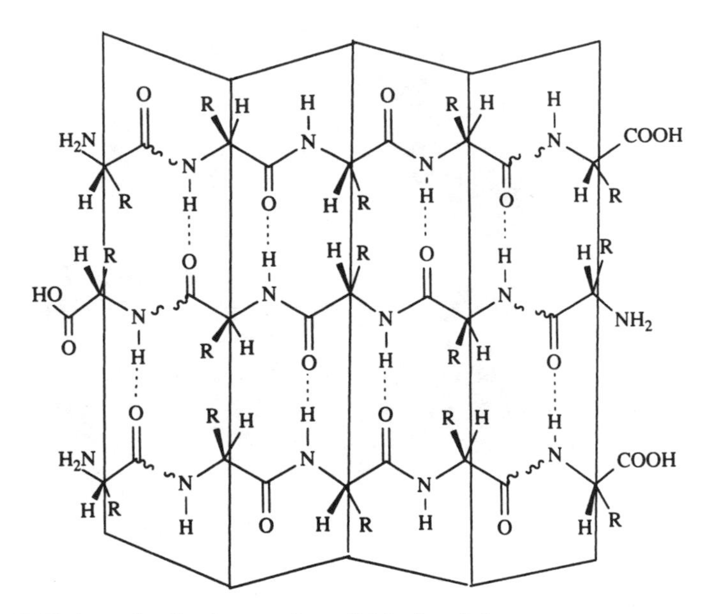

Figure 4.7 *Hydrogen bonding in an anti-parallel* β*-pleated sheet*

4.3.2.3 Conformations of Globular Proteins. Most proteins are water soluble and, having a roughly spherical shape, are termed globular proteins. Their properties do not vary smoothly with external influences, such as temperature or pH; only slight change is observed until a point is reached, often little removed from normal conditions, where an abrupt alteration accompanied by loss of biological function is seen. This sudden change is called denaturation and is due to a change in conformation rather than breakage of covalent bonds.

The structures of many globular proteins have been determined by *X*-ray

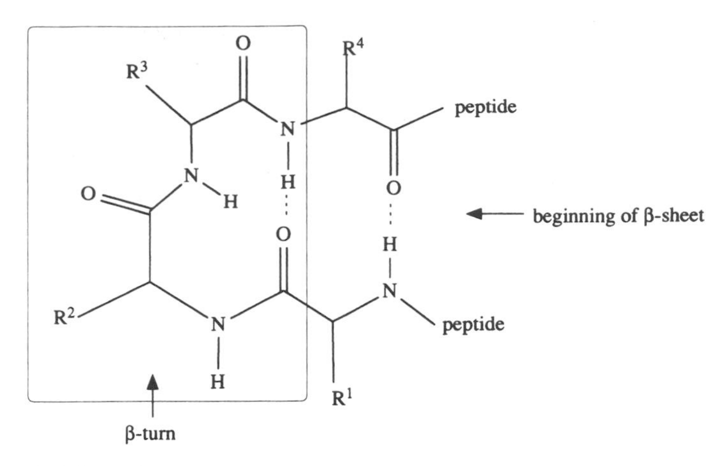

Figure 4.8 *A β-turn in a polypeptide*

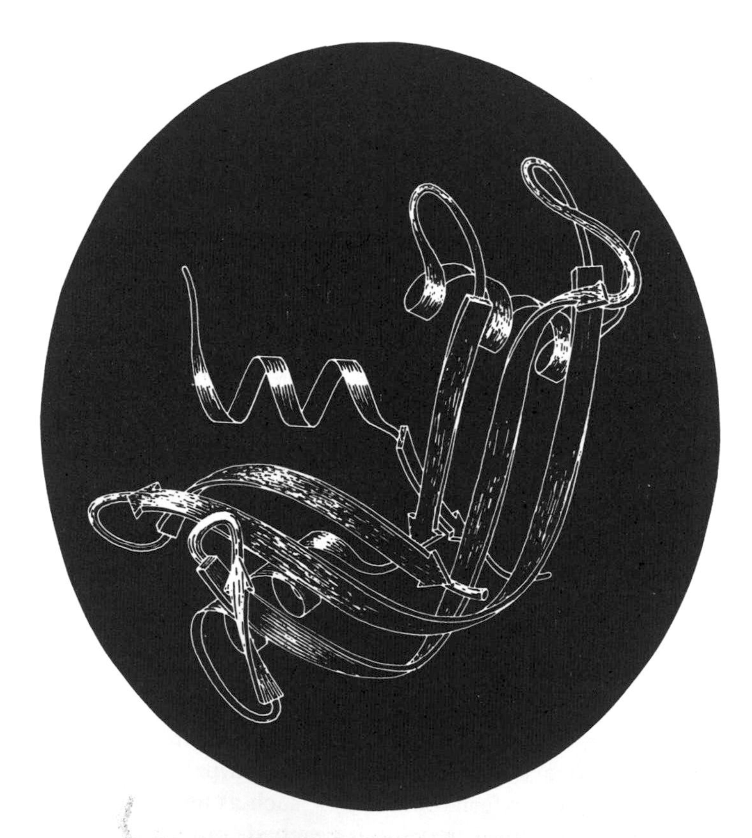

Figure 4.9 *Schematic drawing of the polypeptide backbone of ribonuclease S. Spiral ribbons represent α-helices and arrows represent strands of β-sheet.*
Copyright permission granted by Jane S. Richardson)

crystallography or by NMR spectroscopy in solution. Although their conformations (*e.g.* Figure 4.9) appear irregular it is evident that each globular protein has a specific three-dimensional arrangement which varies little until denatured.

The forces which maintain a protein's conformation are the non-covalent interactions described in Chapter 1 (Section 1.2). Polar groups (ions, amides, OH, *etc.*) occur near other polar groups or on the surface of the protein while non-polar groups cluster together, often near the centre of the molecule. Positive ions such as R_2NH^+ on basic side-chains tend to be close to negative ions such as carboxylate. Many regions show hydrogen bonding and adopt helical or sheet structures described above. In contrast to structural proteins, however, these helices or sheets do not extend far without a change in direction.

4.3.2.4 Enzyme Conformation and Mechanism. Clefts are frequently seen in the structure of globular proteins and in many cases it has been established that these form the active site of the protein. Thus the substrate of an enzyme reaction will enter the cleft and interaction with the specifically orientated groups there causes it to react. After reaction the product generally has less affinity for the active site and is thus released enabling the enzyme to function with fresh substrate.

In a few cases structural studies of an enzyme and its enzyme–substrate complexes have allowed a mechanism to be deduced for the catalytic action of the enzyme. We shall consider one of the best understood group of enzymes, the serine proteases, which cleave amide or ester bonds in peptides and have a serine residue of high reactivity. Serine proteases serve a wide variety of functions including food digestion, blood clotting, hormone production, and fertilization; examples include many and diverse enzymes such as trypsin, chymotrypsin, elastase, and thrombin.

The mechanism followed by such enzyme reactions begins with nucleophilic attack by the OH of the Ser residue on the carbonyl of the ester or amide substrate (Scheme 4.8) to give a tetrahedral intermediate (**33**). Cleavage of peptide or ester bond (C—X) with release of HX (an amine or alcohol derivative) follows, leaving the other half of the substrate bound to the enzyme as an ester.

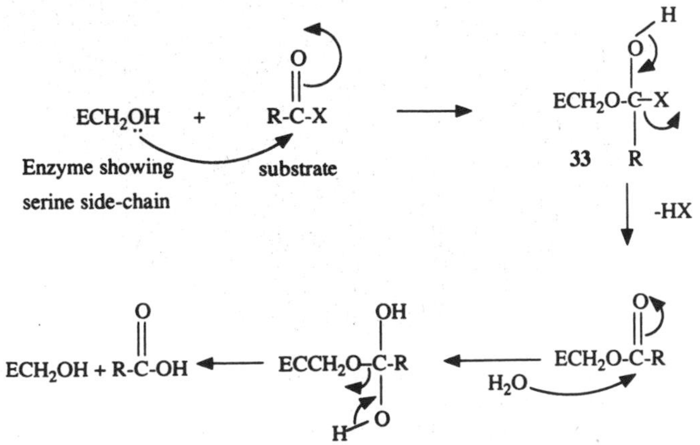

ECH_2OH + R-C(=O)-X → ECH_2O-C(OH)(R)-X (**33**)
Enzyme showing serine side-chain; substrate

↓ -HX

ECH_2OH + R-C(=O)-OH ← $ECCH_2O$-C(OH)(R)-O-H ← ECH_2O-C(=O)-R (H_2O)

Scheme 4.8 *Hydrolysis of an amide or ester substrate by a serine protease enzyme*

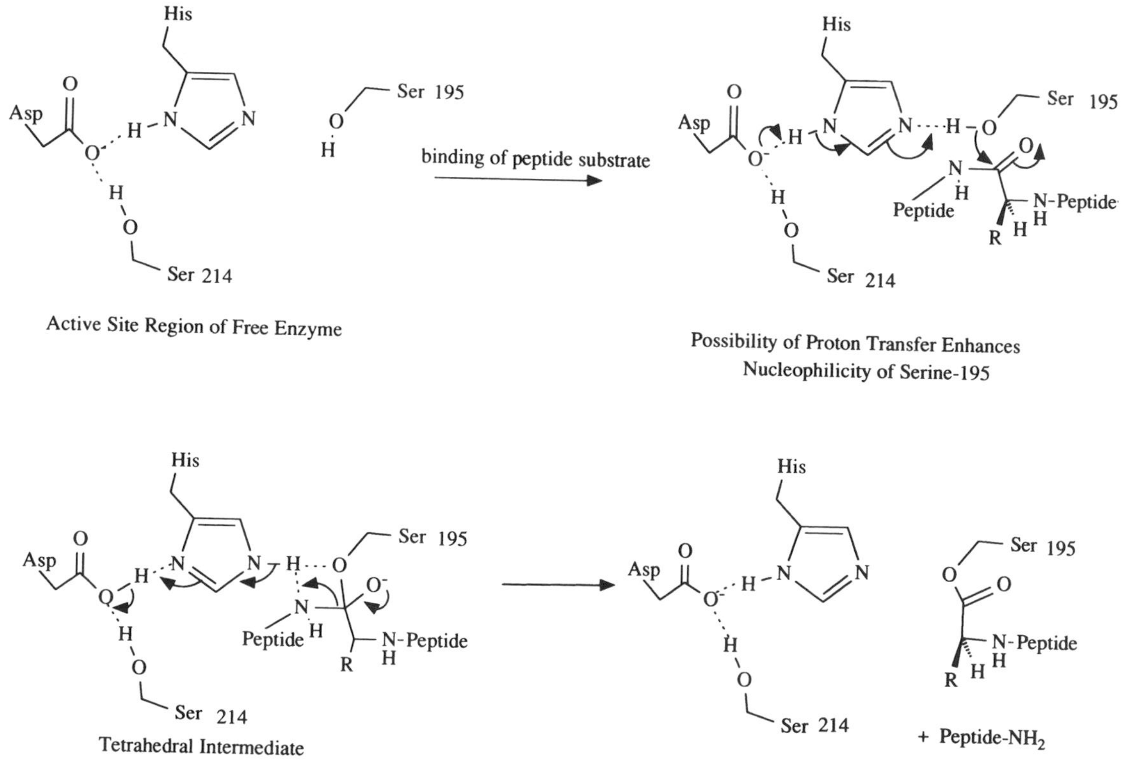
His
Asp
Ser 195
Ser 214
binding of peptide substrate
Active Site Region of Free Enzyme
Peptide
N-Peptide
Possibility of Proton Transfer Enhances
Nucleophilicity of Serine-195
Tetrahedral Intermediate
+ Peptide-NH2

Scheme 4.9 *Neighbouring group participation during chymotrypsin-catalysed hydrolysis*

Attack by a water molecule then regenerates the enzyme and releases the carboxylic acid.

Serine itself would not do this; the activity of the serine in the enzyme is greatly enhanced by initial binding of the substrate in a position ready for attack, and by neighbouring group participation of correctly orientated His and Asp residues. Scheme 4.9 (the residue numbers refer to chymotrypsin) illustrates how participation is thought to occur in the initial steps. The basicity of His-57 is increased by the neighbouring aspartate residue, and the histidine's basic nitrogen is well placed to remove the proton from Ser-195. This increases the nucleophilicity of the serine's oxygen for attack on the substrate carbonyl. The possibility of proton transfer through histidine to or from aspartate also catalyses subsequent steps. Ser-214 is important for holding the aspartate near to His-57.

The manner of binding between substrate and enzyme is dependent on the particular enzyme. In chymotrypsin there is a narrow hydrophobic pocket formed mainly from the loop

```
                    Met(192)-Cys(191)
                              |
                              S
                              |
                              S
                              |
-Gly(216)-Ser-Ser-Thr-Cys(220)
```

into which aromatic residues will fit. This corresponds with chymotrypsin's relative specificity for peptide bonds having Phe, Tyr, or Trp on their carbonyl sides.

4.4 ANALYSIS OF PEPTIDES AND PROTEINS

The steps taken to determine the sequence of a protein by chemical analysis are purification, chain cleavage to polypeptides of less than about 30 residues, successive determination of the amino acids from one end (usually the *N*-terminal end) for each fragment, and finally piecing together the information to give the whole sequence. We will first consider purification then structural elucidation of small peptides, and gradually work up to a full protein structure determination. Finally methods of peptide structure determination using modern physical methods—NMR and mass spectrometry—will be briefly covered.

4.4.1 Purification

Peptides and proteins from biological sources are normally contaminated with many other peptides and proteins. They may be separated using adsorption column chromatography or electrophoresis which exploit differences in pK_a and solubility, or by gel chromatography which exploits differences in molecular weight.

Adsorption chromatography relies on compounds having different extents of adsorption on a solid material when passed over it in solution. For peptides the degree of retention on the solid is most affected by the net charge that the peptide or protein carries at a particular pH.

This is governed by the nature of its many side-chains and since no two proteins have an identical set of side-chains their retention by a solid phase will be different so allowing separation by ion-exchange chromatography or electrophoresis in the same way as previously outlined for amino acids (Sections 4.1.2–3). Ion-exchange columns are suitable for fractionation of unrefined samples since fairly large samples may be separated and they are also used for small samples in high pressure (or performance) liquid chromatography (HPLC). 'Reverse-phase' HPLC columns are also widely employed in the analysis and purification of peptides by HPLC and allow separation of many peptides using a single polar eluant (solvent). Reverse-phase columns contain small particles of silica to which long-chain alkyl groups are covalently attached, thus presenting a hydrophobic surface. Polar molecules have little attraction for this surface and are eluted rapidly while more lipophilic ones are retained longer.

Electrophoresis is normally carried out on a polymer gel soaked in aqueous buffer. Peptides with a net positive charge migrate towards the cathode with a velocity dependent on the magnitude of that charge while negatively charged species migrate in the opposite direction. Electrophoresis is generally used analytically to check for the presence or purity of peptides but instruments for preparative electrophoresis are available and separation on the basis of molecular weight is also possible using a modification of gel electrophoresis as described below.

Chromatography exploiting different pK_a values as described above is good for separating small peptides but proteins are better separated on the basis of their molecular weights. A crude form of molecular weight separation is dialysis. This employs a semi-permeable membrane which allows small molecules to pass through with solvent but obstructs the passage of large molecules. Being simple to perform it is sometimes used to concentrate proteins from crude material and to separate them from small molecules and ions.

A technique with better selectivity is gel filtration. This is a form of column chromatography using a solid support composed of a polymer gel with many pores of chosen size(s). Large molecules are excluded from the pores and are eluted rapidly while molecules able to penetrate the whole gel volume are retained longer. The order of elution has some dependence on other factors such as polarity and so fractionation is not always sharp but can usually give highly purified proteins.

An extensively used and highly discriminating technique for protein fractionation is electrophoresis on a polyacrylamide gel in the presence of sodium dodecylsulphate [$CH_3(CH_2)_{11}SO_3^- Na^+$, SDS] normally referred to as SDS-PAGE. The SDS is a detergent which binds to the surface of proteins, denaturing them to similarly shaped particles bearing a constant net negative charge. Electrophoresis of these complexes causes migration as a function of the protein's molecular weight alone. Usually the technique is used to analyse samples for certain protein components. The samples are loaded onto the top of a

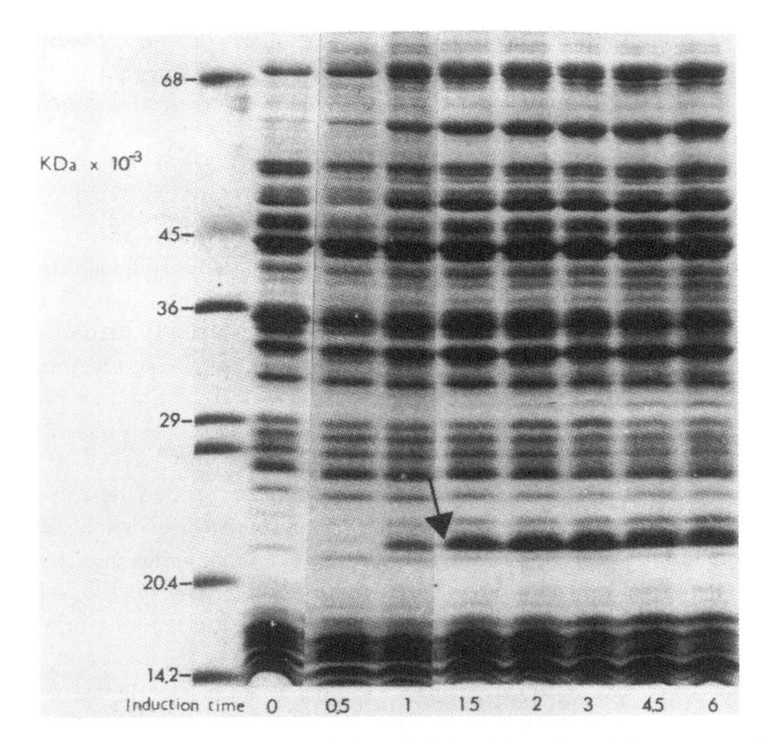

Figure 4.10 *Coomassie blue stained SDS/polyacrylamide gel electrophoresis chromatogram of bacterial proteins. The arrow indicates an extra protein of* 21 kDa *induced by warming the bacterial for the times shown. On the left are marker proteins of known molecular mass* (Courtesy of Dr. M. A. Kaderbhai)

gel and allowed to run downwards in narrow lanes alongside standards. At the end of a run the protein components are stained to reveal individual sharp bands (Figure 4.10) whose distance from the origin is inversely proportional to the logarithm of each protein's molecular mass (larger ones go slower). Molecular masses of proteins are generally given in kilo-Daltons where one kDa is a molecular mass of 1000. Use of thicker gels allows small amounts of (denatured) protein to be isolated by cutting out the desired band and washing the protein off with solvent.

Purification of a naturally occurring peptide or protein is usually a multi-step process. Firstly a crude separation using ion-exchange chromatography is employed to isolate proteins of similar pK_a. This might be followed by gel filtration and/or HPLC for small-scale isolation of pure product whose purity will be checked by SDS-PAGE. Final purification can sometimes be achieved using some biological property such as complexation with an antibody—if the complexing agent can be attached to a solid support in a column, for example, highly selective separations are possible by 'affinity chromatography'.

A final requirement for successful purification is to be able to detect peptides and proteins in solution or on chromatograms. The methods of detection are essentially the same as for amino acids (Section 4.1.4)—UV absorption, and staining. Staining with ninhydrin or fluorescamine is sometimes used but Coomassie Brilliant Blue and particularly silver stains are much more sensitive. In addition highly sensitive and specific biological methods of detecting peptides and proteins are often available.

4.4.2 Determination of Amino Acid Composition

Total hydrolysis of peptides is achieved using 6 M aqueous hydrochloric acid at 110 °C for at least 24 hours in a sealed evacuated tube. Trp is largely destroyed under these conditions and the amido side-chains of Asn and Gln are hydrolysed to acidic side-chains producing Asp and Glu. Cys and cystine residues are also unstable but this problem is usually circumvented by oxidizing them with performic acid to the stable cysteic acid ($R = CH_2SO_3H$) prior to acidic hydrolysis.

Analysis of the liberated amino acids is usually performed by ion-exchange chromatography as previously described. The proportion of each amino acid (except Trp) may thus be ascertained. If only one amino acid is detected and the molecular weight is known the structure may be deduced but in all other cases we still need to know the order in which the amino acids are joined.

4.4.3 *N*-Terminal Residue Determination

The structure of a dipeptide may be found by ascertaining which amino acid is at the *N*-terminal end—that is having its α-amino group free. The first method to determine the *N*-terminal amino acid was to react (Scheme 4.10) the peptide with the Sanger reagent (1-fluoro–2,4-dinitrobenzene). Substitution at a benzene ring normally involves electrophilic attack and loss of H^+. With the Sanger reagent, however, the presence of three strongly electron withdrawing groups allows nucleophilic attack on the aromatic ring by an amino group and displacement of fluoride—a good leaving group. Acidic hydrolysis breaks all amido bonds to give a free amino acid (or many if starting with a polypeptide)

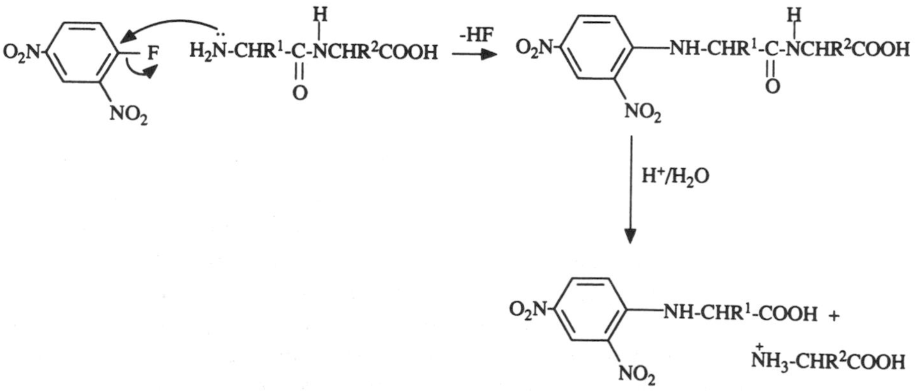

Scheme 4.10 *Reaction of a dipeptide with the Sanger reagent*

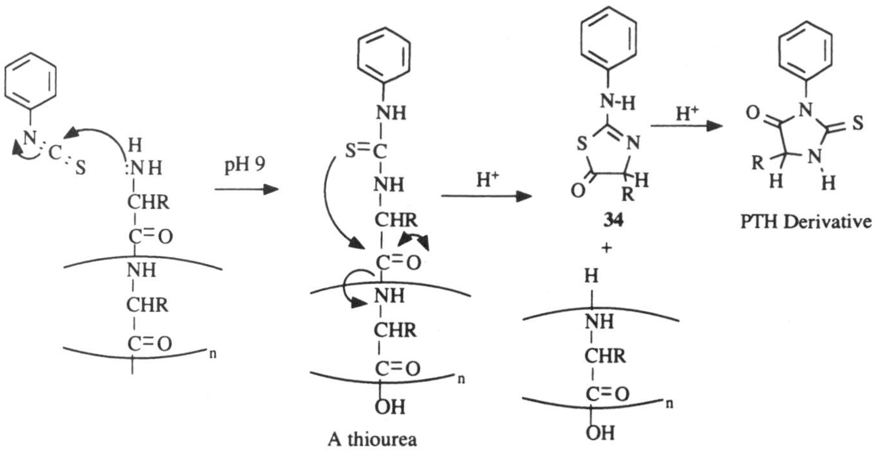

Scheme 4.11 *Edman degradation of a polypeptide*

and one amino acid bearing a dinitrophenyl (DNP) group. Thin layer chromatography then allows rapid identification of the yellow DNP derivative so ascertaining which amino acid was at the *N*-terminal end of the chain. This completes the structural elucidation of a dipeptide but with a longer peptides we still require to know how the remaining amino acids are connected.

Sequential determination of amino acids down a chain may be achieved by the Edman degradation using phenyl isothiocyanate. Again the *N*-terminal amino acid is derivatized but in this case it is possible to hydrolyse selectively to give derivatized *N*-terminal amino acid and leave the rest of the peptide changed only by loss of the single amino acid. The *N*-terminal amino acid is converted into a phenylthiohydantoin (PTH) derivative whose identity may be found by chromatography. The whole cycle, starting with reaction with basic phenyl isothiocyanate, is then repeated to find the new *N*-terminal amino acid. This repetitive process has been automated allowing rapid and reliable determination of the primary structure of peptides up to 30 amino acids; sequences of up 60 amino acids have been established in favourable circumstances.

The initial reaction gives a thiourea which is then hydrolysed to a PTH derivative by a rather complex mechanism. Initially the carbon of phenyl isothiocyanate (Scheme 4.11), which is highly susceptible to nucleophilic attack, reacts with the peptide's free amino group giving a thiourea. Strong acid such as trifluoroacetic acid is then added which rapidly destroys excess phenyl isothiocyanate and promotes nucleophilic attack by the sulphur of the thiourea on the first peptide bond down the chain. This gives the *N*-terminal amino acid as an unstable thiazoline derivative (**34**) which rearranges to a PTH derivative.

Enzymes called exopeptidases may be used to cleave amino acids sequentially from the *N*-terminal (or *C*-terminal) end of a polypeptide. This is a less reliable method of sequencing than the Edman degradation, requiring luck to determine even small structures. The main problem is that if cleavage of the *N*-terminal amino acid is greatly slower than cleavage of the next amino acid, both will be liberated at virtually the same time making it impossible to deduce their order.

4.4.4 Chain Cleavage

The Edman degradation will normally permit structures to be found for polypeptides of up to at least 30 residues, but if longer chains need to be sequenced they should be chopped into smaller portions. Chain cleavage will also be required for cyclic peptides or those in which the *N*-terminal group is blocked by an acetate or other group. Cleavage may be achieved by acid, cyanogen bromide or enzymic means.

Most peptide bonds are stable to 6 M HCl at room temperature but those having an adjacent nucleophilic side-chain are susceptible to hydrolysis. Peptide bonds having aspartic acid on their carboxylate sides are particularly sensitive; the mechanism, involving neighbouring group participation, is shown in Scheme 4.12.

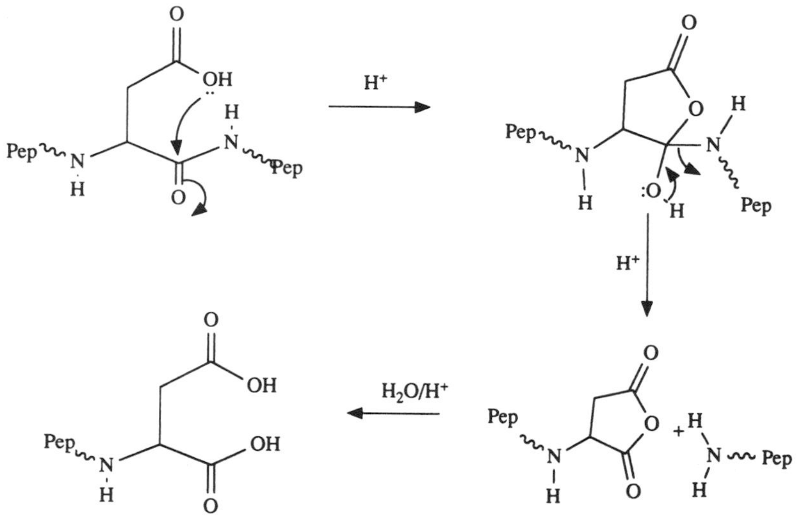

Scheme 4.12 *Peptide cleavage at Asp using* 6M *hydrochloric acid*

The other important chemical method of peptide cleavage, also utilizing neighbouring group participation, is to react with cyanogen bromide (BrCN). Only bonds having methionine residues on their carboxylate sides are cleaved by this treatment and, since Met occurs in about one in sixty residues, fairly long chains are produced. Cyanogen bromide normally reacts with sulphides, R—S—R, giving an alkyl thiocyanate and an alkyl bromide (Scheme 4.13) but in methionyl peptides the carbonyl of the peptide bond acts as a nucleophile instead of the bromide and the peptide bond is transformed into the readily hydrolysable iminium ion (Scheme 4.14).

Enzymes may also be used to cleave peptide bonds and there is a vast range of such peptidases with varying selectivities for different bonds. The two most commonly used are trypsin, which cleaves on the carboxy side of basic amino

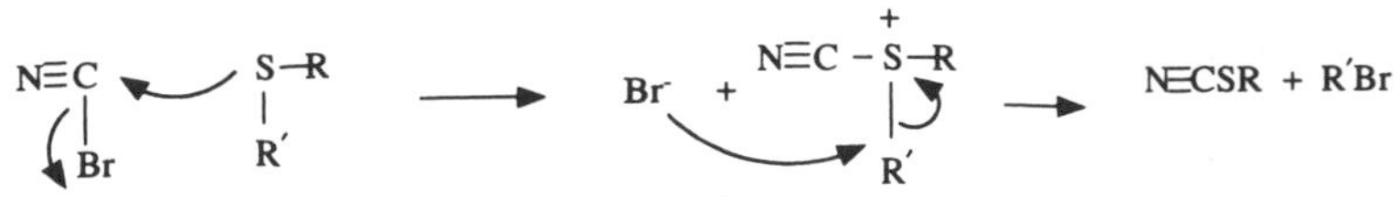

Scheme 4.13 *Reaction of cyanogen bromide with a sulphide*

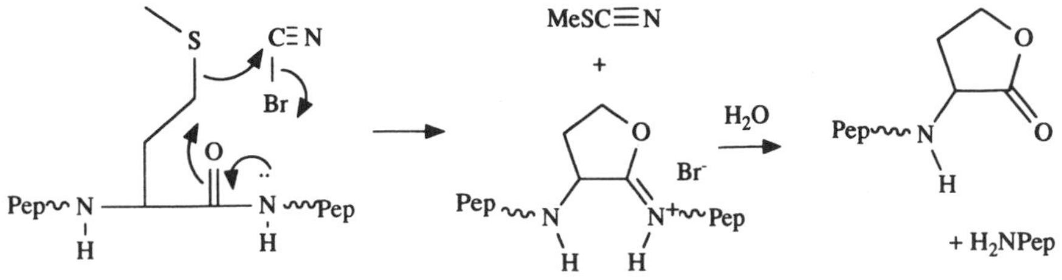

Scheme 4.14 *Peptide cleavage next to met using cyanogen bromide*

acids (Lys, Arg), and chymotrypsin which cleaves on the carboxy side of aromatic residues (Phe, Tyr, Trp).

4.4.5 Mass Spectroscopy

The traditional type of mass spectrometry (electron impact MS) in which electrons are fired at the sample to produce positively charged ions is of little use for peptides since they are insufficiently volatile. However, supporting the sample in a glassy matrix and bombarding it with fast uncharged atoms (FAB-MS) produces large numbers of ions possessing little excess energy, even when using involatile samples.

Peptide molecules protonated on their amido groups are volatilized by FAB-MS and will be accelerated through the spectrometer to be detected as the molecular ions or as smaller fragments. Being low energy ions, they fragment chiefly at peptide bonds to produce stabilized ions (Scheme 4.15) with loss of the *C*-terminal parts as uncharged molecules (which will not be focused onto the detector). The charged *N*-terminal fragments lose carbon monoxide and the resulting stable ions give a mass spectrum having peaks corresponding to the relative molecular masses (−45 for COOH) of each of the constituent peptides. For small peptides the primary structure may usually be deduced directly from a

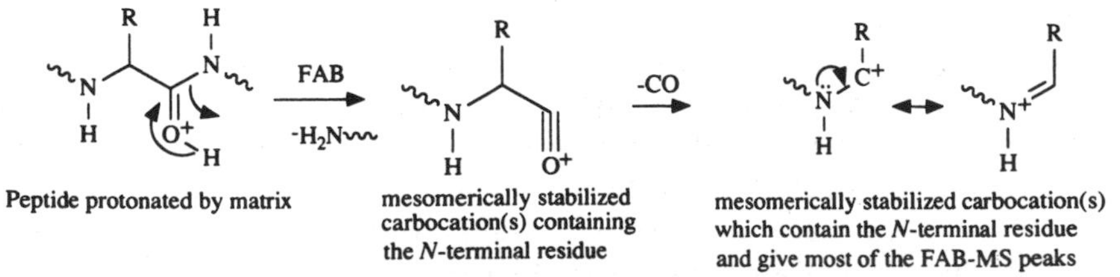

Peptide protonated by matrix

mesomerically stabilized carbocation(s) containing the *N*-terminal residue

mesomerically stabilized carbocation(s) which contain the *N*-terminal residue and give most of the FAB-MS peaks

Scheme 4.15 *Fragmentation during FAB-MS*

single spectrum requiring only a few micrograms of sample. In larger molecules alternative fragmentations are more numerous and will obscure the main peaks of interest allowing only partial assignment of the sequence.

4.4.6 Nuclear Magnetic Resonance

NMR spectroscopy does not provide a simple method of finding the primary structure of a protein but proton NMR is extremely valuable for probing three-dimensional structures. Usually peptides for study will have a known sequence allowing the ^1H NMR peaks to be assigned comparatively easily. Weak irradiation at the resonance frequency of one proton will cause changes in intensity of other peaks from protons close in space. This technique is called NOE spectroscopy and allows the three-dimensional structure of proteins in solution to be deduced.

4.5 SYNTHESIS OF PEPTIDES

The starting materials for peptide synthesis will, of course, be α-amino acids. At first sight one might imagine that merely reacting two amino acids together would give a dipeptide but this approach is excluded by two problems. Firstly the reaction is slow and reversible. Heating an amine and a carboxylic acid with removal of water may produce an amide—as in the production of nylon:

$$\underset{\text{Adipic acid}}{HOOC(CH_2)_4COOH} + \underset{\text{Hexamethylene diamine}}{H_2N(CH_2)_6NH_2} \xrightarrow{280\ °C} \underset{\text{Nylon 6,6}}{HO[CO(CH_2)_4CONH(CH_2)_6]_nH}$$

—but the reaction conditions are incompatible with the stability of the amino acids' side-chains. Secondly each amino acid has both an amino and a carboxylate group and so polymerization and self condensation will compete with dipeptide formation. To synthesize peptides one must increase the reactivity for peptide-bond formation while decreasing the reactivity at competing sites.

A general approach to the synthesis of a dipeptide is shown in Scheme 4.16. The reactivity of the carboxylate group of one amino acid is boosted by an activating group X. To prevent reaction with its own amino group to give a polymer, an amino-protecting group is first attached. The amino-protected carboxylate-activated amino acid may then be reacted with the other amino acid to give a dipeptide.

A problem at this stage is that the free amino acid is zwitterionic and its NH_3^+ group is not nucleophilic. Addition of base will convert it into the free NH_2 group but this carries the risk of racemizing amino acid derivatives (Scheme 4.17). Often a better answer is to protect the carboxylate group of the second amino acid (with Y) allowing its amino group to exist in the absence of base. An additional advantage is that the carboxylate-protected amino acid will be more soluble than the free amino acid in the non-polar solvents required for reaction with the activated amino acid derivative.

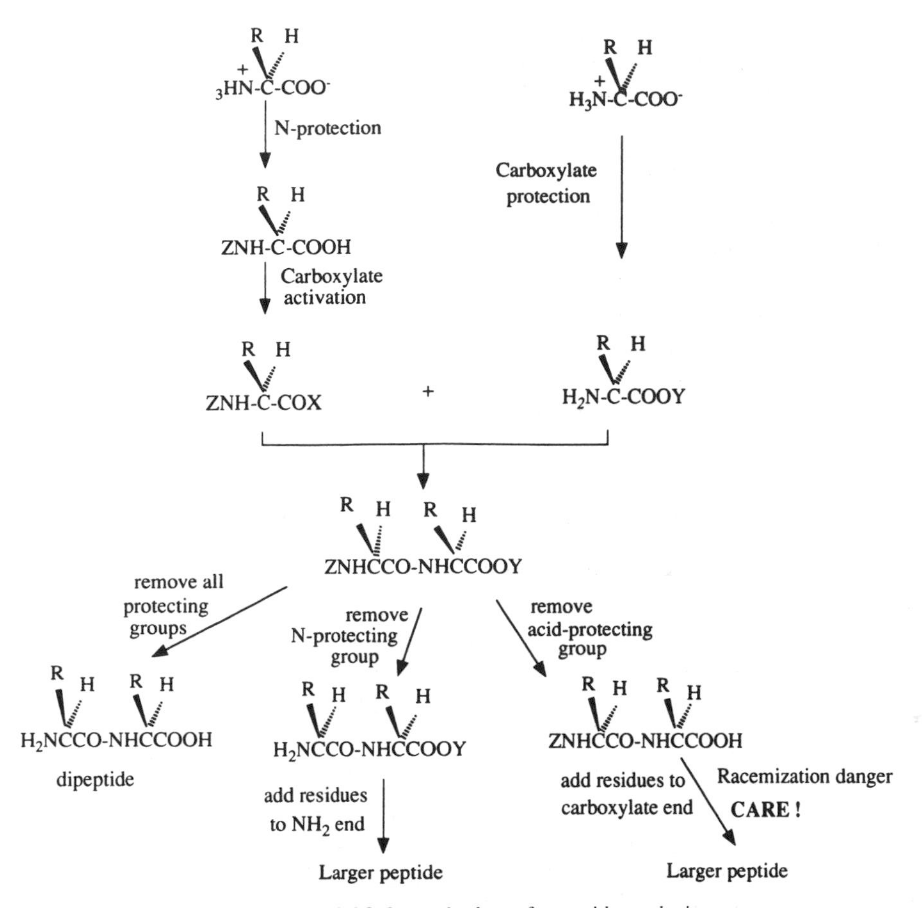

Scheme 4.16 *General scheme for peptide synthesis*

resonance stabilization of anion

-H⁺

+H⁺

Equilibrium shifted to left if X=O⁻ or OH (dianions disfavoured)

Equilibrium shifted to right if
X is electron withdrawing,
base is present, or
acid is present (protonates carbonyl)

racemized

Scheme 4.17 *Racemization of amino acids*

4.5.1 Nitrogen-Protecting Groups

4.5.1.1 Acyl Groups and the Racemization Problem. The simplest way to protect an amino group is to convert it into an amide by reaction with an acidic anhydride or acyl halide. Conjugation of the nitrogen's lone pair with the carbonyl group in an amide such as an *N*-acetate ($RNHCOCH_3$) or *N*-benzoate (RNHCOPh) greatly reduces its nucleophilicity. This method is not usually applicable in peptide synthesis since *N*-acyl groups react with carboxylate-activated amino acids (Scheme 4.18) to give an azlactone. The formation of the azlactone (or oxazolone) does not prevent reaction with a second amino acid—it is another type of activated carboxyl—but it does promote racemization of the amino acid. Its deleterious nature arises from facile interconversion with the aromatic enol which has mirror symmetry. Tautomerization of the enol back to the asymmetric keto form of the azlactone may produce either the *R*- or the *S*-isomer so resulting in partial racemization at this site.

Not only does this preclude the general use of acyl groups to protect the nitrogen during peptide synthesis, but it also means that extending a peptide from the carboxylate end is likely to cause racemization at the activated amino acid residue. This should be obvious by considering the R group in Scheme 4.18 to be the *N*-terminal part of the peptide chain to be extended. Where extension from the carboxylate end is necessary (joining two preformed peptides for example) the amino acid to be activated should be either glycine or proline. Glycine is not chiral and proline is not prone to racemization because its cyclic form with no NH in peptides precludes formation of the oxazolone.

4.5.1.2 Carbobenzoxy, Butyloxycarbonyl, and Fmoc Groups. Protection of the NH_2 of an amino acid with an alkoxycarbonyl group, R^1OCO—, gives a urethane, R^1OCO—NH—CHR—COOH, and these are not easily racemized. It seems that an alkoxy-oxazolone may form but the alkoxy group discourages its deprotonation by mesomeric electron donation.

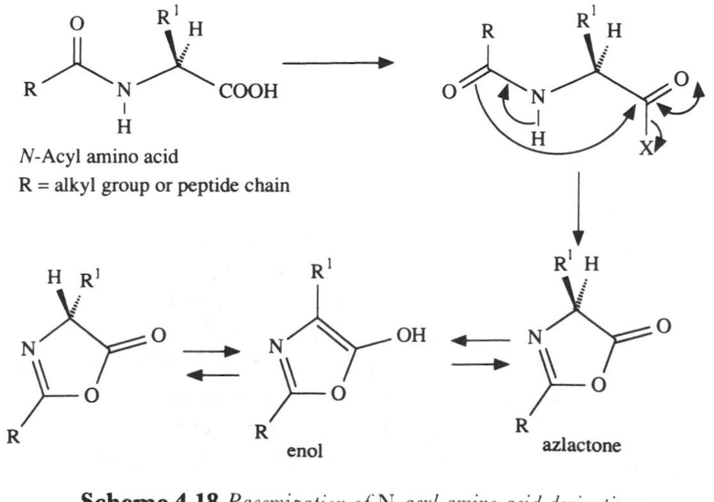

N-Acyl amino acid
R = alkyl group or peptide chain

enol

azlactone

Scheme 4.18 *Racemization of* N-*acyl amino acid derivatives*

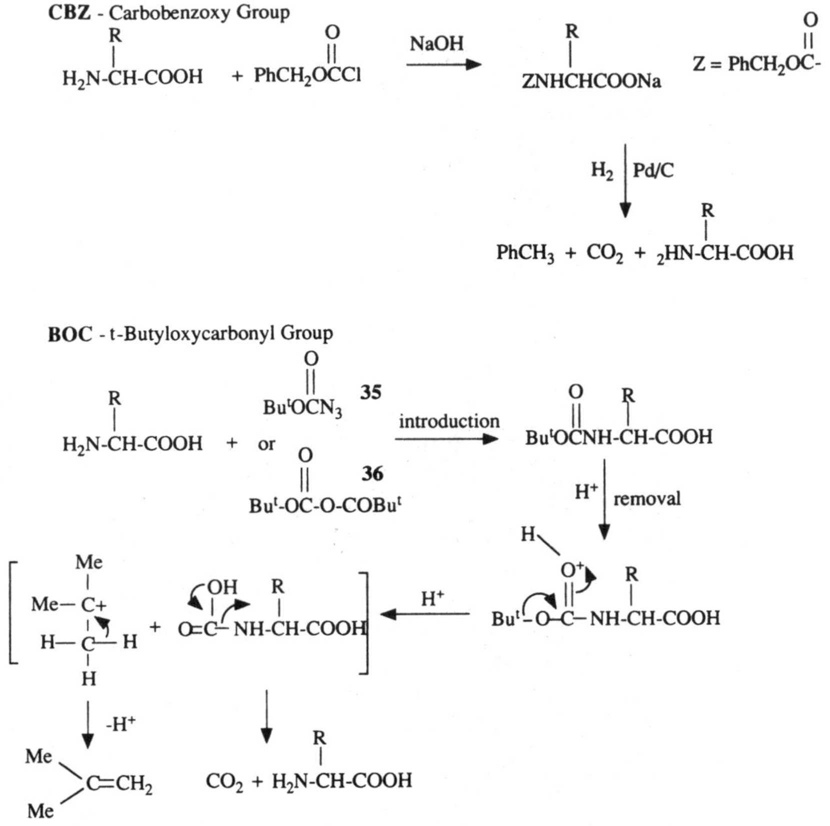

CBZ - Carbobenzoxy Group

$H_2N\text{-}CH(R)\text{-}COOH + PhCH_2OC(O)Cl \xrightarrow{NaOH} ZNHCH(R)COONa$ $\quad Z = PhCH_2OC(O)\text{-}$

$\xrightarrow{H_2\ Pd/C} PhCH_3 + CO_2 + H_2N\text{-}CH(R)\text{-}COOH$

BOC - t-Butyloxycarbonyl Group

$H_2N\text{-}CH(R)\text{-}COOH + Bu^tOC(O)N_3$ (**35**) or $Bu^t\text{-}OC(O)\text{-}O\text{-}C(O)OBu^t$ (**36**) $\xrightarrow{\text{introduction}} Bu^tOC(O)NH\text{-}CH(R)\text{-}COOH$

$\xrightarrow{H^+ \text{ removal}} Bu^t\text{-}O\text{-}C(=O^+H)\text{-}NH\text{-}CH(R)\text{-}COOH \xrightarrow{H^+} [Me_2C^+\text{-}CH_3 + HO\text{-}C(=O)\text{-}NH\text{-}CH(R)\text{-}COOH]$

$Me_2C^+\text{-}CH_3 \xrightarrow{-H^+} Me_2C{=}CH_2$

$HO\text{-}C(=O)\text{-}NH\text{-}CH(R)\text{-}COOH \rightarrow CO_2 + H_2N\text{-}CH(R)\text{-}COOH$

Scheme 4.19 *Introduction and removal of alkoxycarbonyl N-protecting groups*

Amino-protection with a carbobenzoxy group (CBZ, Z, or benzyloxycarbonyl group) is achieved by reaction of the amino acid with benzyl chloroformate and sodium hydroxide (Scheme 4.19). A mild and simple way to remove the group is hydrogenation, for example over palladium on charcoal at normal temperature and pressure. This produces toluene and carbon dioxide as easily removable by-products. The main limitation to hydrogenolysis is that it will often desulphurate methionine residues to α-aminobutyric acid residues. The CBZ group may also be removed using dry hydrogen bromide in acetic acid but this also gives side-reactions with methionine due to attack of the benzyl carbocation intermediate on sulphur. Reductive removal of CBZ using sodium in liquid ammonia is probably the cleanest method when methionine is present.

The *N*-t-butyloxycarbonyl (BOC) group is also commonly used for amino-protection. It is introduced using t-butyl azidoformate (**35**) or the anhydride (**36**) since the chloroformate is explosively unstable. The BOC group is normally removed using acid, such as cold trifluoroacetic acid or even acidic ion-exchange resin. The BOC group's extreme lability towards acids is due to the relative stability of the t-butyl carbocation produced as an intermediate in the reaction (see Scheme 4.19).

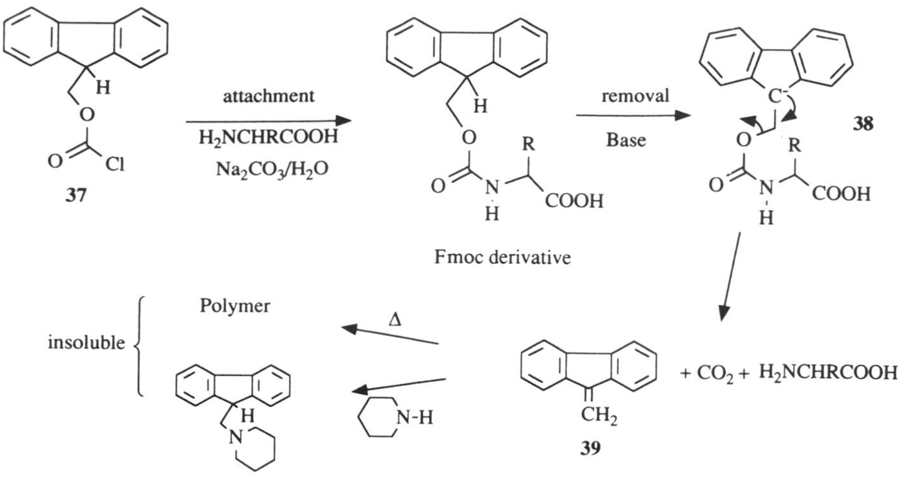

Scheme 4.20 *Attachment and removal of fluorenylmethoxycarbonyl derivatives*

A modern alternative for N-protection is the fluorenylmethoxycarbonyl group, almost always called Fmoc. It is attached (Scheme 4.20) using the chloroformate (Fmoc-chloride, **37**) in aqueous sodium carbonate. Fmoc is easily removed with mild base such as 20% piperidine in DMF to give the free amino acid and dibenzofulvene (**39**) which either polymerizes or reacts with piperidine to give water insoluble by-products. The ease of hydrolysis results from an $E1_{CB}$ elimination from the carbanion (**38**) which has aromatic stability (cyclopentadienyl anion has six π-electrons). Fmoc derivatives strongly absorb light near 280 nm which makes them convenient for use in HPLC analysis of amino acids.

4.5.2 Carboxylate Protection

Carboxylate protection is achieved by conversion into an ester. Generally a mixture of the amino acid and a large excess of an alcohol is treated with thionyl chloride or dry HCl (Scheme 4.21).

All esters may be hydrolysed using alkaline hydrolysis but more selective methods are available for removal of certain esters. Amongst those frequently used are benzyl, and the more acid-resistant 4-nitrobenzyl, esters which are readily removed by hydrogenolysis and t-butyl esters which are usually removed by dry trifluoroacetic acid.

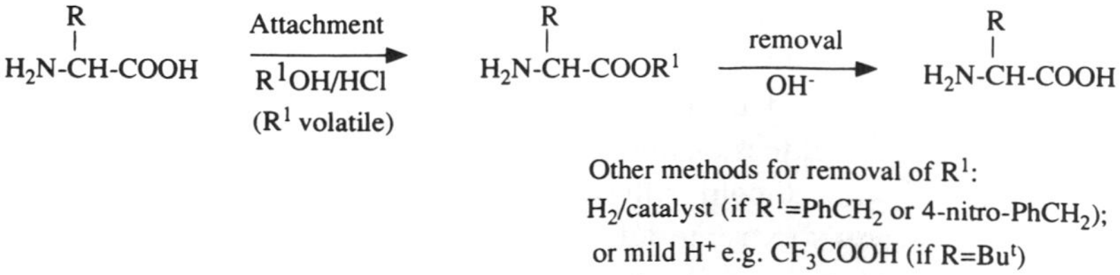

$H_2N\text{-}CH(R)\text{-}COOH \xrightarrow[\text{R}^1\text{OH/HCl (R}^1\text{ volatile)}]{\text{Attachment}} H_2N\text{-}CH(R)\text{-}COOR^1 \xrightarrow[\text{OH}^-]{\text{removal}} H_2N\text{-}CH(R)\text{-}COOH$

Other methods for removal of R^1:
H_2/catalyst (if $R^1=PhCH_2$ or 4-nitro-$PhCH_2$);
or mild H^+ e.g. CF_3COOH (if $R=Bu^t$)

Scheme 4.21 *Carboxylate protection–ester formation*

4.5.3 Side-chain Protection

Side-chains which usually need protection are the amino and guanidino groups of lysine and arginine, the carboxylates of aspartic and glutamic acids, and the nucleophilic thiol and imidazole groups of cysteine and histidine respectively. Methods for protection are described below and are the same or very similar to those already described but must be modified to ensure that the side-chain groups react in preference to the α-amino or *C*-terminal carboxylate groups. Separate protecting groups of different stability to those at the *C*- and *N*-terminal ends are usually employed so that the end groups may be exposed selectively for coupling of further amino acids. The side-chain protecting groups are usually kept in place until the end of the synthesis.

Lysine, like most amino acids, forms a chelated complex (Scheme 4.22) with copper salts. This blocks the α-amino group and so allows the ε-amino group to be selectively protected, by reaction with benzyloxycarbonyl azide for example. Liberation of the side-chain-protected lysine is accomplished by reaction with hydrogen sulphide. Arginine's side-chain may be protected similarly but, being much more basic, just maintaining the pH below about 12 keeps it protonated and is often sufficient for protection.

The side-chain carboxyl groups of Asp and Glu are further from the electron withdrawing amino group than the other carboxyls. Consequently they are more easily protonated, allowing selective esterification under acidic catalysis. Thus reaction of CBZ-Glu with one molar equivalent of t-butyl alcohol and anhydrous HCl gives CBZ-Glu.
|
BOC

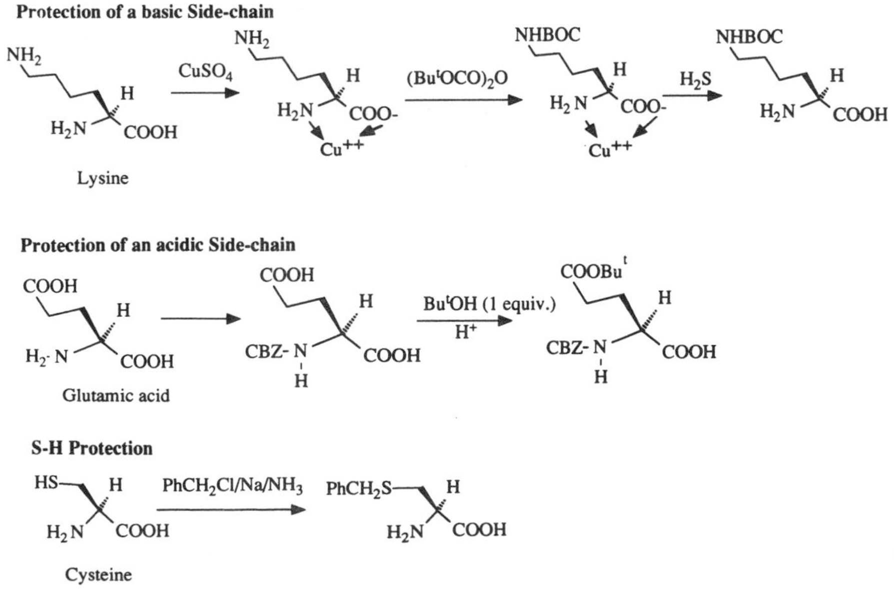

Protection of a basic Side-chain

NH_2, H, H_2N, COOH (Lysine) —$CuSO_4$→ NH_2, H, H_2N, COO-, Cu^{++} —$(Bu^tOCO)_2O$→ NHBOC, H, H_2N, COO-, Cu^{++} —H_2S→ NHBOC, H, H_2N, COOH

Protection of an acidic Side-chain

COOH, H, H_2N, COOH (Glutamic acid) → COOH, H, CBZ-NH, COOH —Bu^tOH (1 equiv.), H^+→ $COOBu^t$, H, CBZ-NH, COOH

S-H Protection

HS, H, H_2N, COOH (Cysteine) —$PhCH_2Cl/Na/NH_3$→ $PhCH_2S$, H, H_2N, COOH

Scheme 4.22 *Side-chain protection*

The nucleophilic SH group of cysteine and NH of histidine may be blocked with benzyl by reaction (Scheme 4.22) with sodium and benzyl chloride in liquid ammonia. It is possible to remove benzyloxycarbonyl groups on a peptide in the presence of *S*-benzylcysteine residues by hydrogenation over palladium in liquid ammonia.

4.5.4 Carboxylate Activation

There are many methods of carboxylate activation. A high level of activation gives rapid peptide bond formation (often known as coupling) but all reactive side-chains must be protected. Since a good carboxylate-activating group is one able to stabilize a negative charge it will also encourage deprotonation (Scheme 4.17) of the amino acid and so increase the danger of racemization. If the carboxylate end of a peptide is being activated oxazolone formation will also be encouraged with additional racemization. Using mild activation racemization is unlikely, and carboxylate and side-chain protection may be unnecessary, but coupling may be slow or incomplete.

4.5.4.1 N,N-*Dicyclohexylcarbodi-imide (DCC or DCCD)*. DCC provides one of the highest levels of activation and so all amino and carboxylate groups not involved in the peptide bond formation must be protected. The free carboxylate group reacts to form an *O*-acylisourea (the activated derivative; Scheme 4.23, **40**) which reacts with the free amino group to give the peptide and insoluble dicyclohexylurea. There is no need to prepare the activated derivative separately; the two protected amino acid derivatives are mixed in an aprotic solvent such as dichloromethane and DCC is added.

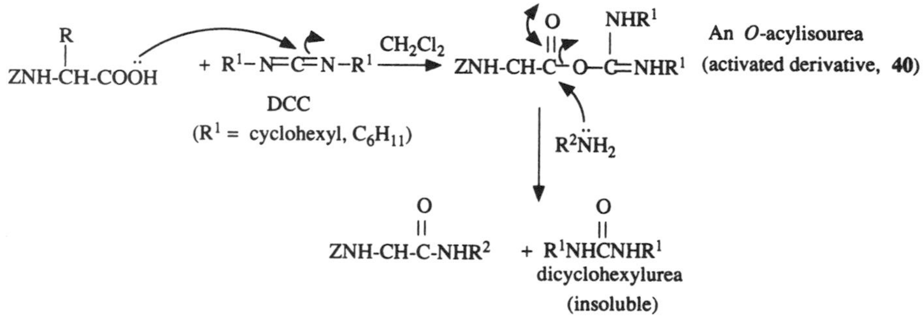

ZNH-CH(R)-COOH + R^1–N=C=N–R^1 (DCC, R^1 = cyclohexyl, C_6H_{11}) $\xrightarrow{CH_2Cl_2}$ ZNH-CH-C(=O)–O–C(=NR^1)–NHR^1 (An *O*-acylisourea (activated derivative, **40**))

$\downarrow$ R^2NH_2

ZNH-CH-C(=O)-NHR^2 + R^1NHC(=O)NHR^1 dicyclohexylurea (insoluble)

Scheme 4.23 *Peptide synthesis activated by dicyclohexylcarbodi-imide (DCC)*

4.5.4.2 Activated Esters. Most esters are only weakly reactive towards nucleophiles. Electron withdrawing groups on the alkoxy part of the ester will boost reactivity because the corresponding alkoxide ion will be stabilized thus becoming a good leaving group.

4-Nitrophenyl esters (**41a**, Scheme 4.24) are frequently used to give a fairly mild degree of activation to carboxylate groups for peptide formation. They may be

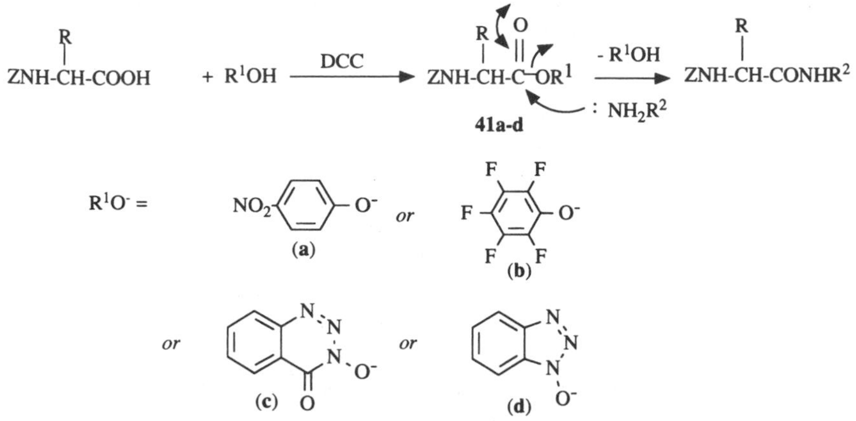

ZNH-CH(R)-COOH + R^1OH —DCC→ ZNH-CH(R)-C(=O)-OR^1 (**41a-d**) —($-R^1OH$; : NH_2R^2)→ ZNH-CH(R)-$CONHR^2$

R^1O^- = NO_2-C_6H_4-O^- **(a)** *or* C_6F_5-O^- **(b)** *or* **(c)** *or* **(d)**

Scheme 4.24 *Peptide coupling using activated esters*

prepared by DCC-promoted condensation of the amino acid with 4-nitrophenol. Their reaction with a free amino group in another amino acid or peptide causes phenoxide ion to be liberated and its bright yellow colour in slightly basic solution allows a simple method of monitoring the extent of reaction.

There are many alternative activated esters in common use including the pentafluorophenyl, dihydro-4-oxo-1,2,3-benzotriazin-3-yl (Dhbt), and benzotriazol-1-yl esters (**41b–d**).

4.5.4.3 Anhydrides. Symmetrical anhydrides (**42a**, Scheme 4.25) of *N*-protected amino acids may be used for synthesizing peptides but one half of the anhydride acts as the leaving group and will not be incorporated into the peptide. To avoid this waste one may use an unsymmetrical or mixed anhydride designed so that the nucleophilic attack is on the amino acid portion. Attack on the other carbonyl of the anhydride may be disfavoured by steric hindrance, as in diphenylacetic (**42b**) and pivolyl (**42c**) mixed anhydrides, or by electron donation, as in the carbonic acid anhydrides (**42d**). Phosphite derivatives (**42e**) are another type of mixed anhydride which sustain nucleophilic attack at the amino acid carbonyl.

4.5.4.4 Azides. The method of activation which is reputed to be the mildest and least likely to cause any racemization is formation of the acid azide. It is too slow for general use but is the method of choice for couplings having especially high risk of racemization such as the coupling of two peptide fragments.

Introduction of the acid azide is a multi-step sequence via the hydrazide (**43**) but does not always require isolation of each intermediate (Scheme 4.26). The preparation and subsequent coupling reaction of an acid azide must not involve elevated temperatures since it will undergo the Curtius rearrangement to give the isocyanate (**44**) which will react with amino groups to give ureas.

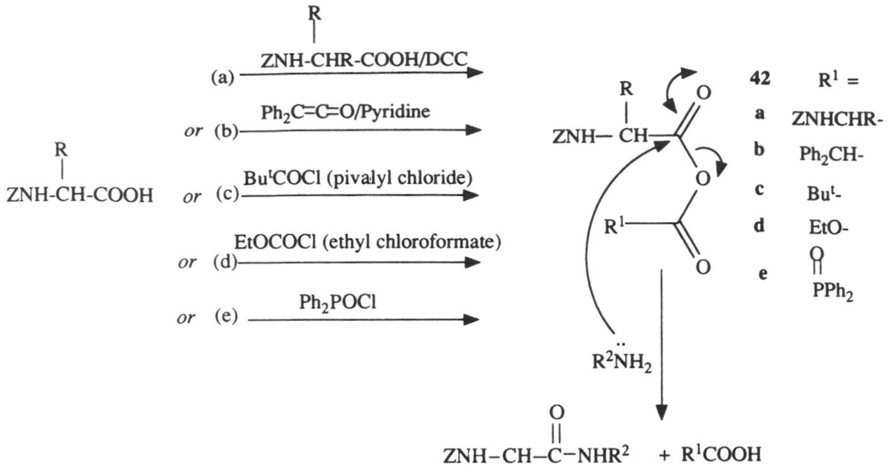

$$\text{ZNH-CH(R)-COOH} \xrightarrow[\text{or (b) } Ph_2C{=}C{=}O/\text{Pyridine; or (c) Bu}^tCOCl \text{ (pivalyl chloride); or (d) EtOCOCl (ethyl chloroformate); or (e) } Ph_2POCl]{\text{(a) ZNH-CHR-COOH/DCC}} \text{ZNH-CH(R)-C(=O)-O-C(=O)-R}^1 \ \mathbf{42} \xrightarrow{R^2\ddot{N}H_2} \text{ZNH-CH-C(=O)-NHR}^2 + R^1COOH$$

42	R^1 =
a	ZNHCHR-
b	Ph_2CH-
c	Bu^t-
d	EtO-
e	$P(=O)Ph_2$

Scheme 4.25 *Peptide synthesis* via *symmetrical* (a) *or unsymmetrical* (b–e) *anhydrides*

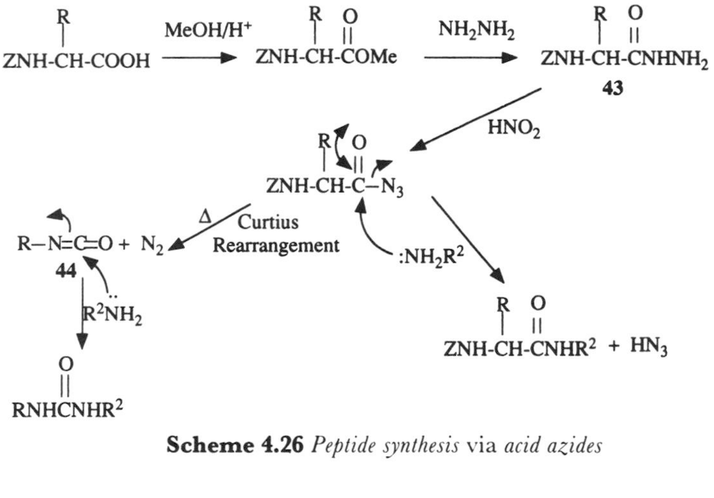

$$\text{ZNH-CH(R)-COOH} \xrightarrow{MeOH/H^+} \text{ZNH-CH(R)-C(=O)OMe} \xrightarrow{NH_2NH_2} \text{ZNH-CH(R)-C(=O)NHNH}_2 \ \mathbf{43} \xrightarrow{HNO_2} \text{ZNH-CH(R)-C(=O)-N}_3$$

$$\text{ZNH-CH(R)-C(=O)-N}_3 \xrightarrow{:NH_2R^2} \text{ZNH-CH(R)-C(=O)NHR}^2 + HN_3$$

$$\text{ZNH-CH(R)-C(=O)-N}_3 \xrightarrow[\text{Curtius Rearrangement}]{\Delta} \text{R-N=C=O} \ \mathbf{44} + N_2 \xrightarrow{R^2\ddot{N}H_2} \text{RNHC(=O)NHR}^2$$

Scheme 4.26 *Peptide synthesis* via *acid azides*

4.5.5 Synthesis of the *C*-Terminal Tetrapeptide of Gastrin

Gastrins are natural hormones which stimulate gastric secretion and contain around 17 amino acid residues. Testing of many synthetic peptide analogues has showed that just the *C*-terminal tetrapeptide amide (the *C*-terminal end is $CONH_2$ rather than COOH) is required to give the physiological action of the natural hormones. This tetrapeptide amide (**46**) is fairly challenging to make since it contains Asp (requiring side-chain protection) and Met (hydrogenolysis of protecting groups unsatisfactory). One of the best methods for its synthesis is shown in Scheme 4.27. Retention of aspartic acid side-chain protection (as the t-butyl ester) throughout DCC-promoted couplings led to poor yields and so

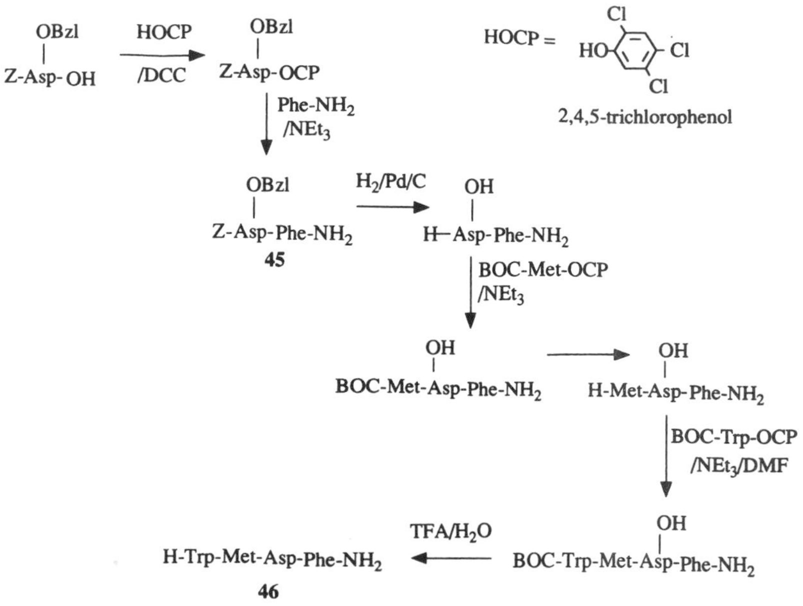

OBzl | Z-Asp-OH —HOCP/DCC→ OBzl | Z-Asp-OCP

HOCP = 2,4,5-trichlorophenol (HO, Cl, Cl, Cl)

↓ Phe-NH_2/NEt_3

OBzl | Z-Asp-Phe-NH_2 **45** —H_2/Pd/C→ OH | H-Asp-Phe-NH_2

↓ BOC-Met-OCP/NEt_3

OH | BOC-Met-Asp-Phe-NH_2 ——→ OH | H-Met-Asp-Phe-NH_2

↓ BOC-Trp-OCP/NEt_3/DMF

H-Trp-Met-Asp-Phe-NH_2 **46** ←TFA/H_2O— OH | BOC-Trp-Met-Asp-Phe-NH_2

Scheme 4.27 *Synthesis of the C-terminal tetrapeptide of gastrin*

activated ester couplings were used allowing no protection of the β-carboxylate
OBzl
|
of Asp for most steps. Thus Z-Asp was activated as its trichlorophenyl (CP) ester and reacted with phenylalanine amide (Phe-NH_2) to give the dipeptide derivative (**45**). Hydrogenation removed Asp's side-chain-protecting benzyl group and the benzyloxycarbonyl (Z) group. *N*-t-Butyloxycarbonyl-L-methionine (BOC-Met) was activated as its trichlorophenyl ester and reacted with the dipeptide derivative to give the N-protected tripeptide amide. The BOC group was removed using trifluoroacetic acid to reveal the free amino group which was coupled with the trichlorophenyl ester of BOC-tryptophan. Treatment with trifluoroacetic acid dislodged the BOC group to give the desired product.

4.5.6 Solid-phase Synthesis

Custom designed synthesis of peptides using solution chemistry, as above, is very laborious, especially the purification of each intermediate by crystallization or otherwise. Another approach is to attach the growing peptide to a polymer which renders it insoluble and therefore easy to isolate even when excess reagents are used to encourage complete reaction at each stage. An early version of such solid-phase synthesis on polystyrene is shown in Scheme 4.28. The α-amino groups are protected with BOC, side-chains with benzyl derivatives, and activation is by DCC. It was used to synthesize bovine pancreatic ribonuclease having 124 amino acids in 17% overall yield. The great strength of the method is the excellent yields rapidly obtained without the need for top-class synthetic skills. A

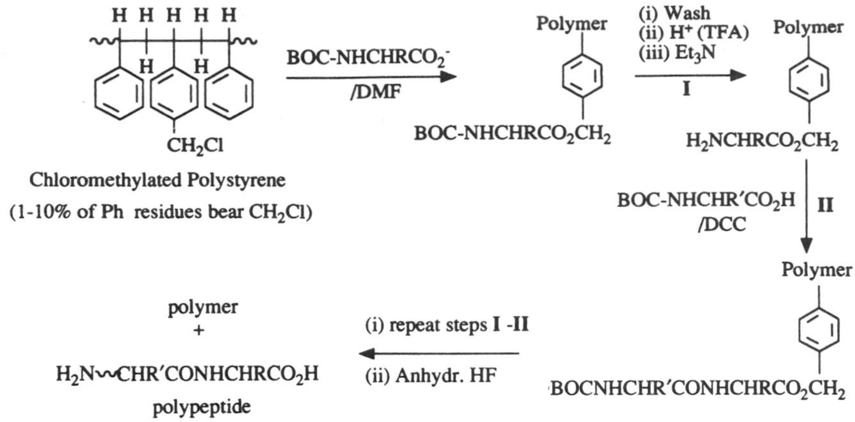

Scheme 4.28 *Solid-phase synthesis of a polypeptide*

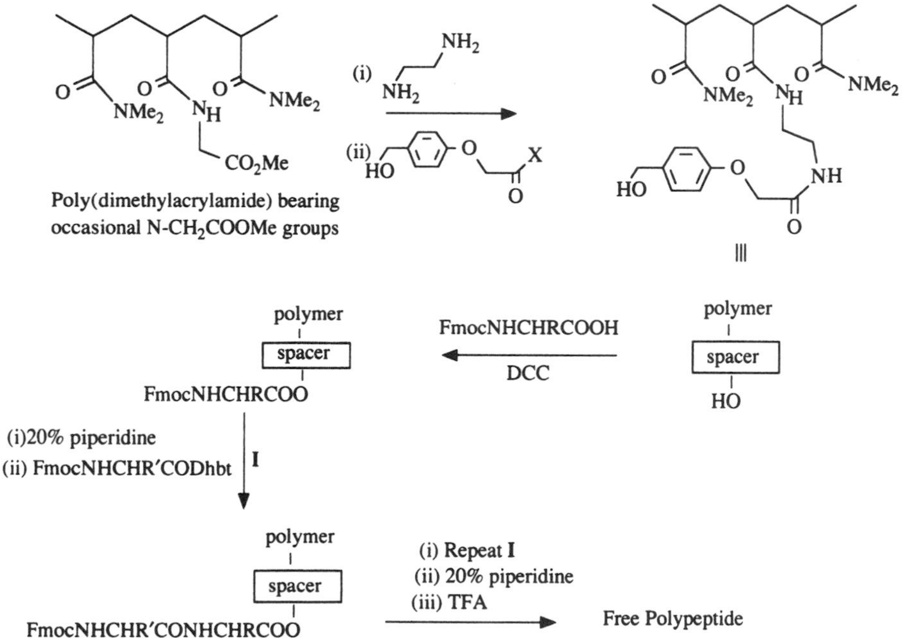

Scheme 4.29 *Solid-phase synthesis using Fmoc methodology*

weakness is the use of the rather harsh conditions—DCC as coupling agent and HF for final release from the polymer—which may cause some racemization or degradation of the peptide.

A more modern version (Scheme 4.29) of solid-phase peptide synthesis links the growing peptide via a spacer group to a polyacrylamide resin which is readily permeable by solvents like DMF and may be detached by mild acid treatment. The *N*-terminal protecting groups must then be removed under non-acidic conditions and the base-labile Fmoc group (Section 4.5.1.2) is commonly employed. Side-chain protection may be by t-butyl groups which will be detached on the final acid treatment. Activation is usually by Dhbt esters (Section 4.5.5.2), which is less harsh than using DCC and the yellow colour produced on coupling allows simple monitoring of the reaction.

Solid-phase peptide synthesis is so simple, fast, and repetitive that automated peptide synthesizers have been made and are available commercially. A remaining limitation is that only small quantities of polypeptide (0.01–0.1 g) may be made. Synthesis of moderate amounts of polypeptide is normally tackled by solution chemistry methods. Large amounts of polypeptide (such as human insulin for drug treatment) require the techniques of genetic engineering, described in the next chapter, to modify microbial cells to make the desired protein.

CHAPTER 5

Nucleic Acids

Nucleic acids were first isolated from cell nuclei over 100 years ago. In 1944 it was found that altering the DNA of bacteria could transform a non-virulent species into a virulent one in an irreversible and inheritable manner. Studies directed towards establishing the detailed structures of the two types of nucleic acid, DNA and RNA, came to fruition shortly after and the double helix structure for DNA was proposed in 1953. It was apparent that this structure could account for the self-reproducing (replicating) properties of DNA necessary for genetic inheritance and that the sequence of nucleotide bases could provide all the information necessary to specify the diversity of biomolecules essential to cells. The code by which nucleotide sequences were translated into amino acid sequences was determined in the 1960s along with the mechanism of how RNA molecules (much smaller than DNA) were intermediaries in the process. In 1973 recombinant DNA was made outside cells by combining DNA from two different sources and shown to be able to replicate and express itself in the common bacterium *E. coli.* This was the start of genetic engineering and now permits gene-therapy—treating disease by the insertion of missing genes into human cells. These methods have been made possible by development in the 1970s and 1980s of the analysis and synthesis of nucleic acids by methods combining enzymology and chemistry. It is now a routine operation to establish the base sequence of a gene or to synthesize fairly long nucleic acids and even genes.

5.1 COMPONENTS OF NUCLEIC ACIDS

5.1.1 Heterocyclic Bases

Nucleic acids are high molecular weight (2×10^4–3×10^{10}) polymers. Hydrolysis of RNA (ribonucleic acid) by heating with aqueous acid gives D-ribose, phosphoric acid, and the heterocyclic bases adenine, guanine, cytosine, and uracil (t-RNA also yields some unusual bases, see Section 5.2.1). DNA (deoxyribonucleic acid) may be hydrolysed under forcing acidic conditions to the same heterocyclic bases except that thymine replaces uracil. Typical conditions—12 M perchloric acid at 100 °C—degrade the sugar component but it may be shown to be 2-deoxy-D-ribose rather than D-ribose present in RNA.

The monocyclic bases cytosine, thymine, and uracil, being derivatives of pyrimidine, are known as pyrimidines while the bicyclic bases, being derivatives of purine, are known as purines. An important difference from the parent compounds is that all those containing exocyclic oxygens exist primarily (>99.99%) in the keto and amino forms shown (Scheme 5.1) rather than any enolic or imino forms even where the latter has greater apparent aromatic character. Scheme 5.2 shows tautomeric forms of cytosine.

Purines Pyrimidines

Adenine Guanine Cytosine Thymine Uracil

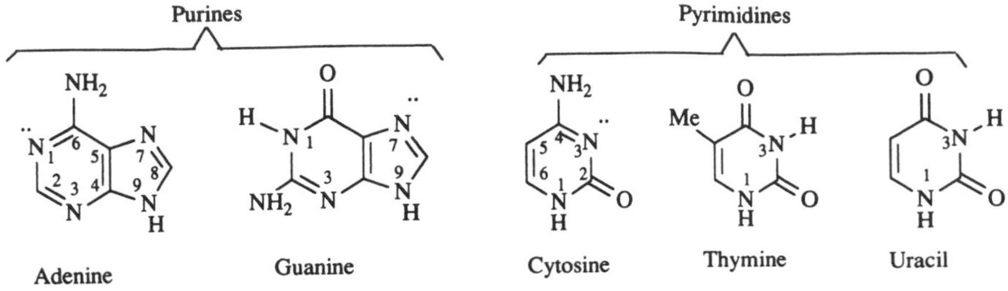

Scheme 5.1 *Structures of the major purine and pyrimidine bases of nucleic acids in their dominant tautomeric forms. The most acidic NH are denoted by N-H, and the most basic N are shown as N:*

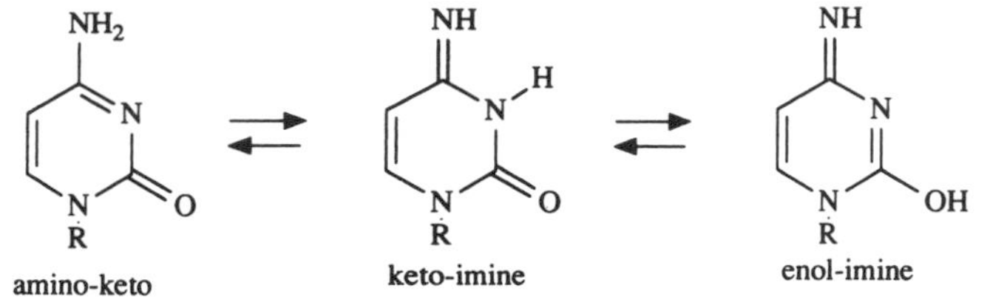

amino-keto keto-imine enol-imine

Scheme 5.2 *Tautomeric forms of cytosine (R = H) and its nucleosides (R = ribose or deoxyribose)*

Purines have conjugate bases with pK_a 2–5 and thus they are normally quite basic. This is because (Scheme 5.3) the NH of the five-membered ring feeds its lone pair into the ring's π-system causing the other nitrogens to gain electron density and basicity. In pyrimidines the nitrogens can not contribute their lone pairs to the ring's π-system but tend to withdraw electrons inductively. Strongly acidic (pH >1) conditions are thus required to protonate thymine or uracil. However, in cytosine the exocyclic amino group can donate its lone pair to the ring in much the same way as for the purines thus raising the basicity of the ring nitrogens; it is protonated below pH 4.4.

The heterocyclic bases of nucleic acids can act as acids or bases but under normal physiological pH's none of them is ionized. It requires a pH below 3.3 to

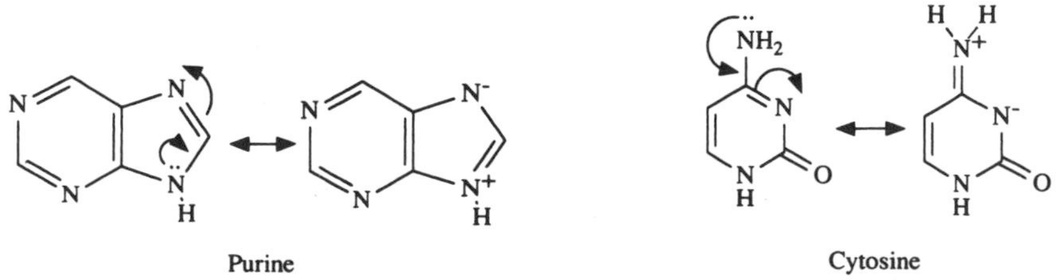

Purine Cytosine

Scheme 5.3 *How conjugation accounts for the basicity of purine and cytosine*

protonate the purines (A and G) or the aminopyrimidine (C) while thymine is hardly basic at all. Under stongly basic conditions (above pH 10), purines and pyrimidines having amido (—CONH—) groups are deprotonated to give negatively charged species.

Chemical manipulation of purine and pyrimidine derivatives has gained importance recently with the discovery of acyclonucleosides with potent biological activity. Acyclonucleosides are similar to nucleosides (see below) except that there is not a complete sugar ring. The synthesis of 'Acyclovir', an effective drug against genital herpes and AIDS, is shown in Scheme 5.4. 2,6-Dichloropurine is used as starting material since its N9 may be selectively alkylated. Ammonia is then used to replace the chlorines and to remove the benzoate blocking group. Diazotization and hydrolysis then selectively converts one of the NH_2 groups into OH which tautomerizes giving acyclovir.

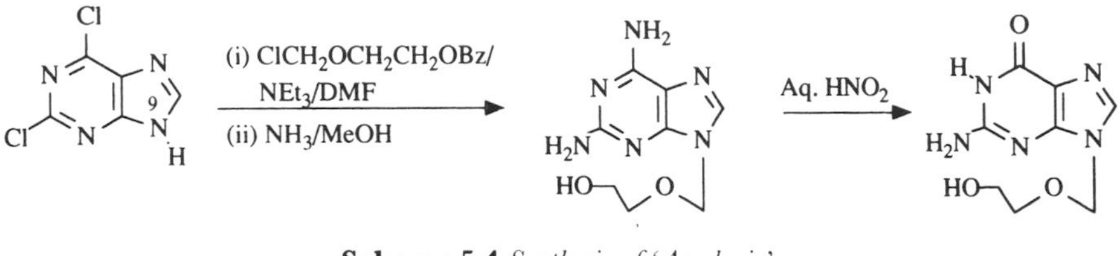

(i) $ClCH_2OCH_2CH_2OBz/NEt_3/DMF$ (ii) $NH_3/MeOH$; Aq. HNO_2

Scheme 5.4 *Synthesis of 'Acyclovir'*

5.1.2 Nucleosides and Nucleotides

Gentle hydrolysis of RNA, using aqueous ammonia, gives the nucleosides, adenosine (A), guanosine (G), cytidine (C), and uridine (U) shown in Scheme 5.5, in which the heterocyclic bases are joined to the anomeric position of D-ribofuranose with the β-configuration. Similarly the 2-deoxy-β-D-ribofuranosyl derivatives, deoxyadenosine (dA), deoxycytidine (dC), deoxyguanosine (dG), and deoxythymidine (dT or T), known as deoxynucleosides, may be obtained from DNA by enzymic hydrolysis to nucleotides (see below) followed by alkaline hydrolysis.

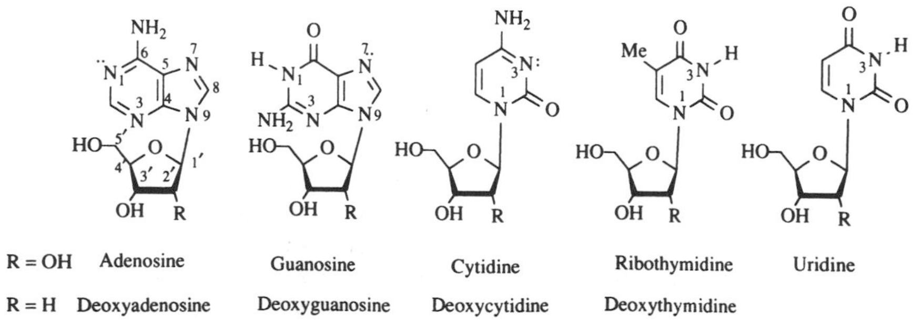

R = OH Adenosine Guanosine Cytidine Ribothymidine Uridine

R = H Deoxyadenosine Deoxyguanosine Deoxycytidine Deoxythymidine

Scheme 5.5 *Structures of the major nucleosides (R = OH) and deoxynucleosides (R = H). Relatively basic nitrogens are indicated by a lone pair (:) and relatively acidic hydrogens by –H*

To distinguish the numbering of the heterocyclic base from that of the sugar, the latter uses primed numbers; position-1 thus refers to a heterocyclic nitrogen whereas position-1′ is the anomeric carbon of ribose.

For ease of drawing and economy in space the nucleosides have been drawn in high energy conformations. In reality an *anti*-conformation (Scheme 5.6) in which the heterocyclic base is as far as possible from C5′ is generally much more favoured than the *syn*-conformations shown. Furthermore the sugar ring is not flat as depicted in Scheme 5.6 but in a puckered half-chair conformation (Scheme 5.7) in which C2′ and C3′ are above and below the planar C1′—O4′—C4′. Which atom is above the plane and which below changes the angle of attachment of groups but, in unsubstituted nucleosides, this makes little energetic difference and the two forms readily interconvert.

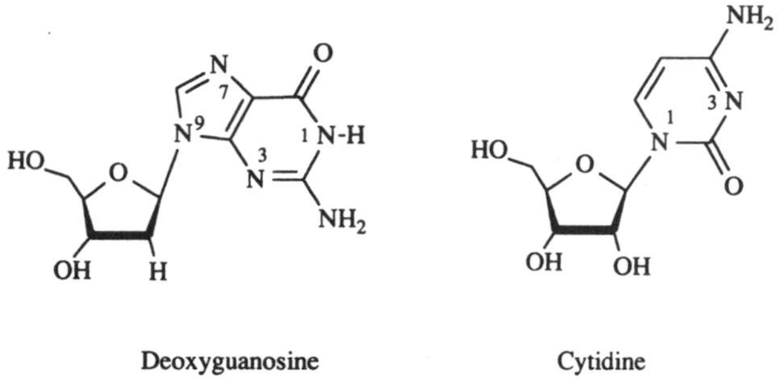

Deoxyguanosine Cytidine

Scheme 5.6 Anti-*Conformations of deoxyguanosine and cytidine*

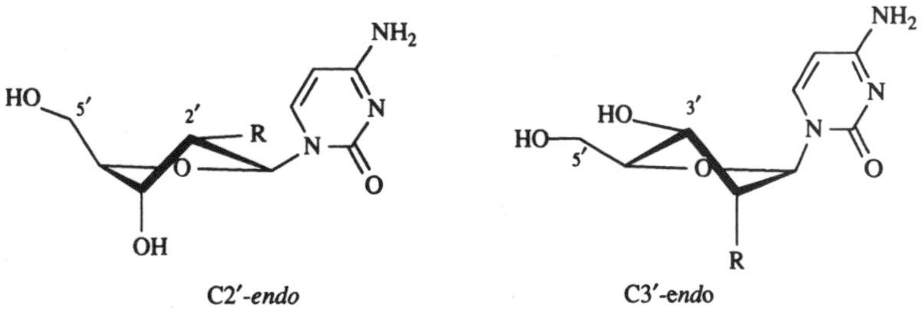

C2′-*endo* C3′-*endo*

Scheme 5.7 anti-*Conformations of cytidine showing the favoured half-chair shapes of (deoxy)ribose*

Other products from the alkaline hydrolysis of RNA are nucleoside 5′- or 3′-phosphoesters; these are called nucleoside 5′- or 3′-phosphates or, more simply, 5′- or 3′-nucleotides. Enzymic hydrolysis of DNA with exonuclease enzymes similarly yields deoxynucleoside 5′- or 3′-phosphates, known as deoxynucleotides. The phosphate groups ($ROPO_3^{2-}$) of nucleotides (and deoxynucleotides) have pK_as of *ca.* 7 for initial protonation and so bear between 0 and 1 protons and a charge of −2 to −1 at physiological pH. Double protonation of the phosphate ester group [to $ROP(O)(OH)_2$] generally requires a pH below 1. The presence of the phosphate group thus makes nucleotides acidic and they are sometimes named as such—5′-adenylic acid is adenosine-5′-phosphate, for example.

5.1.2.1 Phosphate Esters. Phosphate esters are derivatives of phosphoric acid (or more precisely orthophosphoric acid), $O{=}P(OH)_3$. Phosphoric acid is a strong acid; it loses the first proton above pH 2.1, the second at pH 7.2, and all three by pH 12.7. Since the phosphorus is in its highest oxidation state, P^V, and is not easily reduced, the properties of phosphoric acid and its derivatives (phosphorus halides, anhydrides, amides, and esters) are similar to those of carboxylic acids and their derivatives. Esters are produced on reacting phosphoric acid with alcohols in the presence of condensing agents. Replacement of one or two hydrogens by alkyl groups gives monoalkyl and dialkyl phosphates respectively which are still strong acids. Neutral trialkyl phosphates do not form using DCC as the condensing agent but may be produced using 1-(mesitylenesulphonyl)-3-nitro-1,2,4-tetrazole (Scheme 5.8, MSNT). Reaction of MSNT with a dialkyl phosphate, $(RO)_2POOH$, gives a mixed anhydride which may phosphorylate alcohols. The nitrotetrazolide portion is chosen for being a good leaving group with low nucleophilicity and the arylsulphonyl part is bulky to prevent its reacting directly with the alcohol to give sulphonate ester by-products.

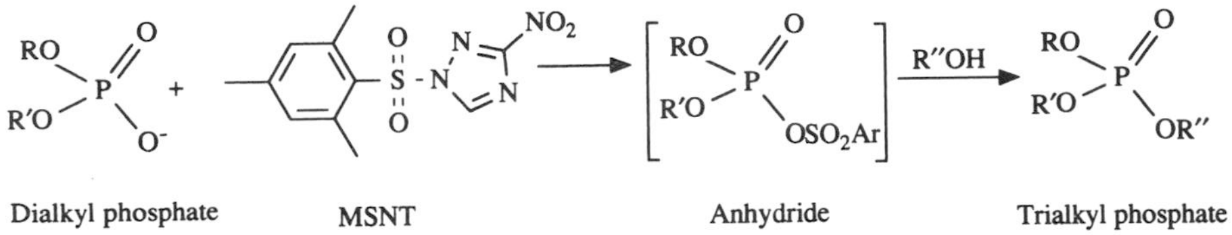

Dialkyl phosphate MSNT Anhydride Trialkyl phosphate

Scheme 5.8 *Use of MSNT in the synthesis of trialkyl phosphates*

Trialkyl phosphates (also called alkyl triesters) are hydrolysed slowly in alkaline solution to alkyl diesters. Further hydrolysis is deterred by the negative charge on the diester and requires heating with strong alkali or acid. Monoalkyl phosphates are also resistant to alkaline hydrolysis but hydrolyse more readily under neutral or acidic conditions. The hydrolysis of phosphate esters of 1,2-diols (such as RNA) is accelerated by the neighbouring hydroxyl group which participates in the reaction. This is shown in Scheme 5.9, where Nu represents a nucleophile (OH^- or enzyme), and accounts for RNA being easier to hydrolyse than DNA both chemically and enzymatically.

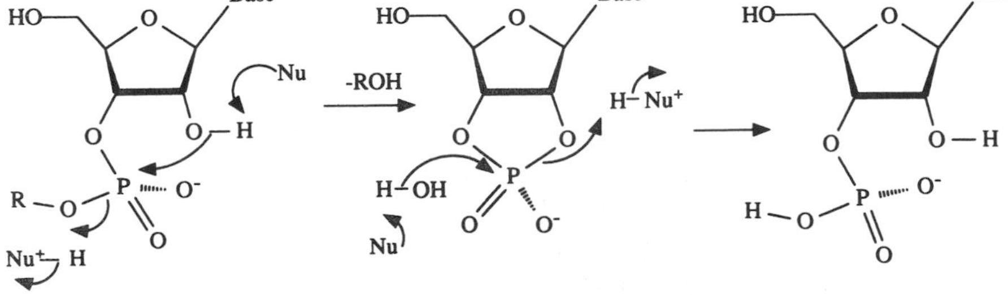

Scheme 5.9 *Neighbouring group participation in the hydrolysis of ribosyl alkyl phosphates*

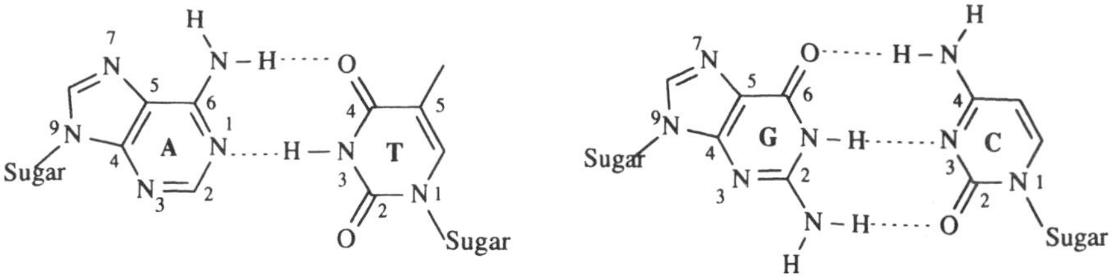

Scheme 5.10 *Normal hydrogen bonding (Watson–Crick base pairing) for A:T and C:G*

5.1.2.2 Hydrogen Bonding between Base Pairs. There are many possibilities for hydrogen bonding between pairs of heterocyclic bases. Most important of these are the hydrogen-bonded base-pairs A:T, A:U, and C:G (Scheme 5.10) proposed by Watson and Crick as part of their double helix structure for DNA (the symbols A, C, and G may refer to nucleosides, deoxynucleosides, or even nucleotides when specifying base-pairings or lengthy nucleotide sequences). The inter-nucleoside attraction is strong for these pairs because their geometries allow optimum hydrogen bonding between two or three pairs of atoms. Although such Watson–Crick pairing between *anti*-conformations is the norm, other pairings have been observed; Hoogsteen pairs (A:T and G:C) and Crick wobble pairs (Scheme 5.11) (G:U, G:A, G:T, and A:C) are important variations involving both *syn*- and *anti*-forms of the nucleosides.

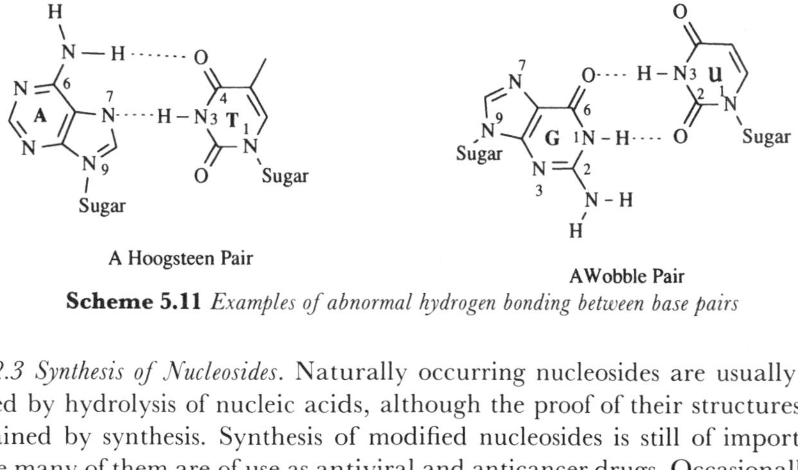

A Hoogsteen Pair

A Wobble Pair

Scheme 5.11 *Examples of abnormal hydrogen bonding between base pairs*

5.1.2.3 Synthesis of Nucleosides. Naturally occurring nucleosides are usually prepared by hydrolysis of nucleic acids, although the proof of their structures was obtained by synthesis. Synthesis of modified nucleosides is still of importance since many of them are of use as antiviral and anticancer drugs. Occasionally, as for the anti-AIDS drug AZT (3′-azido-2′,3′-dideoxythymidine (Scheme 5.12), the starting material can be the related natural nucleoside. More commonly ribose or deoxyribose will be used as the starting material and the heterocyclic base attached.

The first method of nucleoside (*N*-glycoside) synthesis was to react ribofuranosyl halides with purines or pyrimidines in the presence of a heavy metal [mercury(II) or silver(I)] salt catalyst. As in *O*-glycoside synthesis, one generally obtains the required β-anomer if there is a participating group at C2′. Thus the ribosyl bromide (**1**, Scheme 5.13) reacts with the chloromercuri salt (**2**) (from

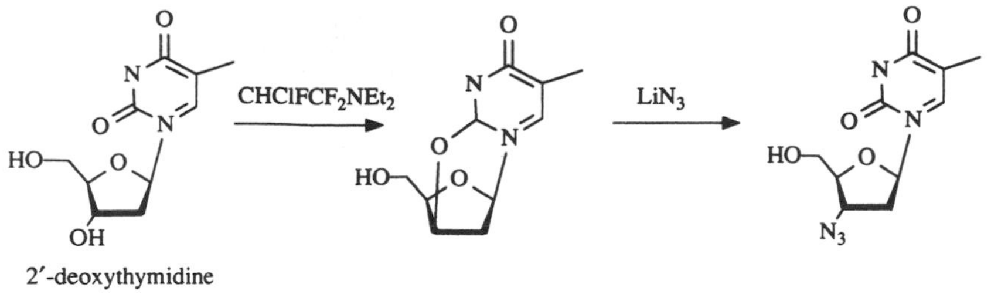

Scheme 5.12 *Synthesis of the anti-AIDS drug AZT*

6-chloropurine and mercuric chloride) to give the *N*-glycoside (**3**) via participation of the 2′-benzoate. Adenosine may then be produced by displacement of the chlorine by ammonia followed by debenzoylation. It would not have been wise to use adenine (6-aminopurine) in the initial coupling reaction since this has other nucleophilic nitrogens which could react with the ribosyl bromide instead of N9.

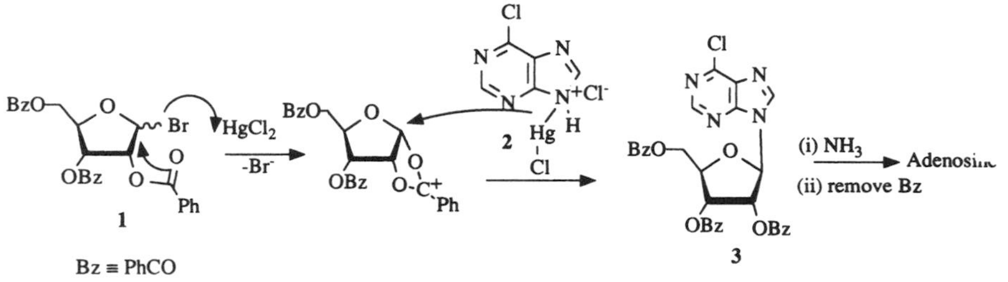

Scheme 5.13 *Heavy metal procedure for nucleoside synthesis*

Substituted pyrimidines, such as the di-enol ether (**4**), are sufficiently nucleophilic to react with ribofuranosyl halides without heavy metal salt catalysis and the resulting quaternary salts are easily converted into nucleosides. This is the Hilbert–Johnson procedure illustrated (Scheme 5.14) for uridine and cytidine syntheses.

A modern variation (Scheme 5.15), known as the silyl base procedure, uses silylated purines and pyrimidines generated by reaction with bis(trimethylsilyl)-acetamide. The silylated bases are sufficiently nucleophilic to react with either 1-*O*-acetyl-ribosyl derivatives or ribosyl halides. A strong Lewis acid, such as the trimethylsilyl triflate shown, is used as catalyst.

Selectivity during formation of the *N*-glycosidic bond is rather variable. Reaction of the wrong nitrogen of the heterocyclic compound is often encountered and 2′-deoxyribosyl halides normally produce a mixture of α- and β-deoxynucleosides. The more modern methods described above have better selectivity but may still give mixtures. When small amounts of a new nucleoside are required chromatographic separation of the reaction products is usually employed but for large quantities a selective route must be devised, largely by trial and error.

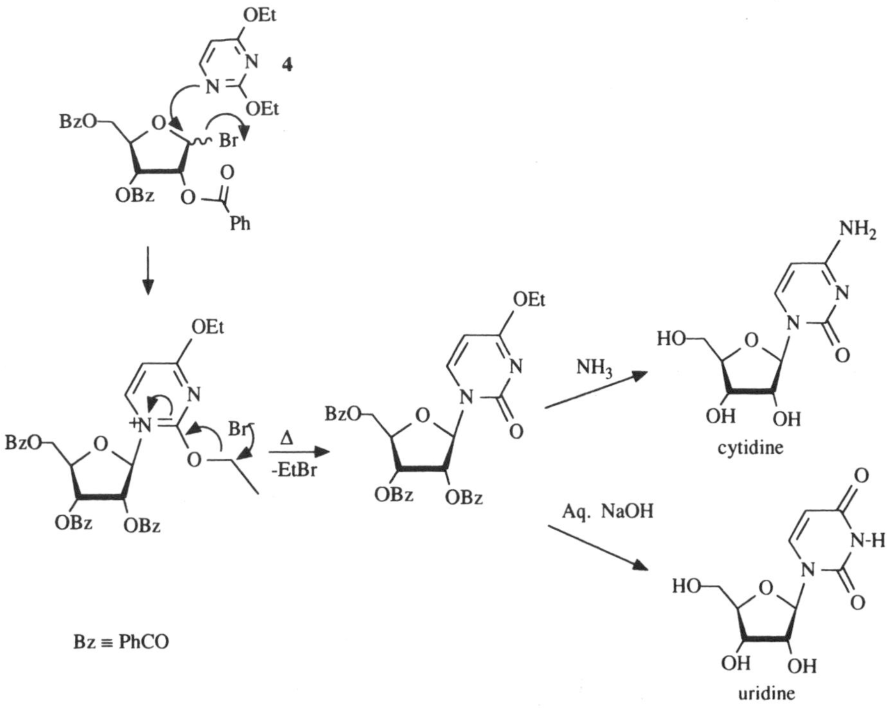

Scheme 5.14 *Syntheses of uridine and cytidine (Hilbert–Johnson procedure)*

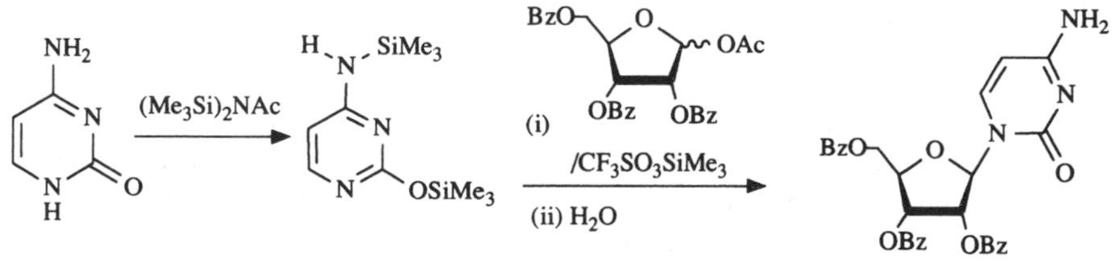

Scheme 5.15 *Silyl base method to synthesize cytidine tribenzoate*

Building the heterocyclic base onto a reactive substituent at the sugar's anomeric position is an alternative approach which largely avoids problems of regio- and stereo-selectivity. An outline of the synthesis of the unusual nucleoside wyosine, which occurs naturally in some tRNA, is shown in Scheme 5.16 starting from a ribosyl isocyanate. Such methods tend to be rather lengthy but are particularly valuable for nucleosides in which the heterocyclic component is not readily available or for *C*-nucleosides—that is C1′ of the sugar is joined to a carbon of the heterocyclic base.

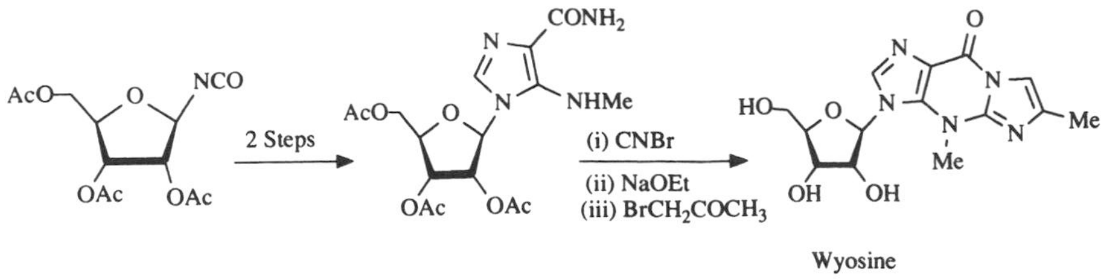

Scheme 5.16 *Synthesis of the unusual nucleoside wyosine*

5.1.2.4 Nucleotide Synthesis. Attaching a phosphate to a nucleoside is complicated by the presence of more than one reactive site in both the nucleoside and the phosphorylating agent. Thus condensation of phosphoric acid and a nucleoside with a condensing agent such as DCC is not generally useful. On the other hand DCC is valuable for the production of phosphate diesters such as cyclic phosphates (see Section 2.4.1.3) since further reaction to triesters does not occur.

Normally the phosphorylating agent will be blocked at all but one position which is activated for reaction as the phosphorus halide or by addition of a condensing agent; the sugar will also be protected except for a single OH at the desired reaction site. Removal of the blocking groups after the condensation reaction liberates the nucleotide. If bulky protecting groups (*e.g.* Scheme 5.17) are present on the phosphorylating agent, selectivity for the primary OH at the 5′-position may be observed so rendering sugar protecting groups unnecessary.

Careful adjustment of reaction conditions has resulted in methods for selective direct phosphorylation of the 5′-position of unprotected nucleosides and deoxynucleosides. Thus, reaction of nucleosides with phosphorus oxychloride in aqueous pyridine may give 5′-nucleotides in over 80% yield.

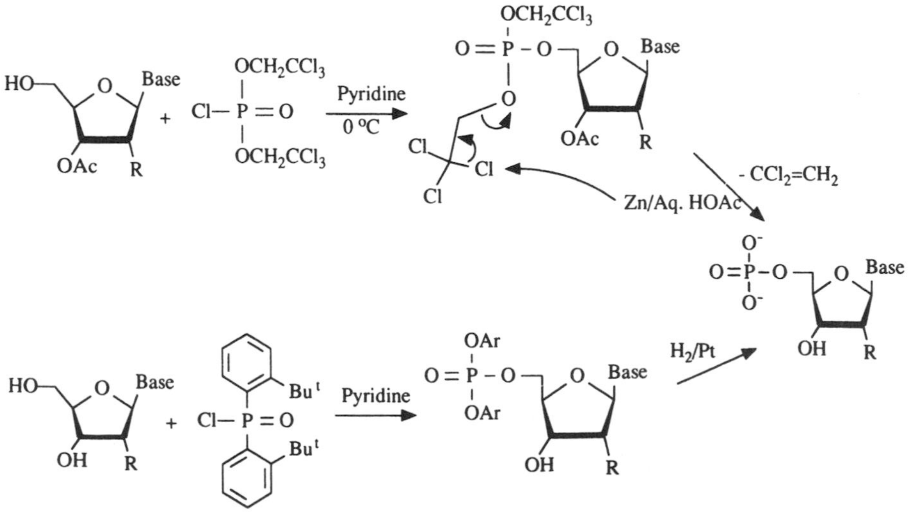

Scheme 5.17 *Nucleotide synthesis using bulky phosphorylating agents*

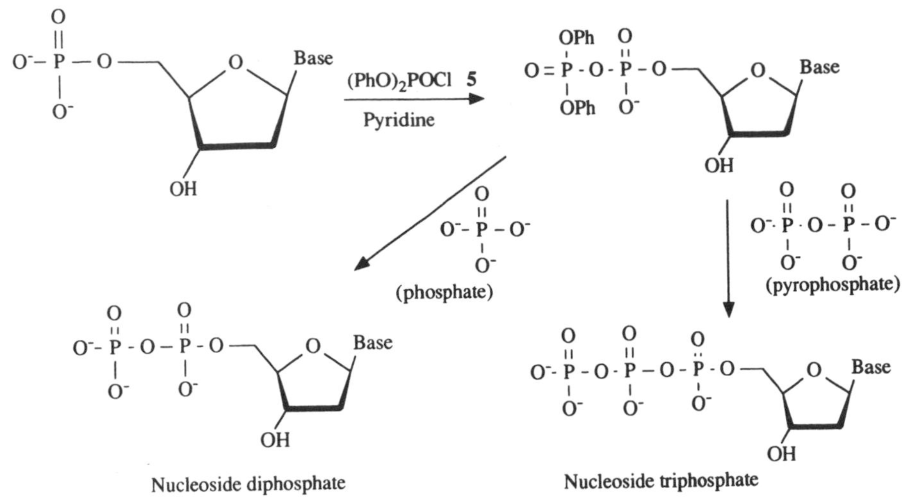

Nucleoside diphosphate

Nucleoside triphosphate

Scheme 5.18 *Synthesis of nucleoside polyphosphates*

Nucleoside polyphosphates participate in a wide variety of biologically important C—O and P—O bond cleavage reactions. They are normally 5′-esters of nucleosides having two or three phosphate groups linked as anhydrides (also called pyrophosphates or condensed phosphates). Although anhydrides, they are fairly stable to hydrolysis because their negative charges repel attacking nucleophiles. They may be synthesized similarly to nucleotides but fully protected condensed phosphate esters have no charge and so are usually too readily hydrolysed to be used as intermediates. One of the most versatile methods (Scheme 5.18) makes use of this lability; the nucleoside is activated by reaction with diphenyl phosphochloridate (**5**) and the resultant pyrophosphate triester is reacted with a nucleophile such as phosphate or pyrophosphate to give the less reactive phosphomono- or di-ester.

5.2 STRUCTURE OF NUCLEIC ACIDS

The primary structures of both RNA (**6**) and DNA (**7**) are comprised of nucleotides linked through the 3′ and 5′ positions as phosphate diesters. This was suggested from the identification of the hydrolysis products from nucleic acids and has been amply proven since. The polymers thus consist of (–sugar-phosphate–)$_n$ chains with heterocyclic bases attached to each anomeric carbon.

The phosphodiester link is not very flexible. The favoured conformation (Figure 5.1) has the atom sequences H4′–C4′–C5′–O5′–P and P–O3′–C3′–C4′ in approximately antiperiplanar (flat zigzag) chains. If the two planes joined at about 180° the ribose rings would be as far apart as possible but the lone pairs on the two phosphate ester oxygen atoms would eclipse each other. *X*-Ray structures of A-DNA and RNA show that these two planes make an angle of about −80° at the tetrahedral phosphorus thus allowing favourable overlap between a non-bonding electron pair on one ester oxygen with the other P–O ester bond.

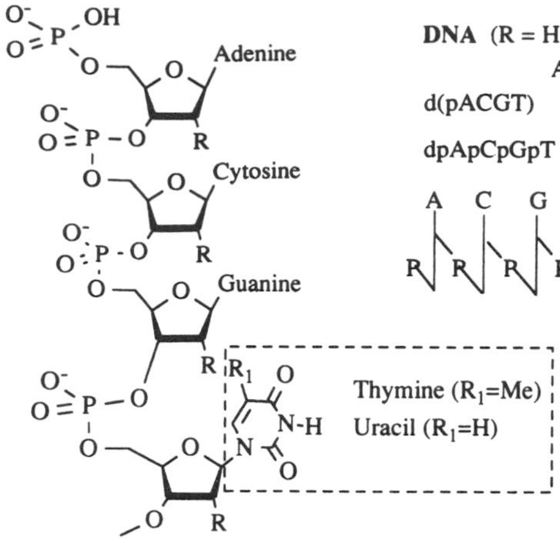
Adenine
Cytosine
Guanine
Thymine (R1=Me)
Uracil (R1=H)

6

DNA (R = H, R_1 = Me)

7

RNA (R = OH, R_1 = H)

Abbreviated forms:

d(pACGT) pACGU

dpApCpGpT pApCpGpU

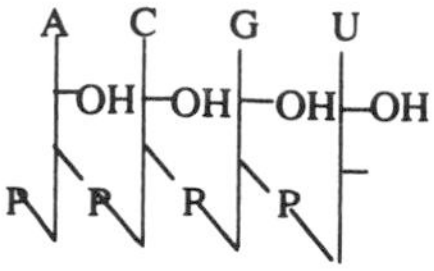
A C G U
OH OH OH OH

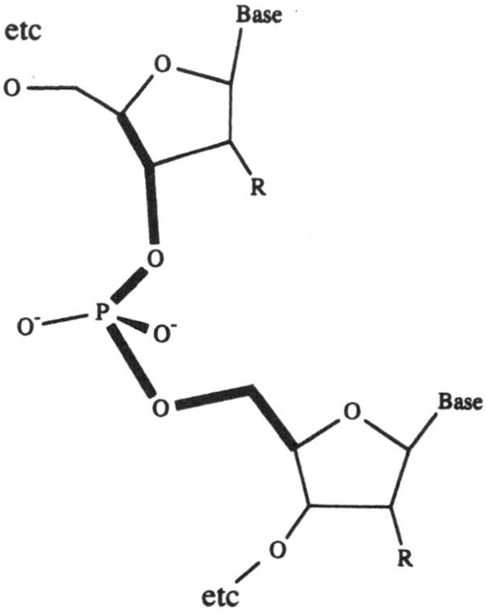
Base
Base
etc
etc

High-lighted bonds (⟶) mark the two sets of coplanar atoms joining at P

Figure 5.1 *Geometry of the phosphodiester link*

5.2.1 RNA

RNAs have relative molecular masses (M) between 10^4 for transfer RNA (tRNA) and 10^7 for ribosomal RNA (rRNA). RNAs are often described by their sedimentation coefficients which are roughly proportional to $M^{1/2}$. Representative examples are 23S RNA, 3700 nucleotides with $M = 1.2 \times 10^6$; 16S RNA, 1700 nucleotides with $M = 0.55 \times 10^6$; and 4S RNA, 75 nucleotides with $M = 2.5 \times 10^4$.

The primary structure of RNA is a long, regular, unbranched chain of ribonucleotides joined by 3′→5′ sugar–phosphodiester bonds. Since there is an OH on C2′, 2′→5′ links are also conceivable but are not normally seen. However, when vertebrate cells are infected by viruses they produce interferon

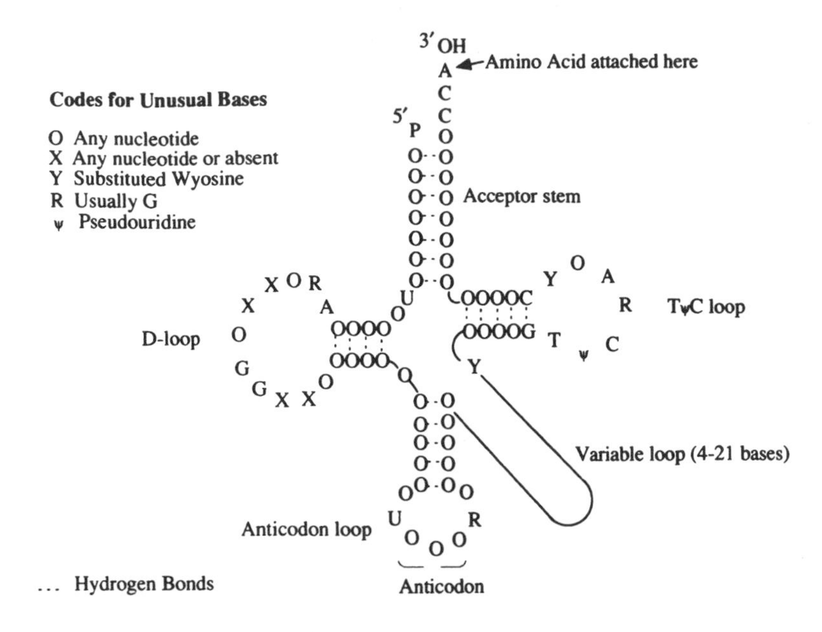

Figure 5.2 *Secondary structure of tRNA*

which causes self-condensation of several adenosine triphosphate (ATP) units to 2′→5′-linked oligonucleotides having antiviral properties.

The long RNA chains adopt well-defined shapes or secondary structures in which sections of the chains double back on themselves and are held together by hydrogen-bonding and lipophilic interactions between the bases. This happens where the sequences are complementary, that is A aligns with U, and G aligns with C (or occasionally U) for several base-pairs, and since the hydrogen bonding is always between one purine and one pyrimidine there is a constant distance between the two backbones. Such regions are generally fairly short and are interspersed with stretches of single-stranded RNA. The hydrogen-bonded sections adopt a right-handed twist which enables the bases to stack one above another with favourable interactions between their π-bonds. The two sugar-phosphate backbones are wound around this stack of bases giving a structure known as a double helix. An example of the ordered structures seen is in the different types of transfer RNA (tRNA) which all adopt a similar clover leaf structure (Figure 5.2). The clover leaf is not obvious in three dimensions and it looks more like a L shape with the anticodon and D loop stems forming the upright and the TψC and acceptor stems making the horizontal piece.

5.2.2 DNA

Natural DNAs are usually much longer than RNAs with relative molecular weights ranging from 1.6×10^6 for DNA from a bacteriophage (virus that parasitizes bacteria) to 10^{11} for a human chromosome. In contrast to RNA, DNA is normally found as a duplex—two non-covalently bound chains held together

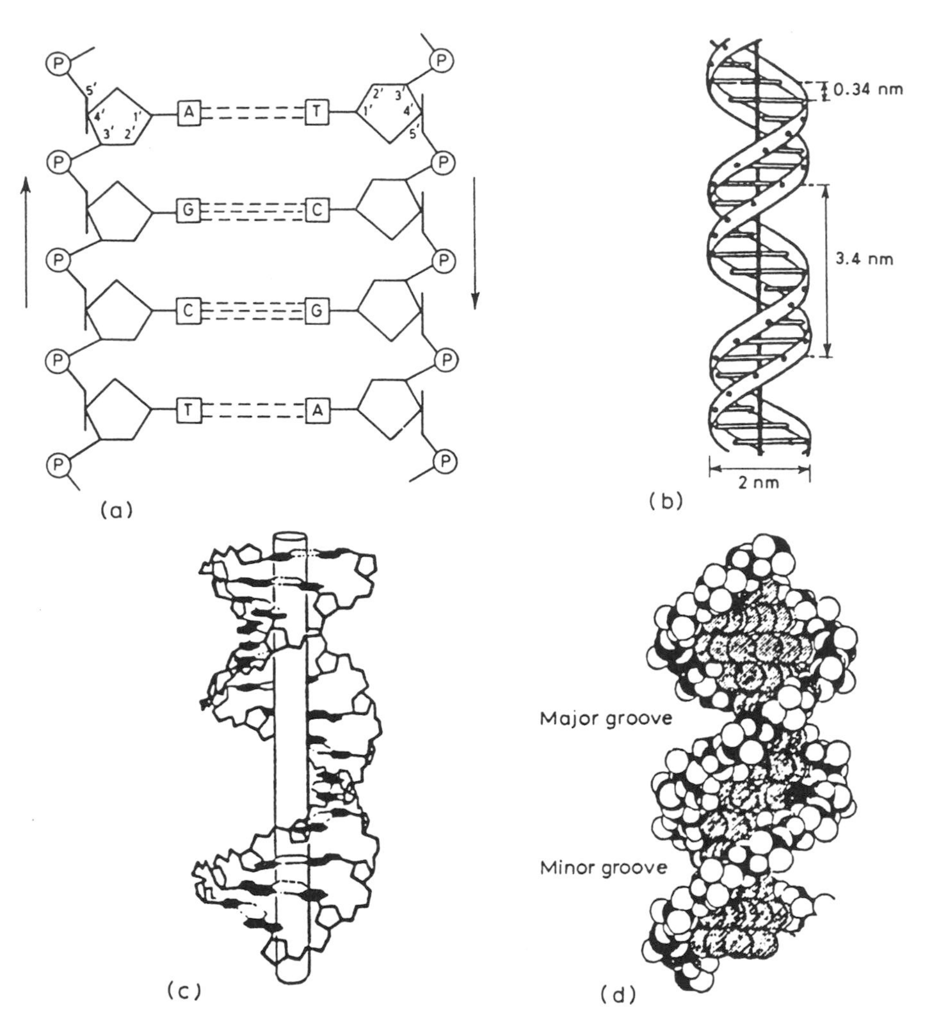

Figure 5.3 *Various diagrammatic ways of representing DNA:* (a) *showing polarity and base pairing but no helical twist;* (b) *showing helical twist and helix parameters but not base pairs;* (c) *showing helix and base pairs;* (d) *space-filling representation showing major and minor grooves*
(Reproduced from 'The Biochemistry of the Nucleic Acids', 10th Edn., R. L. P. Adams, J. T. Knowles and D. P. Leader, with permission from Chapman and Hall)

by a variety of non-covalent interactions, particulary lipophilic 'base-stacking' forces. The size of DNA is often quoted in kilobase pairs; the single molecule comprising a typical human chromosome is *ca.* 10^5 kbp and its helical structure is *ca.* 4 cm long. Most DNA are found to be double-stranded helices—that is they are made from two strands of polynucleotides which are hydrogen-bonded together and wound round the same axis to form a helix with the sugar-phosphate chain on the outside. This is generally similar to that described for RNA except that two chains in opposite directions (or 'polarity') are aligned instead of a single looped chain. Also the base pairs are always A:T and G:C in

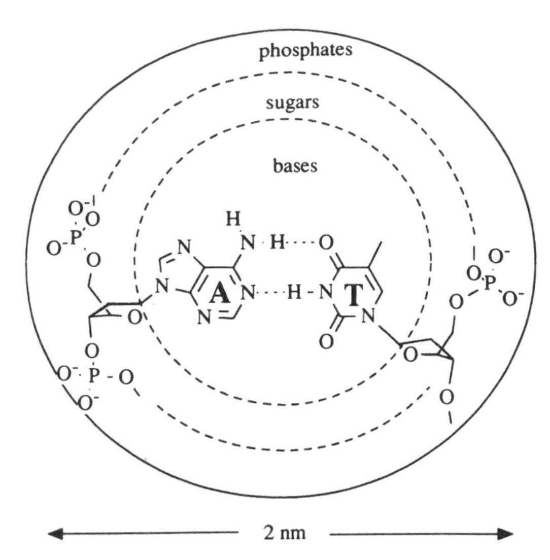

Figure 5.4 *B-DNA: view down helix*

DNA without the occasional variations seen in RNA. In shorthand notation the 5′→3′ chain is placed above its complementary 3′→5′ chain thus:

5′ AGCTTAG 3′
3′ TCGAATC 5′

The helix is right-handed and looks like a tightly twisted spiral staircase (Figure 5.3) with the base-pairs forming the treads. One complete turn takes 10 base pairs and covers 0.34 μm. The base pairs are perpendicular to the axis of the helix and are stacked almost one above the other (Figure 5.4). They are surrounded by the sugar–phosphate backbones which bear one negative charge per phosphate and form two hydrophilic ridges. This leaves two helical spaces known as the major and minor grooves—the minor groove runs close to the 'lower' edge of the base pairs (*i.e.* C1′–N1–-C2 in pyrimidines and C1′–N9–C4–N3 in purines) while the major groove follows the 'upper' edge of the base pairs (*i.e.* C4–C5 in pyrimidines and N7–C5–C6 in purines). These grooves, particularly the minor groove, contain many water molecules which interact favourably with amino and keto groups of the bases.

Natural DNA is thought to exist primarily in the B conformation as described above, but it may vary its conformation according to the local sequence or with changing conditions. When the humidity is reduced to 75% another form known as A-DNA is found. A-DNA has a right-handed helix like B-DNA but it is wider (2.6 nm instead of 2.0 nm) and each turn has more bases (11 instead of 10) but covers less vertical distance (28 nm instead of 34 nm). As in B-DNA all the nucleosides are in their *anti*-conformations but the base pairs are tilted at about 20° to the axis of the helix and leave a 0.6 nm hollow core (Figure 5.5). The conformation of helical RNA has been found to be of the A-type; the 2′-hydroxyl group of RNA hinders formation of B-type helices.

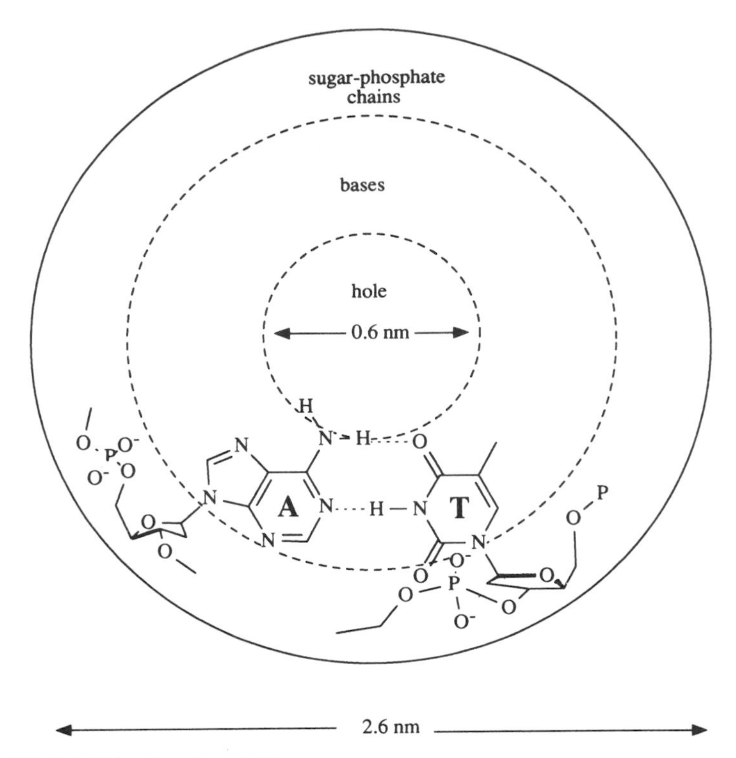

Figure 5.5 *A-DNA: view down helix*

Another type of DNA known as Z-DNA has been found in some synthetic double-stranded oligonucleotides, especially those in which purines alternate with pyrimidines down the chain and under high salt concentration. Z-DNA is a left-handed helix with a minor groove so deep that it reaches the helix axis but it has no major groove thus exposing cytosine-C5 and guanine-N7–C8 at the helix surface (Figure 5.6). The conformation of the nucleosides alternates between *syn* and *anti* down each chain and every base pair also has one nucleoside, normally the purine, in its *syn*-conformation. The sugar pucker of the *syn*-nucleosides is C3′-*endo*—*i.e.* C3′ up (instead of C2′ up for B-DNA)—which causes their 5′-phosphates to be much nearer their sugars' plane. The phosphate backbone adopts a zig-zag path around the helix and it is this that led to the designation Z-DNA. These contortions of the backbone allow Watson–Crick type hydrogen bonding to be retained. Evidence for the existence of Z-DNA in living cells has been obtained but it is not yet clear whether it has any biological role.

5.3 BIOLOGICAL FUNCTIONS OF NUCLEIC ACIDS

The functions of nucleic acids are to provide a self-reproducing source of genetic data and a mechanism for converting this information into biological characteristics. Four processes have been identified which carry out these roles. Replication is the process by which DNA copies itself, recombination is the

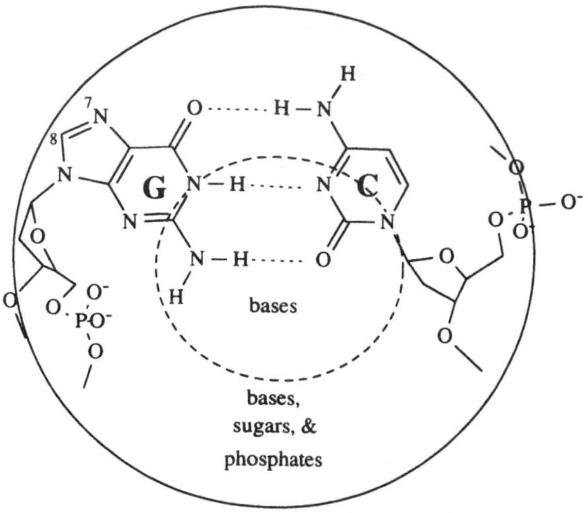

Figure 5.6 *Z-DNA: view down helix axis*

production of new DNA from two parental DNA molecules, transcription is the process by which a molecule of RNA is made complementary to a portion of DNA, and translation is the process by which protein is made to the specifications contained in the RNA sequence.

The number of chromosomes, or DNA duplex molecules, present in every cell to specify the genetic inheritance varies between species. Normally there are two copies of each chromosome in every cell and the number of unique chromosomes, called the haploid number, is half the usual total, or diploid, number. Humans have 23 haploid chromosomes, the fruit fly, drosophila, has four haploid chromosomes, and most bacteria have a single, circular chromosome. Each chromosome is composed of many (usually several thousand) genes. Each gene is that portion of the DNA which specifies the sequence of a protein.

5.3.1 Replication

The structure of DNA as two complementary strands suggested how genetic information might be passed from parents to offspring in multiplying cells. The idea, now proven, is that DNA unwinds into its two non-identical, but complementary, strands. Two new strands, complementary to the old ones, are synthesized onto the templates of the separated strands. On completion two identical DNA molecules result.

The process of replication is rather more complex than might be inferred from the above outline description. The start of DNA replication requires that the DNA is folded into a favourable conformation, and a short stretch of 'primer RNA' is synthesized onto a partially unwound section of DNA. New DNA then begins to be attached to the RNA and later the RNA primer is hydrolysed from the new DNA chain. The unwinding of template DNA is facilitated by strand breaks which are later resealed. Both strands are synthesized simultaneously in the 5′→3′ direction onto partially unwound DNA and this means that one of the new DNA chains must be synthesized in portions which are joined later. The whole process is controlled by numerous enzymes and, at least in certain bacteria, by a regulator RNA as well.

5.3.2 Recombination

Many organisms reproduce asexually in a manner similar to that of individual cells. They simply divide into two and the genetically identical halves grow independently. For a species to survive environmental change the genetic make-up of its individuals must be capable of change. In many species this is achieved by sexual reproduction giving offspring which are not clones of their parent but have chromosomes comprised of an assortment of genes from two parents. This is an example of genetic recombination which occurs in all living organisms.

Genetic recombination during sexual reproduction is initiated by formation of a junction between two similar chromosomes from different parents. The junction is able to move along DNA helices and this branch migration can cause a gene to be swapped between the two attached DNA strands (Figure 5.7). The two DNA molecules resulting will have regions of their double-stranded structures which do not wholly match (A:T and G:C). These regions are recognized and repaired by DNA repair enzymes. The newly inserted sequence may be deleted by replacement with a sequence complementary to the old one or the newly inserted sequence may be 'fixed' by generation of its complementary sequence to supplant the old. The control of junction formation and DNA repair is very accurate and involves recognition of portions of DNA by proteins.

Recombination may also occur where the genes involved are not homologous (very similar). An important example is transposition which involves movement of a genetic element (such as a copy of a DNA segment) to a new site on the same DNA molecule or to another chromosome. Certain viruses, such as the retroviruses responsible for AIDS and some forms of cancer, are specialized forms of

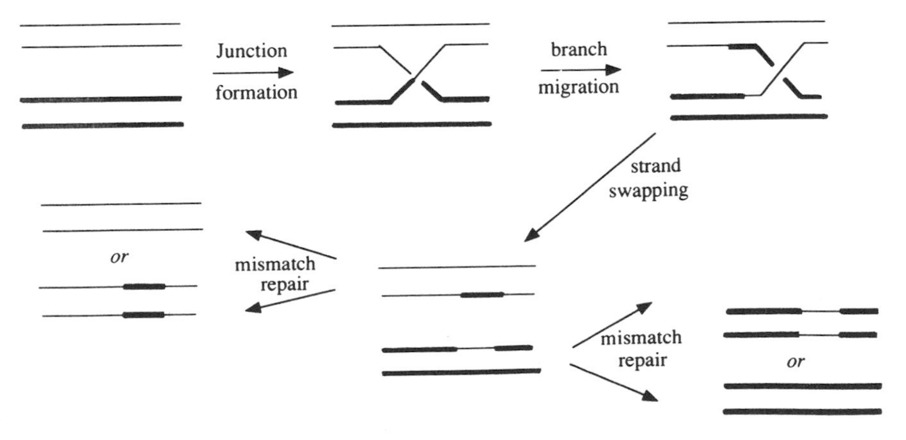

Figure 5.7 *Recombination between pairs of similar chromosomes*

transposons. They consist of little more than a short DNA or RNA chain which reproduce themselves by insertion into another organism's DNA.

Plasmids are another class of mobile genetic elements. They are circular duplex DNA molecules which are able to replicate autonomously of the host chromosome and may be considered as small supplementary chromosomes. Plasmids are able to pass between bacteria and transmission of *R*-factor plasmids which contain genes bestowing resistance to several different antibiotics has led to rapid establishment of bacterial strains possessing multi-drug resistance. Plasmids have recently been turned to man's advantage by their use in genetic engineering. New genetic material can be readily inserted into plasmid 'vectors' and then the recombinant DNA molecules are introduced into suitable cells where they are amplified many times and, given the appropriate conditions, the genes they contain expressed. Large quantities of the inserted DNA or of the protein for which it codes can be isolated from cultures of cells containing recombinant DNA.

5.3.3 Transcription

Those organisms which have cell nuclei—eukaryotes—store the genetic information in their cell nuclei but it must control protein synthesis which occurs on ribosomes away from the nucleus. The information is transferred to the ribosomes by messenger RNA (mRNA). Transcription is the process in which mRNA is synthesized with a sequence complementary to the DNA of the gene to be expressed (*i.e.* its protein is to be made). Transcription is very similar to replication with Watson–Crick hydrogen bonding of base pairs dictating that A gives U, T gives A, G gives C, and C gives G. The correct start and end points of each gene are identified by a promoter sequence in the DNA upstream of the gene and a terminator sequence downstream. Only the $3' \rightarrow 5'$ strand of DNA is used as a template and so gives mRNA with a sequence identical (except T = U) to the $5' \rightarrow 3'$ or 'coding strand' of DNA.

5.3.4 Translation

Translation is the synthesis of protein on ribosomes under the direction of mRNA. Ribosomes have a total mass of $2\text{–}5 \times 10^6$ Da and are composed of many different proteins and several different rRNA molecules which form into one small and one large ribosomal subunit. Ribosomes become attached to the mRNA like beads on a string and the mRNA determines the protein that is made by the ribosome. Protein synthesis is initiated at the mRNA's 5′-end and requires an opening sequence of an AUG or GUG nucleotide triplet. The rest of the mRNA is composed of a succession of 'codons', each comprised of three nucleotides, coding for a single amino acid.

Amino acids for protein synthesis are brought to the site of translation joined to a transfer RNA (tRNA) molecule specific for that amino acid. With the exceptions of Met and Trp, there are at least two possible codons for each amino acid (tabulated in the 'genetic code', Table 5.1) but such degenerate codes usually

Table 5.1 *The standard genetic code*

First letter (5')	*Second letter*								*Third letter (3')*
	U		*C*		*A*		*G*		
U	UUU	Phe	UCU	Ser	UAU	Tyr	UGU	Cys	U
	UUC		UCC		UAC		UGC		C
	UUA	Leu	UCA		UAA	Stop†	UGA	Stop†	A
	UUG		UCG		UAG		UGG	Trp	G
C	CUU	Leu	CCU	Pro	CAU	His	CGU	Arg	U
	CUC		CCC		CAC		CGC		C
	CUA		CCA		CAA	Gin	CGA		A
	CUG		CCG		CAG		CGG		G
A	AUU	Ile	ACU	Thr	AAU	Asn	AGU	Ser	U
	AUC		ACC		AAC		ACG		C
	AUA		ACA		AAA	Lys	AGA	Arg	A
	AUG	Met*	ACG		AAG		AGG		G
G	GUU	Val	GCU	Ala	GAU	Asp	GU	Gly	U
	GUC		CGG		GAC		GGC		C
	GUA		GCA		GAA	Glu	GGA		A
	GUG		GCG		GAG		GGG		G

† Stop = termination codon.
* Also usual initiation codon.

differ only slightly and each amino acid has only one tRNA. The anticodon part of the tRNA (Figure 5.2) binds to a codon of the mRNA and the tRNA is able to recognize each of the codons specific for its amino acid, by using both Watson–Crick and wobble pairing of the bases. Peptide synthesis proceeds simultaneously at several ribosomes but amino acids are always added to the *C*-terminal end of the peptide so that the *N*-terminal protein end corresponds to the 5′ mRNA end.

5.4 NUCLEIC ACID STRUCTURE DETERMINATION

Detailed three-dimensional information about nucleic acids has come from *X*-ray data but such studies are very time-consuming and require crystalline material. Usually just the primary structure, *i.e.* the order of the nucleotides, is required. This is achieved by production of all possible fragments, their separation according to size by electrophoresis, and identification of the terminal nucleotide in each. Two different approaches have been developed using mainly chemical or mainly enzymic techniques—we shall concentrate on the former.

Natural DNA is extremely long and so the first step in its sequencing is to chop it into manageable fragments. This is accomplished using enzymes called restriction endonucleases which cleave DNA at specific points defined by a series of four to seven nucleotides. Normally the DNA will be incubated with one restriction enzyme to give many restriction fragments of 100–600 nucleotides. Fresh DNA is then incubated with a different restriction enzyme to produce other pieces whose sequences overlap those produced from the first endonuclease. Sequencing of both sets of restriction fragments should allow the full sequence to be deduced from a 'restriction map' in the same manner as for proteins.

5.4.1 Chemical Cleavage Method

Once a set of restriction fragments suitable for sequencing is available, the next step is to attach covalently a radioactive or fluorescent marker to an end of each fragment to allow easy detection of tiny amounts. This is normally achieved by an enzymic reaction; the 5′-ends of DNA strands may have a ^{32}P-labelled phosphate attached to a 5′-hydroxyl by reaction with radiolabelled adenosine triphosphate (ATP) and polynucleotide kinase, for example. The various labelled fragments are separated by electrophoresis (see Section 4.4.1) on a denaturing gel to yield the individual single-stranded labelled DNA fragments. The denaturing gels cause strand separation of double-stranded DNA fragments due to the presence of sodium dodecyl sulphate; they are similar in nature and mode of action to those used to uncoil and separate proteins according to mass.

All of the separated restriction fragments are then split into four portions, each of which is subjected to a mild cleavage reaction with selectivity for the four different nucleotides. Dimethyl sulphate, acid, or hydrazine are used to modify the heterocyclic bases and then the modified base is displaced and the strand split on both sides of the 1′-unsubstituted sugar using basic hydrolysis. The chosen conditions are such that each DNA molecule is likely to be cleaved at only a few sites but since there are millions of molecules every possible cleavage will be encountered.

The four different base-modification reaction conditions may be summarized as follows:

(1) At G only: Methylation on N7 using dimethyl sulphate in aqueous buffer. Adenine methylates only slowly and on N3 under these conditions.

(2) At A and G: Protonation of purine ring nitrogens using aqueous formic acid (adjusted to pH 2 with piperidine).

(3) At T and C: Aqueous hydrazine (NH_2NH_2) ring-opens pyrimidines by conjugate nucleophilic addition to a carbonyl group.

(4) At C only: Hydrazine in 5 M sodium chloride solution only ring-opens cytosines.

In each case displacement of the modified heterocyclic bases and chain splitting is accomplished quantitatively by heating with aqueous piperidine. Mechanisms for the cleavage reactions at guanine and thymine are shown in Schemes 5.19 and 5.20.

The reaction products from the four cleavage reactions are then separated by gel electrophoresis in four parallel lanes and their positions visualized by exposing the gel to a photographic plate. Only radioactively labelled fragments will show (Figure 5.8) and these appear in order of molecular weight. Assuming that the 5′-position is labelled, the fastest band (usually shown at the bottom) will be due to the 5′-terminal mononucleotide which can not be identified; however, the lane(s) in which it appears establishes the identity of the base (second in the original DNA fragment) to which it was attached. The next fastest band will be a

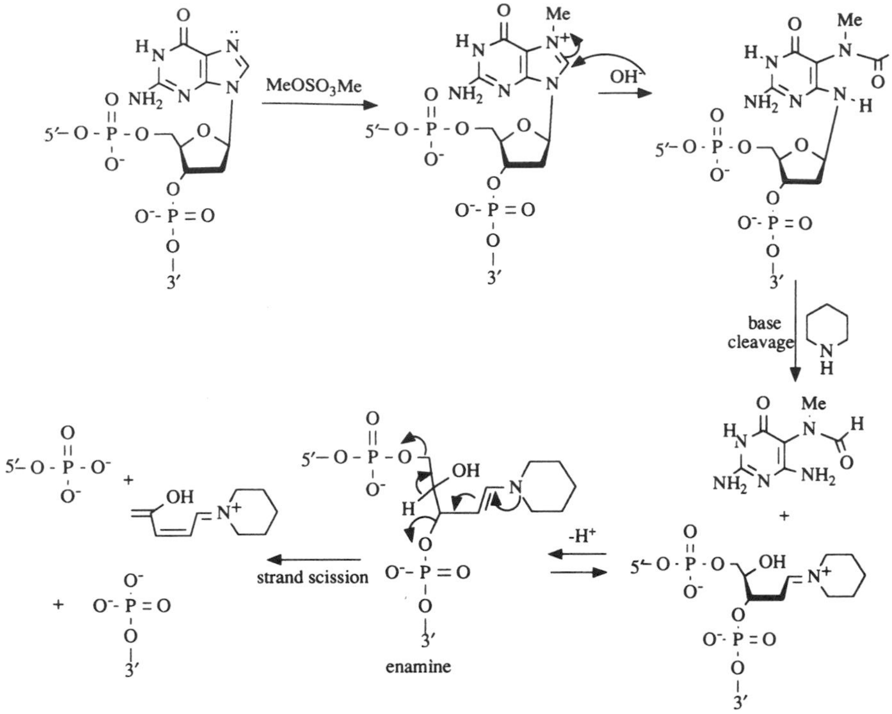

Scheme 5.19 *Mechanism of DNA cleavage at deoxyguanosine*

dinucleotide and the lane(s) in which it appears will identify the third base in the original DNA fragment. Similarly the lane(s) in which the third band occurs gives the identity of the fourth base and so on up the gel. Separation of polynucleotides on thin polyacrylamide gels can achieve excellent resolution—cleanly resolving nucleotides of up to 600 bases which differ in length by just one unit. Usually around two hundred bands may be distinguished on a single gel and so for longer DNA segments two or more gels would be run to cover different size ranges.

An elegant modification to this procedure is to label the chain ends with four differently coloured fluorescent dyes—one for each base-selective reaction. Only one lane is then used on the gel electrophoresis, eliminating error due to gel variation, and the sequence may be read from the colour of the bands down the gel. The bands may be read during electrophoresis as they pass a detector near the bottom of the gel thus allowing detection of each nucleotide fragment from long DNA segments of up to 600 nucleotides.

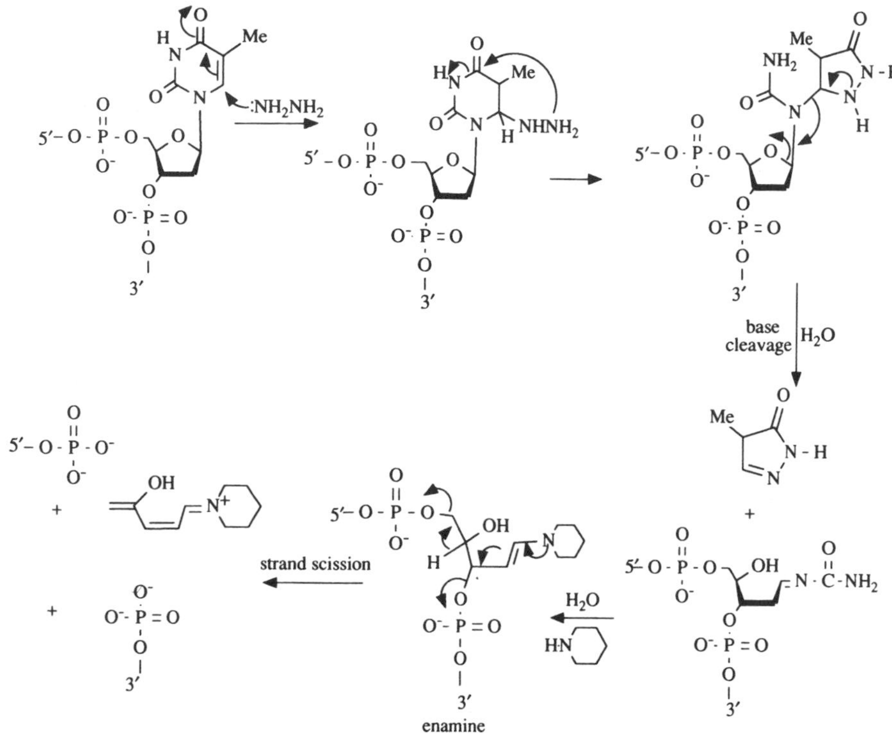

Scheme 5.20 *Mechanism of DNA cleavage at deoxythymidine*

5.4.2 Enzymic or Chain-termination Sequencing

In this method single-stranded DNA fragments of suitable length to be sequenced are produced as described above and are copied four times using a test-tube version of natural DNA replication. In each enzymic copying one nucleotide of the four feed-stock nucleotides is present partially as a 2′,3′-dideoxy form. The new DNA chain grows in the 5′ to 3′ direction but when one of the dideoxy nucleotides is incorporated, extension will cease since there is no 3′-hydroxyl to accept the next nucleotide. The resulting DNA molecules are the complete set of fragments which terminate with the nucleotide that is present in modified form. The fragments are obtained radiolabelled by supplying the enzymic copying reaction with dATP (deoxyadenosine triphosphate) having $O{=}^{32}PO_2$— or $^{35}S{=}PO_2$— in place of its α-phosphate; this radiolabelled phosphate is incorporated into the growing oligonucleotides. The enzyme used is DNA polymerase from *E. coli* but since the natural enzyme may cause loss of nucleotides from the 5′-end of DNA it is cleaved into two pieces and the larger 'Klenow fragment', which possesses the polymerase activity without unwanted nuclease activity, is used.

Once the four nested sets of labelled DNA fragments have been produced one separates them according to size on a slab gel and reads the sequence from an

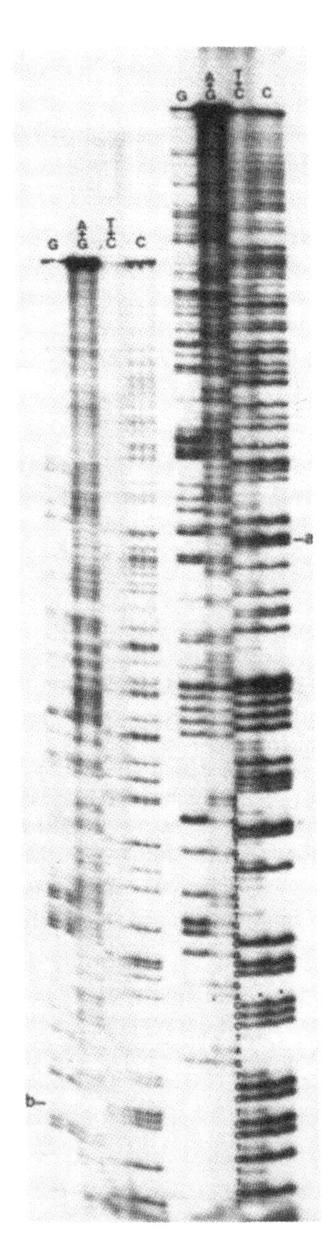

Figure 5.8 *Autoradiograph of a sequencing gel. Portions of double-stranded DNA, labelled with* ^{32}P *at one 5′ end, were partially cleaved using the four nucleoside-selective reactions described on p. 133. The cleaved DNA was then electrophoresed on a* 0.3 × 400 mm *gel as follows: half was loaded (left) and electrophoresed until a xylene cyanol marker dye moved* 300 mm, *at which time the other half was loaded (right) and electrophoresis continued until the new market dye migrated* 300 mm. *To derive the sequence begin at the bottom of the right pattern and read upward until the bands are too closely spaced to continue (arrow* a*), find the corresponding position in the left pattern (arrow* b*), and again read upward* (Reprinted from *Methods in Enzymology*, 1980. **65**, 544, with permission from W. Gilbert)

autoradiogram or a fluorescence detector in the same way as in the chemical method.

5.5 CHEMICAL SYNTHESIS OF NUCLEIC ACIDS

The chemical synthesis of oligonucleotides of specified sequence is an indispensable part of the powerful technology of genetic engineering. Basically, in genetic engineering, one inserts a gene into an organism in order to confer some desired characteristic. Fortunately it is not necessary to synthesize the entire gene. Normally a natural gene may be selected with a similar sequence and 'site-directed mutagenesis' is used to modify it. An oligonucleotide with a short gene segment (say 15 nucleotides) containing the desired alteration is used as a primer in polymerase catalysed DNA replication of the natural gene. Provided that it has only a few base mismatches with the natural gene it will be incorporated and the gene produced in modified form for insertion into the organism.

There are several different methods of oligonucleotide synthesis and they are classified according to how the phosphate is introduced. The earliest was the phosphodiester method (Scheme 5.21) in which the nucleotide to be coupled to

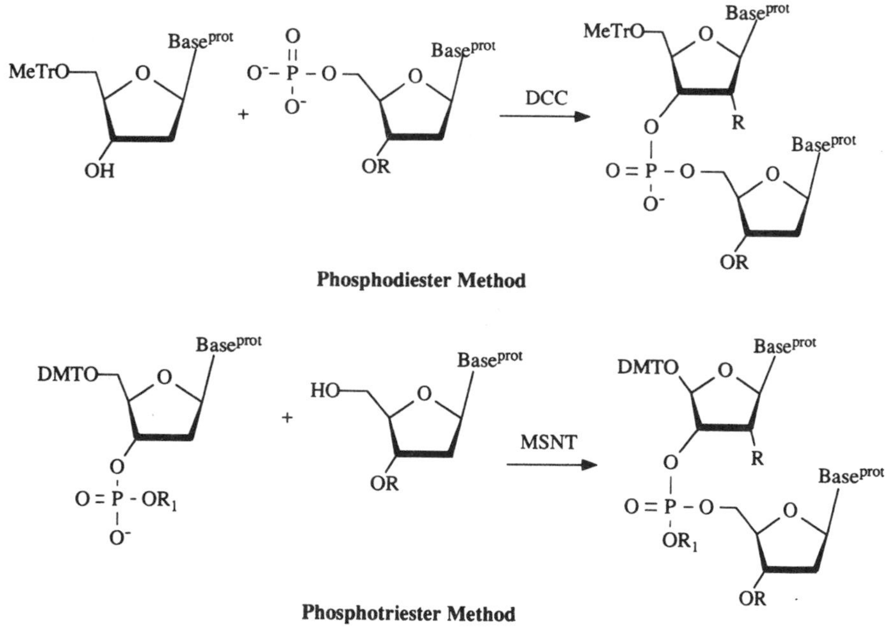

Phosphodiester Method

Phosphotriester Method

MeTr & DMT are acid-labile protective groups;
R is an alkali-labile protective group;
R_1 is a phosphate-protecting group;
$Base^{prot}$ is one of the four DNA bases with exocyclic nitrogens protected - e.g. Thymine, N^2-isobutyryldeoxyguanine, N^6-benzoyldeoxadenine, N^4-benzoyldeoxy-cytosine

DCC & MSNT are condensing agents.

Scheme 5.21 *General methods of coupling in oligonucleotide synthesis*

the growing oligonucleotide is protected on reactive exocyclic nitrogens and at O3′ but not at the phosphate. Coupling is induced with an activating agent to give the oligonucleotide as a phosphodiester and then a 3′-OH is deprotected ready for addition of the next nucleotide. This has been superseded by the more selective and rapid phosphotriester approaches (Scheme 5.21) especially the phosphite triester or phosphoramidite variant in which a permanent P—OH protecting group is introduced and removed only at the end of the synthesis. Originally oligonucleotides were synthesized using solution chemistry but now solid phase synthesis (*i.e.* the growing nucleic acid is attached to a solid support) is predominant.

5.5.1 Protection of Deoxyribonucleosides

Although many alternative protective groups for the exocyclic nitrogens on the heterocyclic bases have been tested, the most popular protecting groups are isobutyryl for guanine, and benzoyl for adenine or cytosine. Thymine has no exocyclic nitrogen and does not need protection. Cytidine's exocyclic nitrogen is sufficiently nucleophilic so that it may be selectively benzoylated. For adenosine and guanosine, however, the ribosyl hydroxyl groups are of similar reactivity and benzoylation occurs at O3′ and O5′ as well as at nitrogen. To obtain *N*-benzoyl-2′-deoxyadenosine (Scheme 5.22) full benzoylation using benzoyl chloride in pyridine may be followed by alkaline hydrolysis which deprotects the hydroxyl groups and hydrolyses *N*,*N*-dibenzoyl derivatives to amides but leaves the amides themselves. An alternative route (Scheme 5.22), making use of the greater thermodynamic stability of O—Si bonds over N—Si bonds (compare Scheme 5.15), starts with transient protection of the hydroxyls with trimethylsilyl groups. The exocyclic nitrogen may then be benzoylated and the silyl groups hydrolysed under mild conditions to the desired *N*-acyl-purine. Introduction of an isobutyryl group onto N2 of deoxyguanosine is similar except that isobutyric anhydride replaces benzoyl chloride.

Since the primary (5′) sugar hydroxyl group is a stronger nucleophile than the secondary (3′) hydroxyl group, coupling proceeds more easily if the phosphate is first attached to the 3′-position; this strategy has been standard since the original diester approach was supplanted. Protection of the 5′-position is usually by the dimethoxytrityl group. Like trityl (Section 2.4.3.4) this bulky group is selective for primary hydroxyls. Its mesomerically electron-donating methoxy groups make it even more acid-labile than trityl itself since they stabilize the triarylmethyl carbocation, Ar_3C^+, which is an intermediate in the decomposition. The DMT group is easily introduced using dimethoxytrityl chloride with a mildly basic catalyst such as pyridine (Scheme 5.23).

5.5.2 Introduction and Protection of Phosphate

Some methods of nucleotide synthesis have been covered in Section 5.1.2.4 and these may provide starting materials for the older phosphodiester approach. In

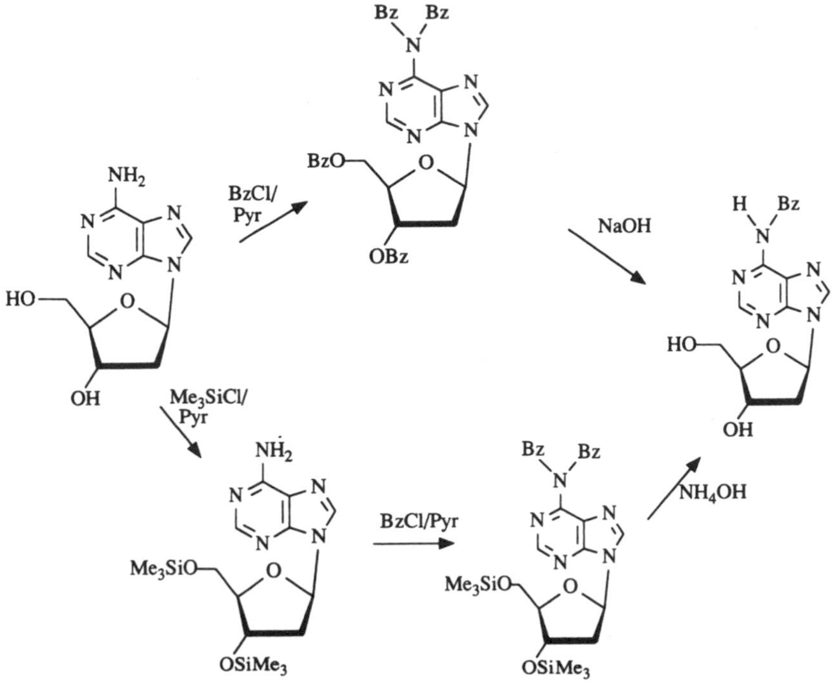

Scheme 5.22 *Routes to* N*-benzoyl-2′-deoxyadenosine*

the more modern methods the phosphate is joined first to the 3′-position and is protected, thus requiring other synthetic methods for the starting materials.

For the phosphotriester methods phosphate protection is by an aryl group, generally 2-chlorophenyl or dichlorophenyl, because alkyl phosphodiester nucleotides are insufficiently reactive. Using 2-chlorophenyl ditriazolophosphate (**9**, Scheme 5.23), only one triazolide is displaced by the nucleoside leaving the other to be hydrolysed.

In the phosphite triester and phosphoramidite oligonucleotide coupling methods the reactive 3′-nucleotide derivatives are alkyl phosphochloridites [nucleoside–P(OR)Cl] or alkyl phosphoramidites [nucleoside–P(OR)NR_2], having phosphorus in its P^{III} oxidation state. These are sufficiently reactive as either alkyl or aryl derivatives and methyl or 2-cyanoethyl are the favoured protecting groups (R). The original attachment of the phosphite derivative is by nucleophilic displacement at P—Cl or (more slowly) P—NR_2 by nucleoside 3′-OH. An example is the formation of a phosphoramidite of deoxyadenosine using 2-cyanoethyl *N,N,N′,N′*-tetraisopropylphosphorodiamidite (**10**) in the presence of the mildly acidic catalyst, tetrazole (Scheme 5.23). The P^{III}—N bonds are of the right reactivity for the reagent to mono-phosphitylate the O3′ of the first nucleoside and for the reasonably stable phosphoramidite product to react later with O5′ of the second nucleoside.

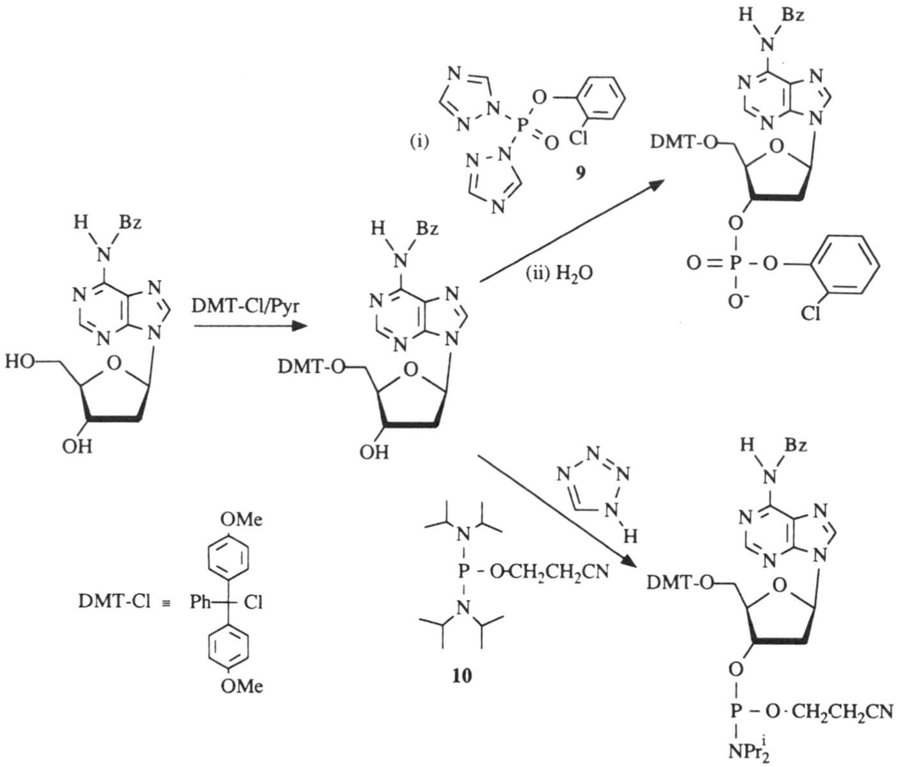

Scheme 5.23 *Introduction of 3′-phosphate (or phosphite) derivatives into deoxyadenosine*

5.5.3 Attachment to a Solid Support

It is standard practice to attach the first nucleotide to a solid support to permit automation and maintenance of excellent yields in the multi-step synthesis of an oligonucleotide. The requirement for consistently excellent yields is illustrated by a 20 step synthesis which gives an overall 36% yield with an average yield in each step of 95%, but only 1% with an average step yield of 80%. A solid support allows simple isolation of product and thus facilitates the use of excess reagents to boost yields. For oligonucleotide synthesis the solid may be silica gel but controlled pore glass beads (CPG) are preferred being rigid, non-swellable, and inert to the reactions of oligonucleotide synthesis.

A long spacer group is used to extend the nucleotide's point of attachment away from the solid surface and thus allow access by all reagents. A 4-nitrophenyl ester of a deoxyribonucleoside-3′-succinate may be reacted with an aminopropyl chain on the glass to give an amide linkage, for example (Scheme 5.24). Even longer spacers are commonly employed. Attachment of many chains to the solid support causes steric hindrance between them and so the loading is kept to 10–50 μmol of nucleotide g^{-1} of solid support. Obviously it is not practical to make large amounts of nucleic acid in this way but this is not a

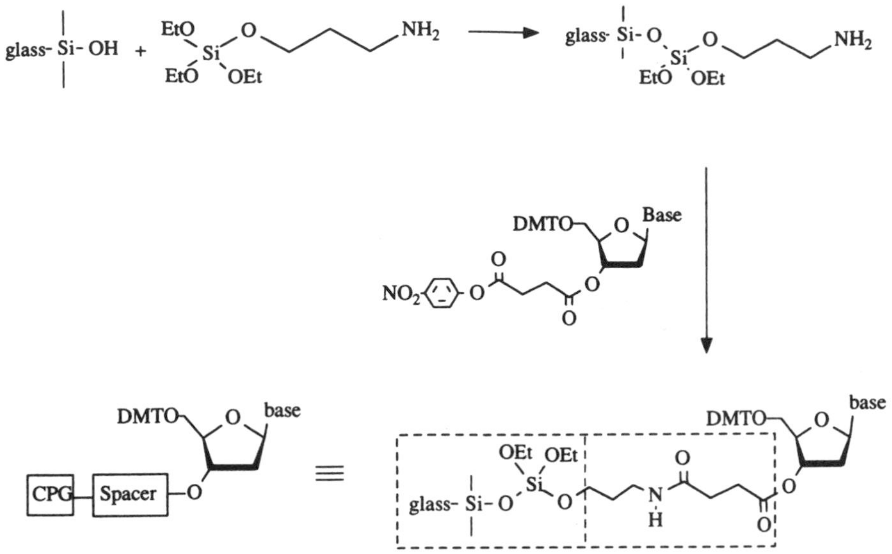

Scheme 5.24 *Attachment of a nucleoside to a solid support*

problem if it is to be incorporated into a gene since it may then be replicated many times.

5.5.4 Nucleotide Coupling

In the solid-phase phosphotriester approach (Scheme 5.25) the protected 3′-nucleotide is reacted with deprotected 5′-OH of the nucleoside in the presence of a powerful condensing agent. DCC is not effective but a range of sulphonyl derivatives have been employed and MSNT (see Section 5.1.2.1) is recognized as one of the best. A problem with the use of strong condensing agents is that deoxyguanosine derivatives may be phosphorylated on O6 and so this position requires protection. A minor side-reaction is sulphonylation of the 5′-hydroxyl group and, although this accounts for 3% of product at most, it limits the

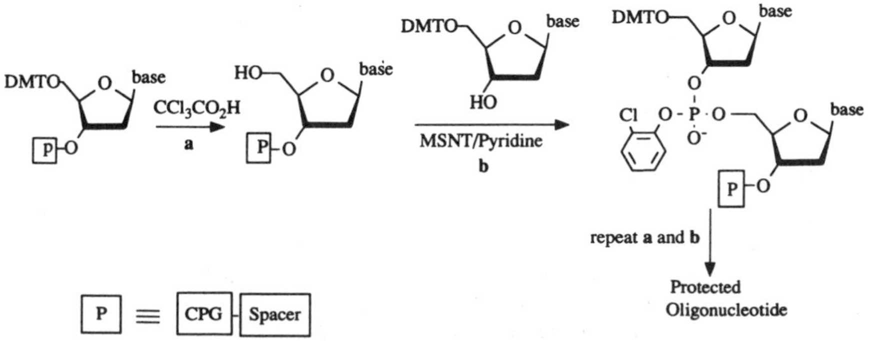

Scheme 5.25 *Nucleotide coupling in phosphotriester oligonucleotide synthesis*

effectiveness of the method. In order to couple the next nucleoside derivative the new 5′-OH is deprotected by removal of the dimethoxytrityl group using trichloroacetic acid in dichloromethane. Repetition of the coupling–deprotection cycle may be repeated as many times as required but build-up of side-products sets a practical limit of about 40 cycles.

The solid-phase phosphoramidite method avoids the wasteful side-reactions of the triester method and has allowed development of efficient automatic DNA synthesizers. The 5′-hydroxyl group of the immobilized nucleotide is reacted with the protected nucleoside 3′-phosphoramidite using tetrazole as weakly acidic catalyst. The dinucleoside phosphite which results is oxidized with iodine to the phosphotriester and the dimethoxytrityl group removed with trichloroacetic acid (CCl_3COOH, TCA) in dichloromethane to complete the assembly cycle (Scheme 5.26) which generally proceeds in better than 98.5% yield. Oligonucleotides of up to 150 residues are prepared by this method.

5.5.5 Capping

If any 5′-hydroxyls remain after coupling they will continue to react in the succeeding cycles and eventually produce oligonucleotides with one too few nucleotide units. Such oligomers are difficult to separate from the intended product and so further growth onto unintentionally free 5′-OH groups is stopped by 'capping' immediately after each coupling. This entails acetylation using acetic anhydride and *N*-methylimidazole. Although the same molar amount of oligonucleotide impurities is still obtained, they are shorter chains and so easier to separate from the desired product. An additional advantage is that any phosphitylation on O6 of guanosine that occurs in the phosphoramidite method of coupling is completely reversed by this treatment.

5.5.6 Deprotection

Once the required number of nucleotides have been coupled all the protecting groups are removed as summarized below:

(1) Dimethoxytrityl is removed in the same way as during coupling cycles—namely trichloroacetic acid in dichloromethane.
(2a) 2-Chlorophenyl groups, used for phosphate protection in the triester method, are removed by treatment with aqueous *syn*-2-nitrobenzaldoximate ion which acts as a nucleophile in a rapid elimination reaction (Scheme 5.27).
(2b) 2-Cyanoethyl groups, used for phosphate protection in the phosphoramidite approach, are removed by aqueous triethylamine or ammonia in a base-catalysed elimination.
(3) Benzoyl and isobutyryl groups, used to protect the heterocyclic bases, are removed using concentrated aqueous ammonia (50 °C, 5 h).
(4) The succinate linkage to the solid support is hydrolysed using aqueous base.

Deprotection is not usually a complicated procedure since several steps can

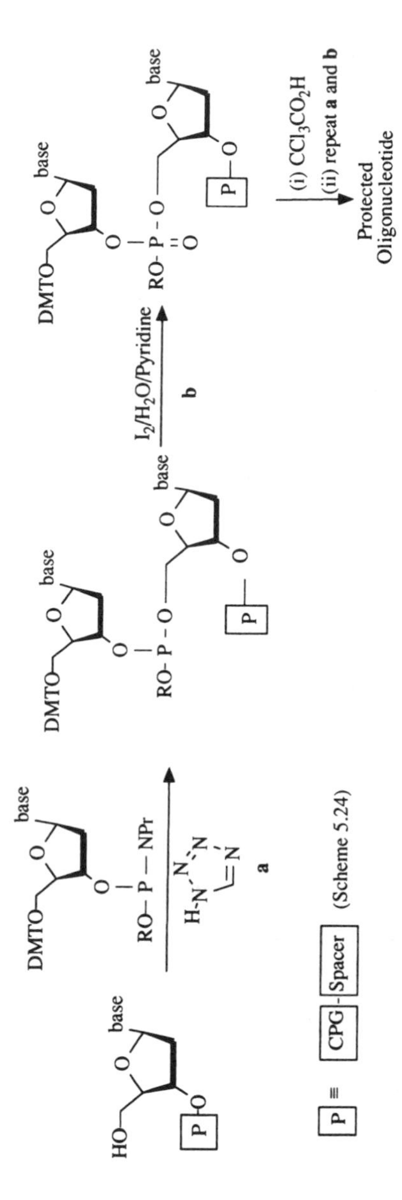

Scheme 5.26 *Nucleotide coupling in the phosphoramidite method of oligonucleotide synthesis*

Scheme 5.27 *Deprotection of 2-chlorophenyl using nitrobenzaldoximate ion*

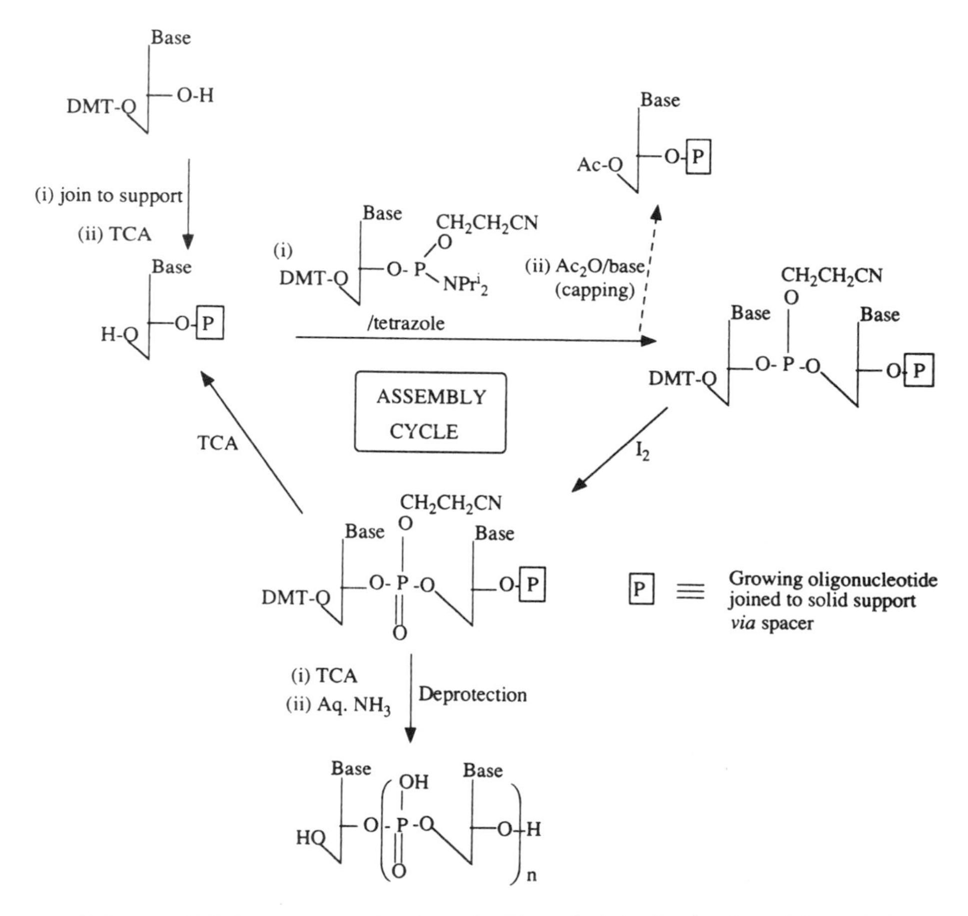

Scheme 5.28 *Overall scheme for oligonucleotide synthesis by the phosphoramidite method*

normally be accomplished simultaneously. Steps 2b, 3, and 4, for example, may be completed in one go using the conditions of 3.

An overall scheme for the synthesis of an oligonucleotide by the phosphoramidite method is given in Scheme 5.28. The final step in oligonucleotide synthesis is purification and this is normally performed using polyacrylamide gel electrophoresis or high performance liquid chromatography.

The advent of genetic engineering, which is already improving agricultural production and health care, was made possible by an understanding of molecular interactions of nucleic acids and by expertise in organic synthesis. Many more wide-ranging benefits are sure to derive from a fuller understanding of the interactions between biomolecules at the molecular level.

CHAPTER 6

Steroids

6.1 LIPIDS

Isolation of biomolecules from hydrolysed aqueous biological fluids (hydrolysed to separate carbohydrates from their aglycones) or from dried plant or animal material often commences with extraction into an organic solvent. Carbohydrates, nucleotides, proteins, and most of the compounds we have considered in previous chapters are very polar and remain in the aqueous or solid phases while those compounds extracted are relatively non-polar and are known as lipids.

The bulk of lipids are fats and oils known as glycerides but since they function mainly as energy stores and have little biological activity, we will study them only briefly. Some lipids, especially phosphoglycerides, are important in cell membranes, helping to control the passage of materials in and out of cells, but the molecular detail of the processes is poorly understood. Others such as steroids, which are the chief concern of this chapter, and the prostaglandins (covered in Chapter 7) are found in very small amounts but are highly potent in the control of activity within cells.

6.1.1 Glycerides and Sphingolipids

Fats, oils, and waxes are used by plants and animals as long-term energy stores. They are esters of long-chain 'fatty' acids which have saturated or unsaturated chains usually composed of an even number of carbon atoms between 12 and 20. Some natural waxes such as beeswax, largely myricyl palmitate, $CH_3(CH_2)_{14}COO(CH_2)_{29}CH_3$, are simple esters of long-chain alcohols, but most fats and oils are glycerides (**1**) (also called triglycerides) which are triesters formed between glycerol, $CHOH(CH_2OH)_2$, and three identical or mixed fatty acid units. Oils are liquid at room temperature and generally have a greater number of double bonds than fats which are solid. Increasing unsaturation causes lowering of melting point because the double bonds are mostly *cis* and give an abrupt change in direction of the carbon chain which will not allow close packing of adjacent molecules. Hydrogenation of animal or vegetable oils saturates the hydrocarbon chains so producing solid margarine. The realization that eating such saturated fat predisposes humans to heart disease has increased the use of fats and oils with greater unsaturation.

Phosphoglycerides (**2**) are found almost exclusively in cell membranes and are

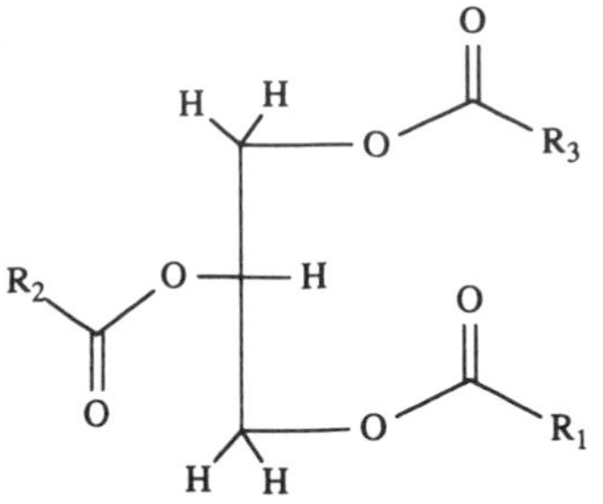

1 A glyceride

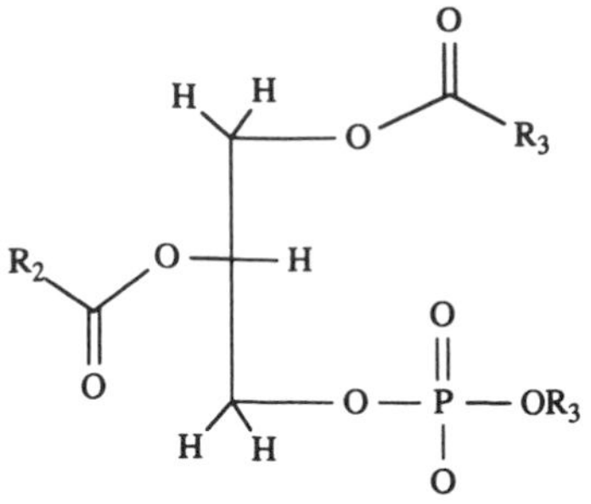

2 A phosphoglyceride

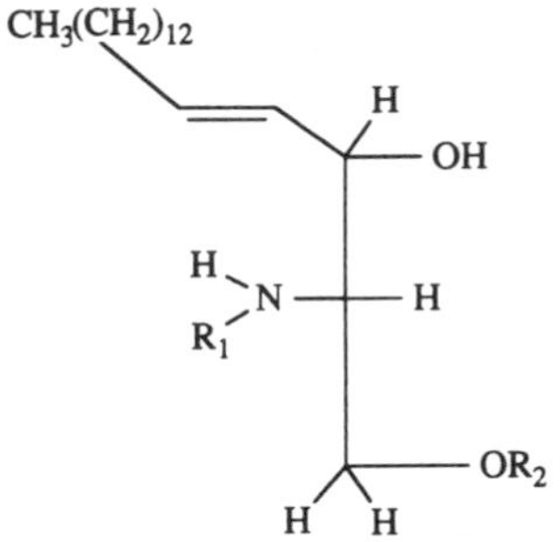

3a R_1, R_2 = H: sphingosine

3b R_1 = COR, R_2 = H : ceramides

3c R_1= COR, R_2 = glycoside: glycosphingolipids

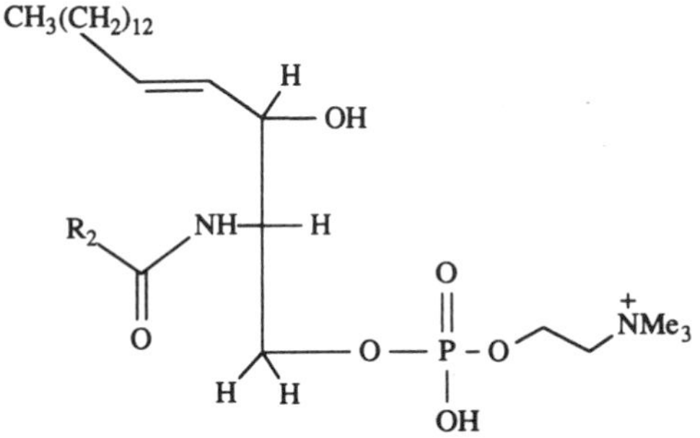

4 A sphingomyelin

similar to the glycerides in fats and oils except one ester branch is formed from a phosphate ester $ROPO_2OH$ rather than a carboxylic acid. The RO- group of the phosphate is derived from a polar alcohol such as choline, $Me_3N^+CH_2CH_2OH$, suiting the phosphate ester branches of triglycerides in their roles as the polar, external, part of membranes while the other two fatty acid chains form the non-polar, internal part of membranes.

Sphingolipids (**3b**, **3c**, and **4**) are structurally similar to glycerides. They are *O*-substituted ceramides; and ceramides (**3b**) are *N*-acyl derivatives of sphingosine (**3a**) or dihydrosphingosine (which has CH_2—CH_2 in place of CH═CH). Sphingolipids bearing a substituted phosphate are called sphingomyelins (**4**) and are components of the myelin sheath which surrounds nerve cells. Cerebrosides have either glucose or galactose at O1 while those having an oligosaccharide of about 2–10 sugar residues are called glycosphingolipids or glycoceramides. If sialic acid is one of the sugars the lipids are known as a gangliosides.

Carbohydrate portions of glycosphingolipids at cell surfaces govern cellular recognition processes of animals including blood typing, immunological defence, and cellular differentiation (the formation of different cell types from an initially homogeneous cell population). Abnormal gangliosides occur in large amounts on cancer cells and have been used as anticancer vaccines since antibodies raised against them bind to whole cancer cells after recognizing the same peculiar carbohydrate sequence. Some gangliosides have been shown to interact with glycoprotein hormones, interferon, opiates, bacterial toxins, and other biomolecules but it is not clear whether they comprise true receptor sites since details of these interactions and their biological consequences are lacking.

6.1.2 Terpenes

Simple lipids such as α-pinene (**5**) (in most conifers), myrcene (**6**) (in bay leaves), limonene (**7**) (in lemons), and zingiberene (**8**) (in ginger) are responsible for the odours of plants and are C_{10}, C_{15}, or C_{20} compounds which appear to be head-to-tail oligomers of isoprene (**9**). The biological building block for these fragrant oils is isopentenyl pyrophosphate (**10**) and the large, diverse group of biomolecules having extended carbon chains assembled from isoprenoid units are known as terpenes (or terpenoids to include those molecules with modified terpene skeletons).

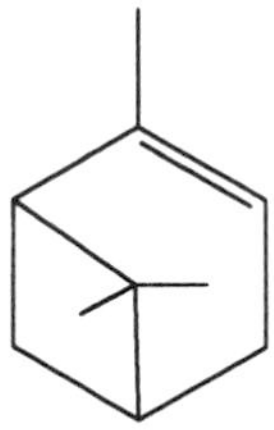

α-pinene
5

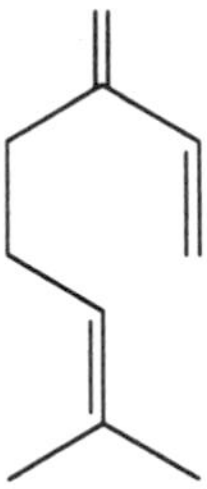

myrcene
6

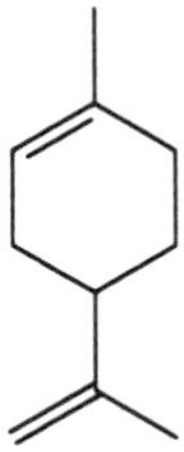

limonene
7

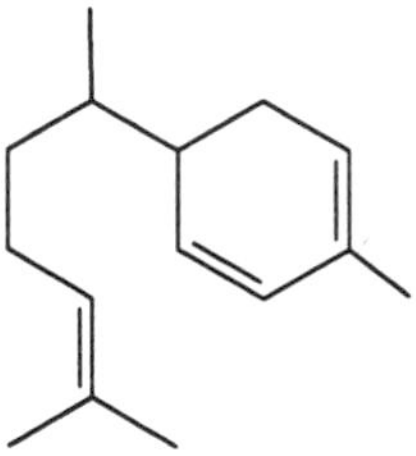

zingiberene
8

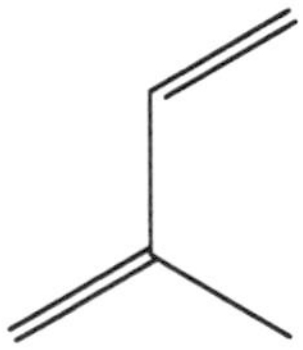

isoprene
9

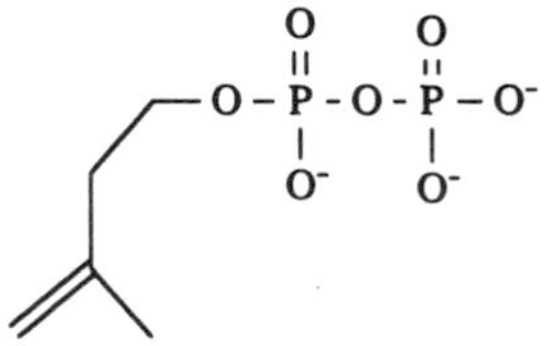

isopentenyl pyrophosphate
10

Terpenes are found in virtually all plants and animals and have a diversity of functions. One important family of tetracyclic diterpenes (*i.e.* four isoprene units) is the gibberellins, such as the C_{20} gibberellin A_{13} (**11**), which function as plant growth hormones; many gibberellins are C_{19} compounds similar to gibberellic acid (**12**) but are still referred to as diterpenes.

The triterpene squalene (**13**) is biosynthesized by dimerization of C_{15} units and after oxidation and cyclization ultimately forms cholesterol, the progenitor of

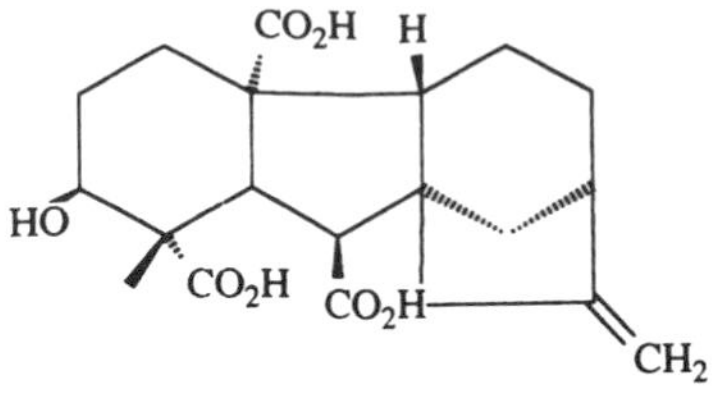

11 Gibberellin A_{13}

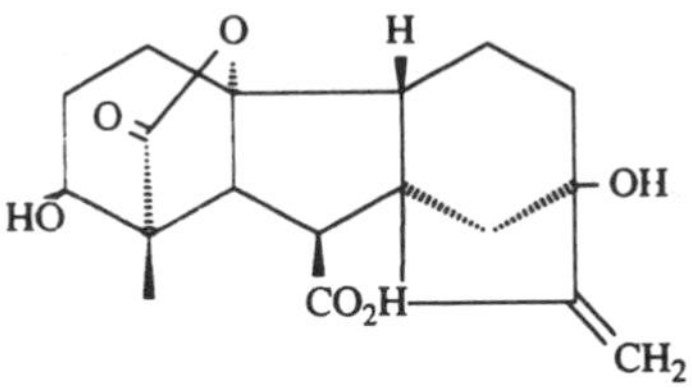

12 Gibberellic acid

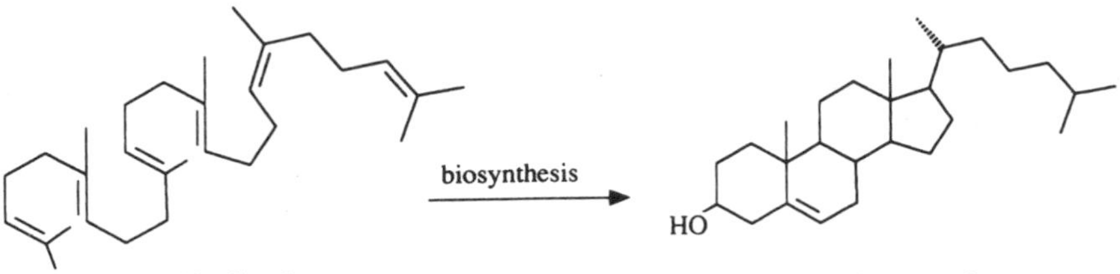

13 Squalene

biosynthesis

Cholesterol

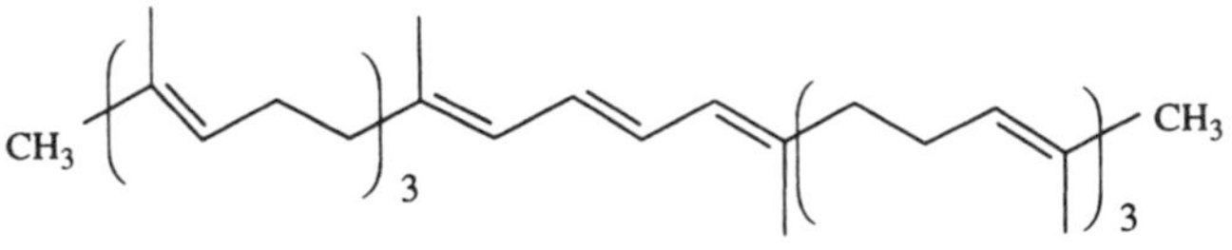

14 Phytoene

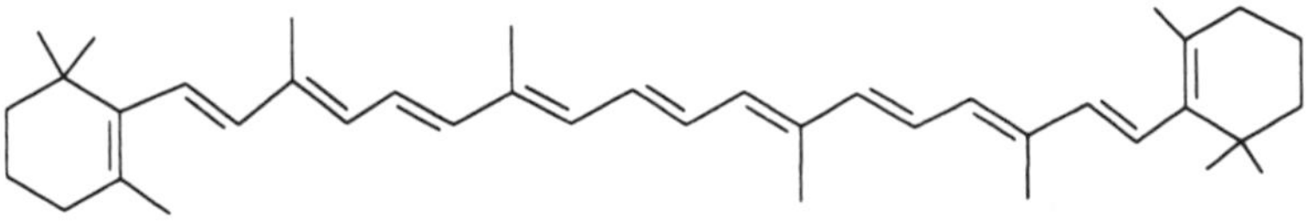

15 β-Carotene

steroidal hormones. Similar dimerization of C_{20} units is employed in the biosynthesis of phytoene (**14**) whose biological dehydrogenation and cyclization gives carotenoids such as β-carotene (**15**), the pigment of carrots used in food colouring. Extended conjugation in the chain of alternating double and single bonds in carotenoids gives a chromophore crucial for absorption of light in the pigments of both vision and photosynthesis.

6.2 STRUCTURE AND NOMENCLATURE OF STEROIDS

Steroids have the perhydrocyclopentanephenanthrene ring structure (**16**) consisting of three six-membered rings (A, B, and C) and one five-membered ring (D) fused together as shown. The three-dimensional structure of steroids varies

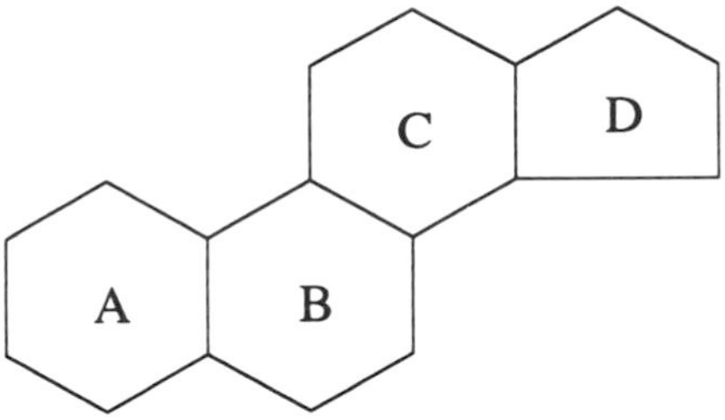

A B C D

16

greatly with the type of ring junction; this will be readily appreciated after a study of the three-dimensional structures of simple decalins.

6.2.1 Decalins

trans-Decalin (**17**) has *trans*-related hydrogens at the ring junction and has a fairly rigid structure in which the rings may deform to boat or skew boat forms on the input of considerable energy but can not flip to alternative chair conformations. Substituents thus have virtually fixed axial or equatorial orientations and those in axial positions are more hindered than their equatorial counterparts since 1,3-diaxial atoms have their centres separated by only 2.50 Å. If, as in steroids and *trans*-9-methyldecalin (**18**) (commonly used decalin numbering

H ← 2.5 Å → H

all-chair

boat-chair

17 *trans*-decalin

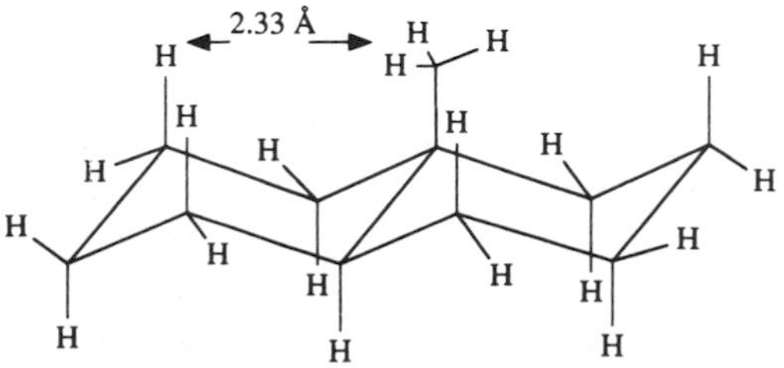

2.33 Å

18 *trans*-9-methyldecalin

differs from both steroid numbering and standard IUPAC recommendations), there is a methyl group at the bridge-head the separation from the 1,3-related axial hydrogens falls to 2.33 Å, which is roughly the sum of their van der Waals radii.

cis-Decalin (**19**) is much more flexible and may easily flip into an alternative all-chair form (see **20**) or into various boat conformations. One CH_2 group in each ring is in an axial position in relation to the other ring and so *cis*-decalin is of higher energy than *trans*-decalin in which all four carbons attached to the bridge-heads are equatorial. In 9-methyldecalins the *trans*-isomer (**18**) is also more stable than the *cis*-isomer (**20**) but the energy difference is quite small since the *trans*-isomer has a destabilizing axial methyl group which partially compensates for the two axial CH_2 groups in the *cis*-isomer.

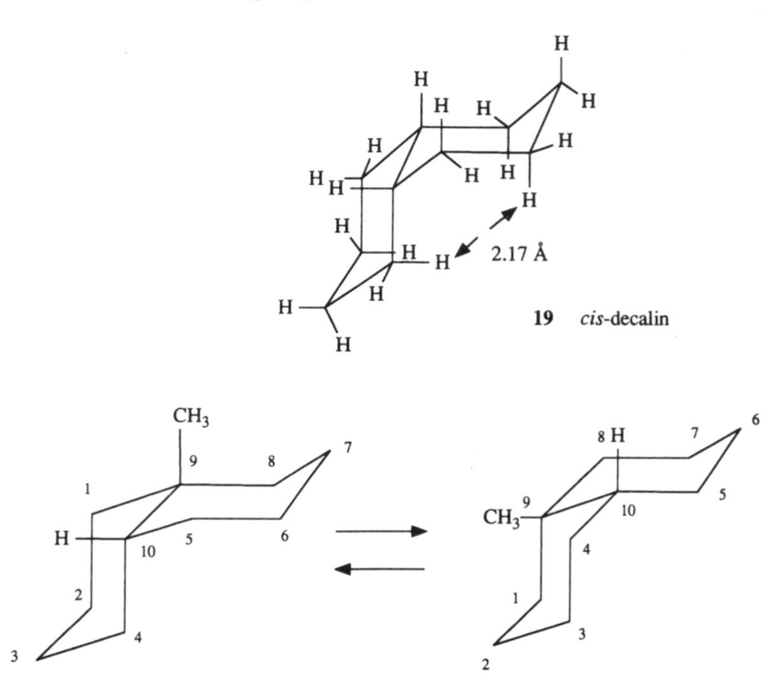

19 *cis*-decalin

Ring-flipping in *cis*-9-methyldecalin **20**

6.2.2 Steroid Nomenclature

The IUPAC numbering of steroids is shown for cholestane (**21**) and derives from historical usage rather than the general rules for cyclic hydrocarbons. Systematic names for steroids are based on the saturated tetracyclic alkane having a similar carbon skeleton. The principal ones are: cholestane (all 27 C's), cholane (C's 1–24), pregnane (C's 1–21), androstane (C's 1–19), and oestrane (C's 1–18). Deviations from one of these basic skeletons is indicated by prefixes; for example 19-norandrostanes lack C19 and an A-homosteroid has a seven-membered ring A. The role of the hydrocarbon names in steroid nomenclature parallels that of simple alkanes: complex structures are named as derivatives by giving the nature

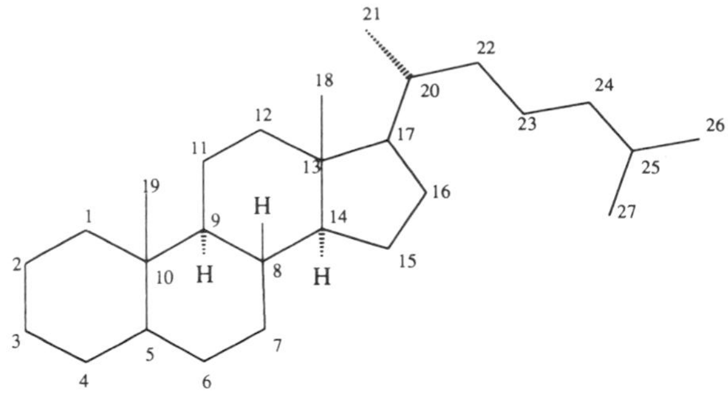

21 Cholestane

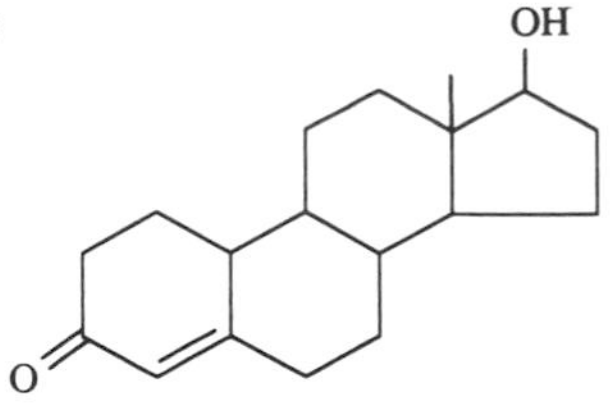

23 Nandrolone

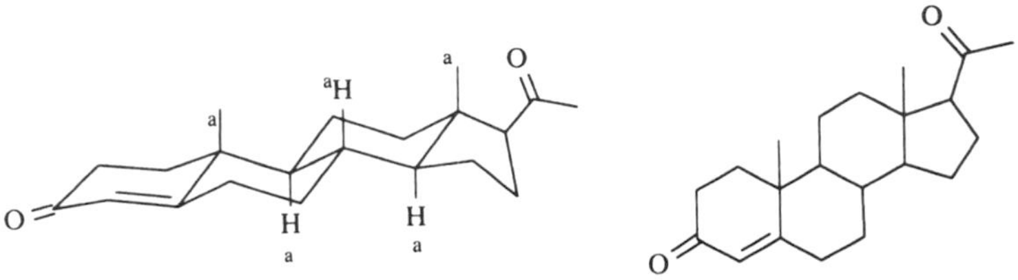

22 Progesterone

and position of all substituents in the molecule and changing the ending 'ane' into 'ene', 'one', or such other suffix as is appropriate. Systematically, progesterone (**22**) is pregn-4-ene-3,20-dione (Δ^4-pregnene-3,20-dione is an earlier version) and such nomenclature is widely used although semi-systematic naming in which steroids are described as modifications of well known trivially named steroids is also commonplace. The anabolic steroid 'nandrolone' (**23**) for example is commonly called 19-nortestosterone rather than its systematic name, 17β-hydroxyoestr-4-en-3-one.

Fusion of the rings in steroids leads to a surface (almost flat in many steroids) having readily discerned upper and lower faces. Substituents on the lower face are designated as α and are shown attached by dashed lines whereas those on the upper face are β and attached by solid lines as for cholesterol, 5-cholesten-3β-ol (**24**). If the configuration is unknown a wavy line is employed with the designation ξ (Xi). The configuration of chiral centres in the side-chain is given by the basic names such as cholestane (**21**) but any additional centres may be specified using the *R* and *S* nomenclature. An older convention is to view the side-chain as a Fischer projection and to define substituents on the right as α- and those on the left β-orientated.

Each ring junction could be either *cis*- or *trans*-fused but BC and CD *trans* forms shown are normal and are implied both in the basic names and in structures where these ring junctions are shown with unspecified stereochemistry as in **25**. Steroids with a *cis* AB ring junction are said to be in the 5β-series (referring to the ring junction hydrogen) and their skeletal names are prefixed by 5β, whilst those which are *trans*-fused are prefixed by 5α. Bile acids like lithocholic acid (**25**), 3α-hydroxy–5β-cholan–24-oic acid, are members of the 5β-series.

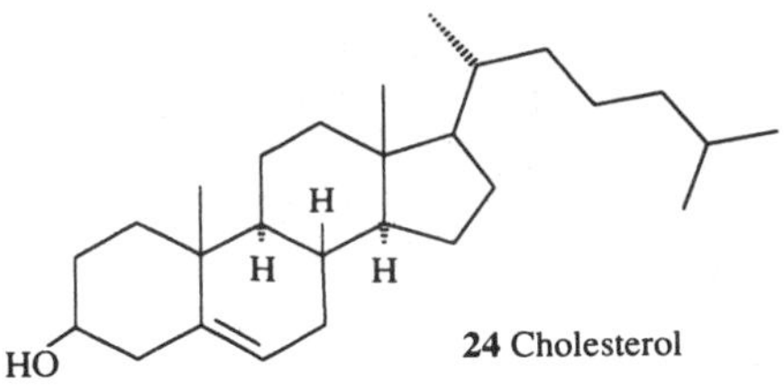

24 Cholesterol

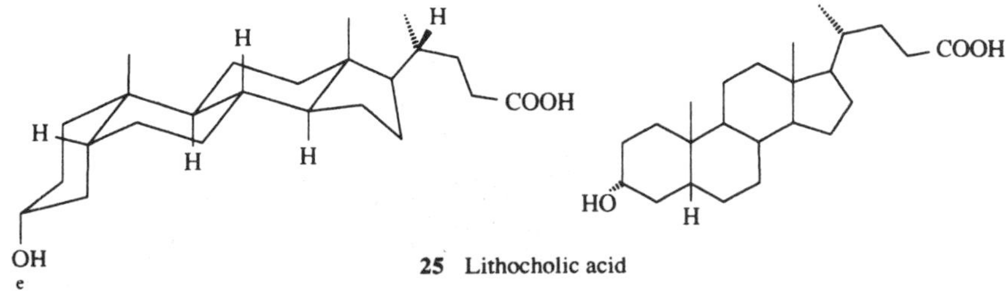

25 Lithocholic acid

6.2.3 Three-dimensional Structure in Steroids

Modern rationalizations of steroid reactions emphasize the steric environment of functional groups and so steroids need to be envisaged as their three-dimensional conformations although, for economy of effort and space, they are not generally shown in this way. All naturally occurring steroids have a *trans* BC ring junction and, with the exception of the cardiac glycosides (**26**) such as the heart stimulant digitoxin, found in foxgloves, they also have a *trans* CD ring junction. The three most commonly encountered shapes thus result from differences in the AB ring junction which may be *trans* as in androsterone (**27**), *cis* as in lithocholic acid (**25**), or unsaturated as in progesterone (**22**).

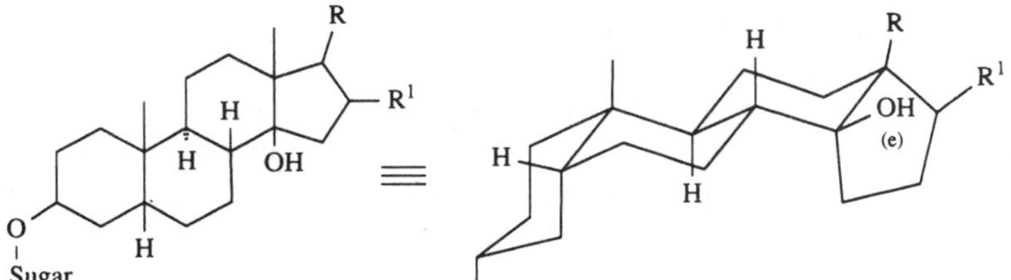

26 Cardiac Glycosides

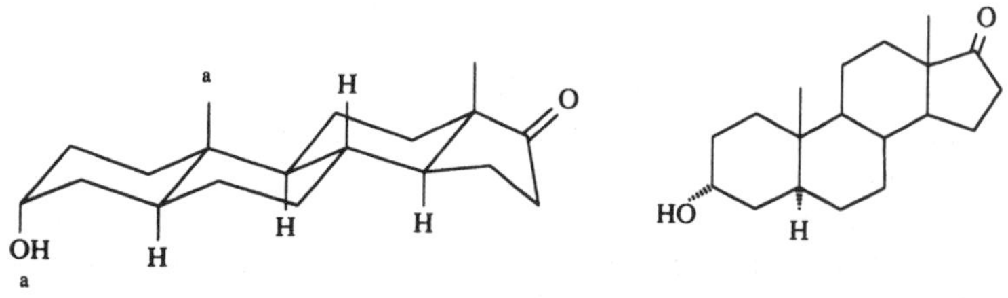

27 Androsterone

The *trans* BC ring junction imposes a rigidity on steroids; no alternative chair forms are attainable by any ring but subtle conformational changes occur easily. Introduction of a bulky substituent causes a deformation of the ring to which it is attached and this change is relayed through the ring junctions to alter each of the other rings. In some cases severe strain in a chair conformation will result in the appearance of a preferred boat conformation; an example is given in Section 6.4.2.3.

In the 5α-series the rings are approximately coplanar, giving an upper β-face which is very hindered by the 'angular' methyl groups and by the side-chain on C17 in cholestanes, pregnanes, and androstanes. In the 5β-series definite upper and lower surfaces are still discernible although now curved. Regio- and stereoselectivity in both series is mainly governed by the large hindrance on the β-face, but whether a substituent is axial or equatorial (indicated by 'a' and 'e' in structures **22** and **25–27**) may also be important, especially in ring A.

6.3 STRUCTURES AND BIOLOGICAL ACTIVITIES OF STEROIDS

Cholesterol (**24**) is the only steroid which is plentiful in vertebrates, but it appears not to have hormonal activity. The hormonal steroids are present in very low concentrations, typically 10^{-9} mol l^{-1} in blood, but they have powerful effects on their target tissues. Being relatively small and non-polar they diffuse easily through membranes into cells of many types but they have little effect except on their target cells which have receptor proteins in the cytoplasm able to complex with the hormone. This complexation activates the receptor protein to bind to certain non-histone proteins on the cell's chromatin thereby stimulating the transcription of certain genes and leading in turn to the synthesis of specific proteins.

The ability of receptor proteins to bind strongly to specific steroids depends principally on small differences in the overall steroidal shape. Although the male sex hormone, testosterone, differs from the female sex hormone, progesterone, merely in the substitution of an acetyl for a hydroxyl group at C17, they bind little to each other's receptor proteins and so elicit entirely different responses. However, specificity is rarely absolute and steroidal therapy often produces undesired hormonal responses as side effects.

Hormonal steroids are classified according to their site of action or of biosynthesis into oestrogens, progestogens, androgens, and corticosteroids.

6.3.1 Androgens

Androgens are male hormones necessary for spermatogenesis and for the development of male sexual characteristics and so a deficiency may lead to a low sperm count and impotence. They may also produce protein anabolic effects such as acceleration of skeletal growth and muscle growth.

Testosterone (**28**, 17β-hydroxyandrost-4-en-3-one) is the most potent natural androgen but if given orally it has little effect since, to reach the blood stream, it must pass through the liver and here it is metabolized (Scheme 6.1) to the weakly

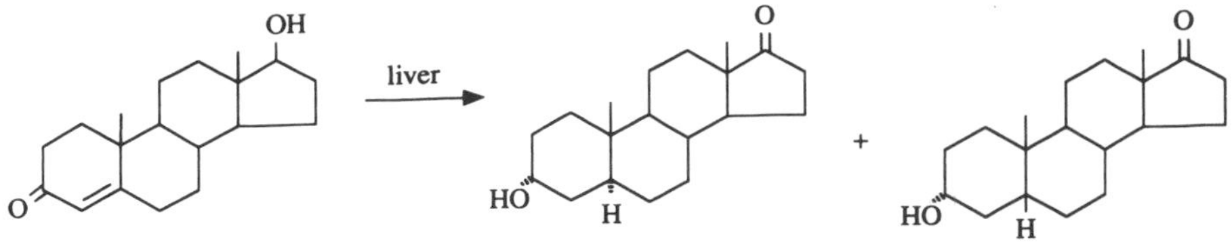

liver

28 Testosterone

29 Androsterone

30 Etiocholanone

Scheme 6.1 *Metabolism of testosterone*

androgenic androsterone (**29**) and the inactive etiocholanone (**30**). For oral activity a substituent at C17 is added to reduce metabolic destruction; 17α-methyltestosterone (**31**) is a widely used orally active androgen. Care is needed with the dosage of androgens since they suppress the release of follicle stimulating hormone (FSH) from the pituitary gland of the brain. FSH stimulates the production of testosterone (principally in the testis) and so its suppression will cause endogenous testosterone levels to fall. Abrupt cessation of steroid therapy may leave abnormally low endogenous steroid concentrations followed by a 'rebound' to abnormally high levels; to avoid serious side-effects withdrawal of steroidal drugs should be gradual.

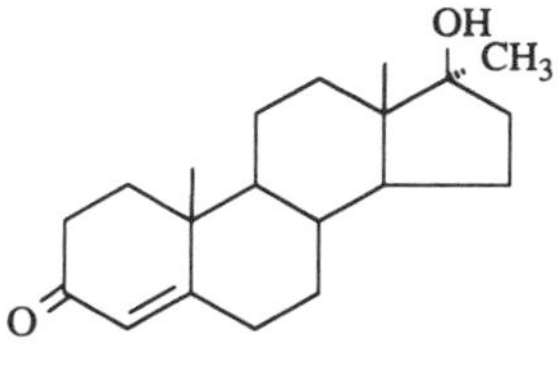

31 17α-methyltestosterone

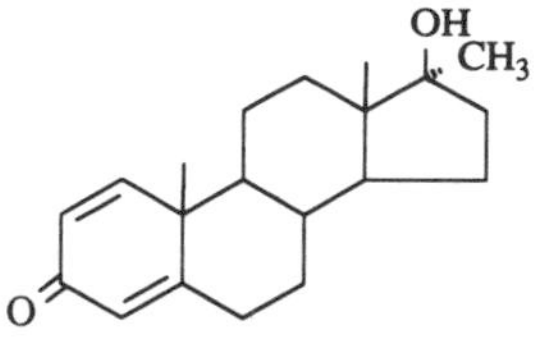

32 Methandrostenolone

Anabolic drugs having low androgenicity are useful in the treatment of underdeveloped children and for patients having debilitating diseases or in convalescence. Methandrostenolone (**32**) is widely used in this way as is 19-nortestosterone (nandrolene **23**) which has the muscle-building power of testosterone with only one tenth of its androgenicity. Women normally have only one fiftieth of the testosterone blood levels of men but can tolerate low doses of androgens without undesirable effects. Large doses of anabolic or androgenic steroids produce low fertility in either sex through suppression of FSH release.

6.3.2 Progestogens

Progestogens are female hormones principally concerned with attainment of conditions suitable for pregnancy. Progesterone (**34**) is the principal one and is biosynthesized from cholesterol via the weak progestogen pregnenolone (**33**). Metabolic reduction or side-chain cleavage, principally in the liver, gives products such as the diol (**35**, Scheme 6.2) with little progestogenic activity. Such metabolism is much slower for derivatives having a 17α-substituent, as in

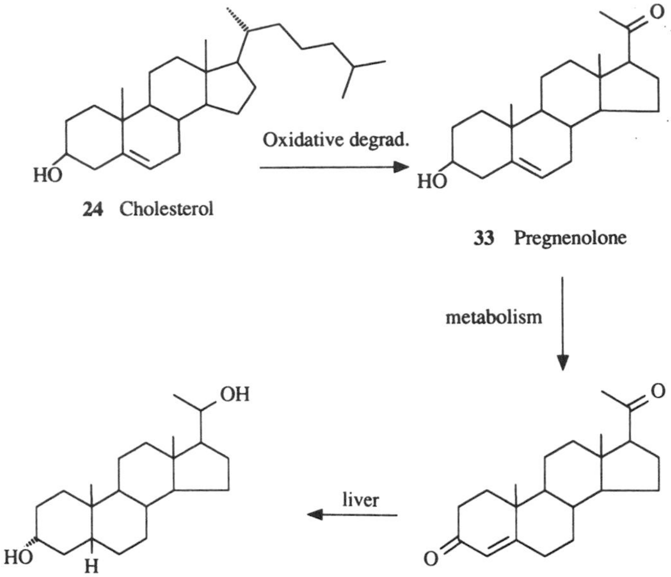

HO
24 Cholesterol

Oxidative degrad.

O
HO
33 Pregnenolone

metabolism

O
O
34 Progesterone

liver

OH
HO H
35 Inactive metabolite

Scheme 6.2 *Biosynthesis and metabolism of progesterone*

17α-hydroxyprogesterone acetate (**36**) having 10 times the oral activity of progesterone or medroxyprogesterone acetate (**37**) with 20 times the oral activity of progesterone. The enlargement of the uterus and fallopian tubes caused by administration of progestogens in conjunction with oestrogens was used to treat infertility in the early 1950s. Of those so treated 15% became pregnant after withdrawal of the hormone therapy and the momentous observation was made that none of the patients became pregnant *during* treatment. An imaginative combined study on the contraception/fertility inducing effects of various steroids followed and led to the development of the contraceptive pill. A composition of 10 mg synthetic progestogen (norethynodrel, **38**) with 0.15 mg synthetic oestrogen (mestranol), known as enovid, was the first, becoming available in 1955 and approved for US use in 1960.

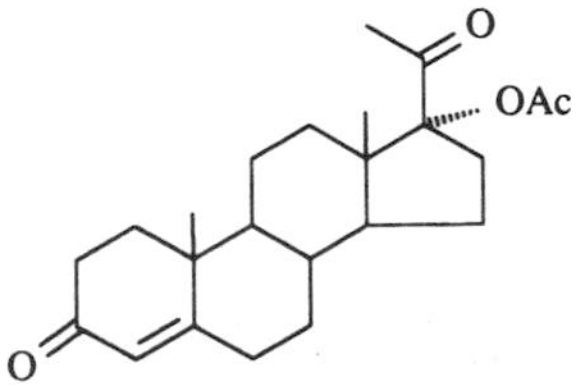

O
OAc
O
36
17α-Hydroxyprogesterone acetate

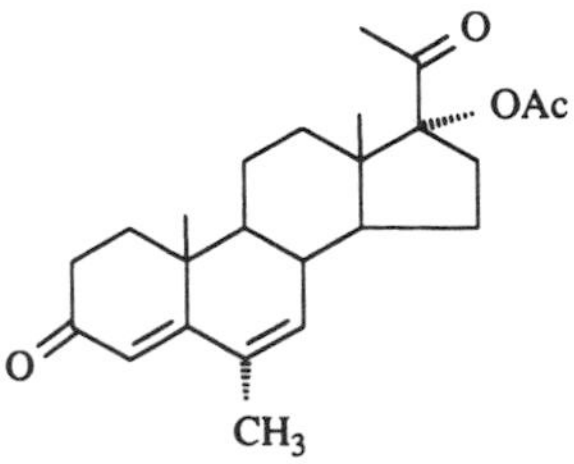

O
OAc
O
CH_3
37
Medroxyprogesterone acetate

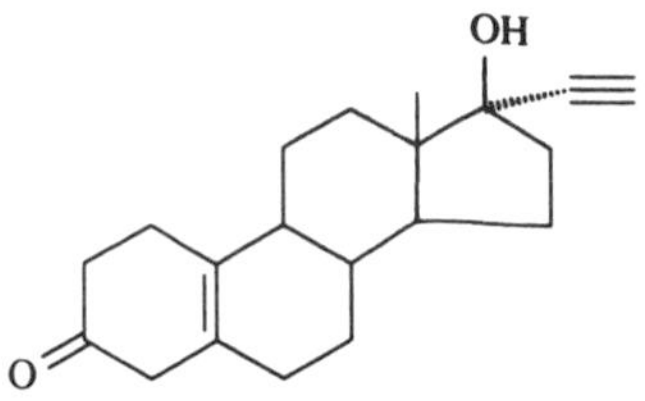

38
Norethynodrel

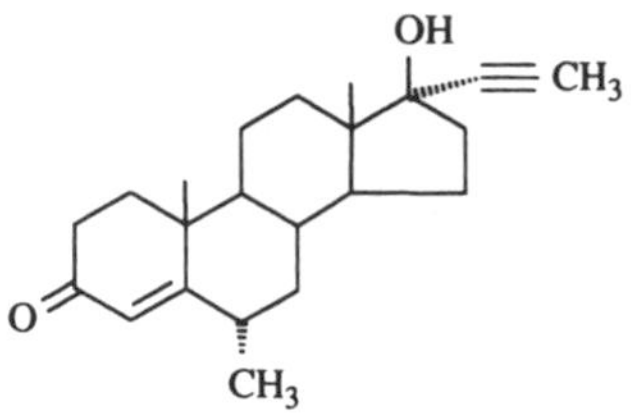

39
Dimethisterone

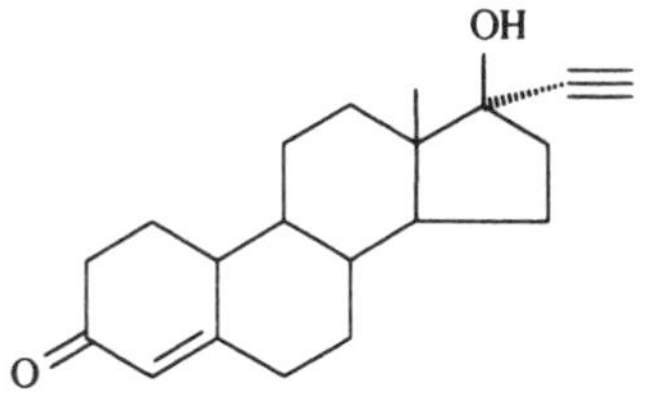

40
Norethisterone

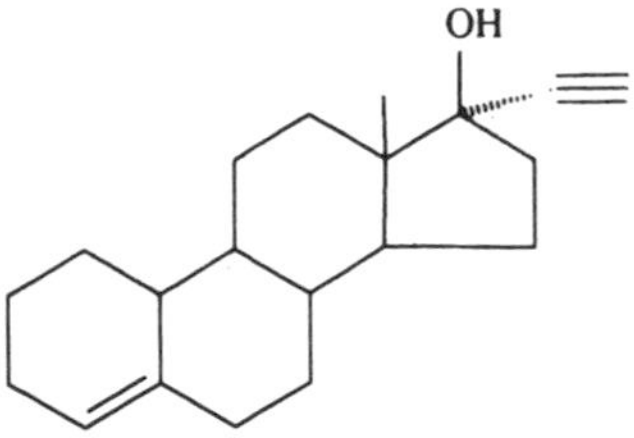

41
Lynestrol

The most commonly used oral progestogens are not derivatives of progesterone but are 20α-ethynyl derivatives of testosterone or 19-nortestosterone. Dimethisterone (**39**), 19-norethisterone (**40**), norethynodrel (**38**), and lynestrol (**41**) are examples of some used extensively in contraceptive pills and for hormone replacement therapy. A low dose of one of these is all that is required for contraception (as in 'mini pills' such as *micronor* containing 0.35 mg norethisterone) but combination with an oestrogen (in 'combination pills') prevents irregular bleeding. Medroxyprogesterone acetate (**37**) administered as a deep intramuscular injection (as in the drug *depo-provera*) provides contraception for up to six months. The low polarity imbued on quingestanol acetate (**42**) by the cyclopentyl ether at position-3 is responsible for its prolonged biological lifetime allowing its use as an oral once-a-month contraceptive.

There is not complete independence of the androgenic and progestogenic

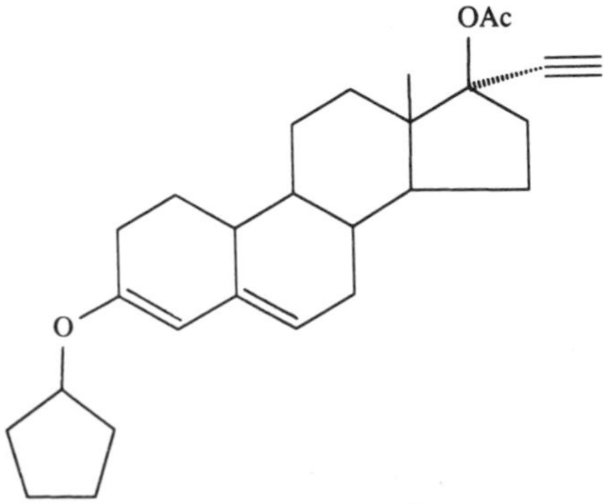

42 Quingestanol Acetate

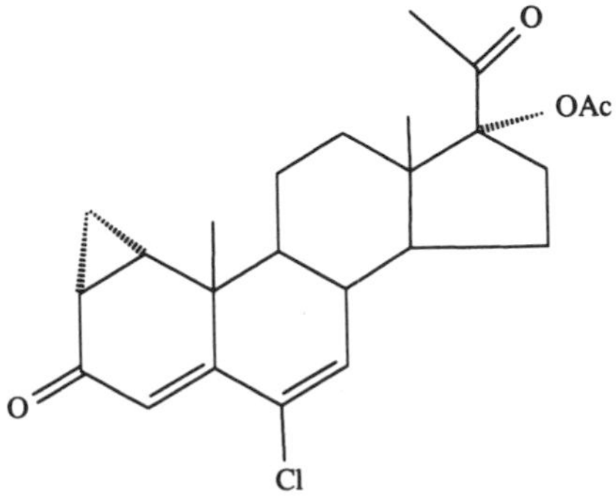

43 Cyproterone Acetate

activities of steroids. Cyproterone acetate (**43**), for example, is a progestogen but also reduces male libido and fertility through competitive inhibition of testosterone and is used to treat male hypersexuality.

6.3.3 Oestrogens

Oestrogens (pronounced as it is spelt in the USA: estrogens) are responsible for female sexual characteristics including the regulation of the menstrual cycle. Their main origin is the ovaries except in pregnancy when the placenta produces large quantities of oestrogens and progestogens.

The most potent natural oestrogen is 17β-oestradiol (**44**) which is biosynthesized from pregnenolone (**33**) in several steps. Oestradiol is converted in the liver (Scheme 6.3) into oestrone (**45**) and thence into oestriol (**46**), both of which have only about one tenth of the activity of oestradiol itself.

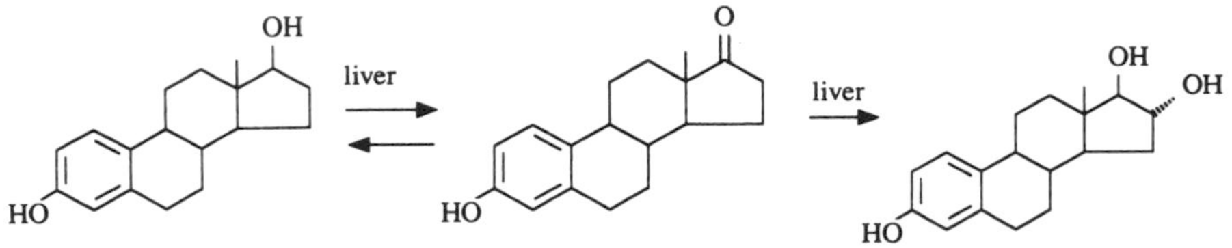

44 17β-oestradiol **45** oestrone **46** oestriol

Scheme 6.3 *Metabolism of oestradiol*

There are two commonly used orally active oestrogens – 17α-ethynyloestradiol (**47**) and its 3-methyl ether mestranol (**48**). The 3-cyclopentyl ether (quinestrol **49**) has been used in conjunction with quingestanol for long-lasting contraceptive action but more commonly quingestanol has been used alone. Other medicinal uses for oestrogens are the treatment of oestrogen deficiency, prostate cancer, and the suppression of lactation.

Modification of the structure of an oestrogen other than by addition of a 3-*O*-carboxyl, 3-*O*-alkyl, or 17α-ethynyl group generally results in removal of oestrogenic activity. For example the benzene ring is essential (analogues with two double bonds in ring A are not oestrogenic), each of the seven stereoisomers of oestrone has much reduced or absent activity, and the addition of alkyl groups or oxygen almost always causes much decreased activity.

A notable exception to this rule is *trans*-diethylstilboestrol, which is a potent orally active oestrogen having a non-steroidal structure (**50**) which appears to owe its activity to having the same spacing between its hydroxyl groups as is

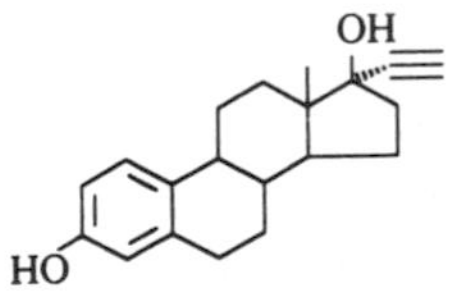

47 Ethynyloestradiol 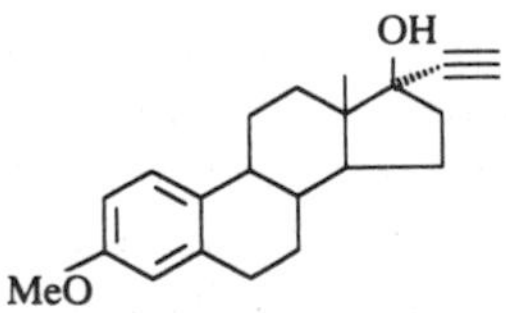 **48** Mestranol 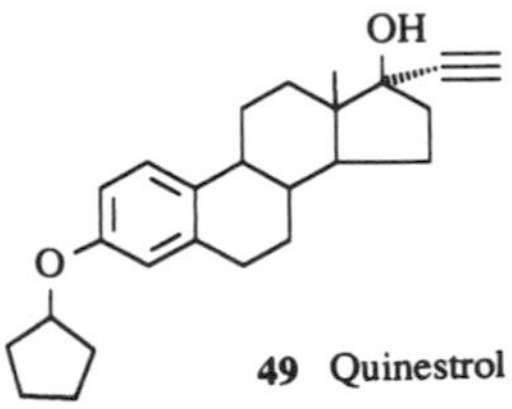 **49** Quinestrol

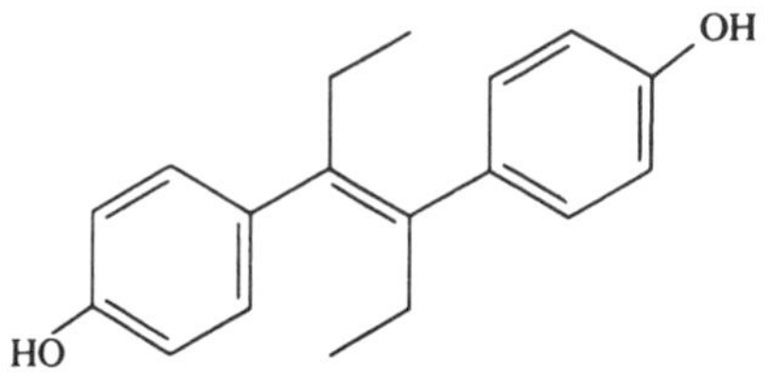

50 *trans*-diethyl stilbestrol

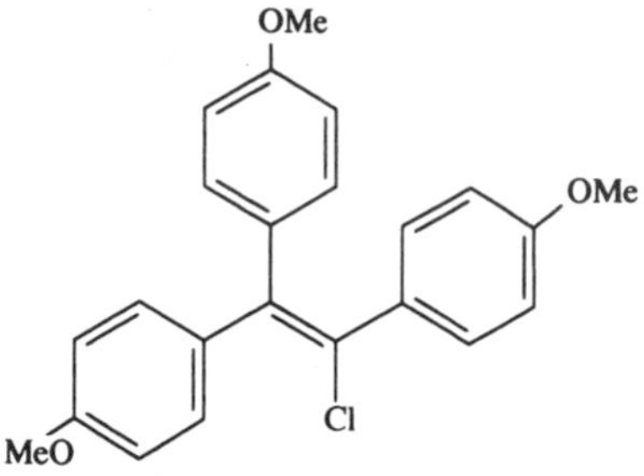

51 Trianisylchloroethylene (TACE)

critical to the physiological activity of oestradiol. Being an inexpensive oestrogen it was widely used but has now been shown to cause human cancer. Structurally similar compounds such as trianisylchloroethene (TACE) (**51**) are used to treat infertility. They are anti-oestrogenic, causing release of gonadotrophins (pituitary hormones eliciting sex-steroid production in the gonads) by interference with the feed-back mechanism.

6.3.4 Corticosteroids

This group of steroids, characterized by an oxygen function on C11, is produced by the adrenal cortex (*i.e.* outer part of adrenal glands near the kidneys) under the influence of corticotrophic hormones from the pituitary. The corticosteroids are split into two main groups: the glucocorticoids, which control fat, protein, calcium, and carbohydrate metabolism, and the mineralocorticoids controlling sodium and potassium levels. Vertebrates inhabiting terrestrial or fresh-water environments produce mainly aldosterone (**52**) as the mineralocorticoid and corticosterone (**53**) as the glucocorticoid. Inhabitants of salt water substitute hydrocortisone (otherwise known as cortisol, **54**) for aldosterone.

Cortisone (**55**), a powerful glucocorticoid, also occurs naturally in mammals and the discovery of its antirheumatic properties (1949) led to its acclaim as a wonder drug. The affect of cortisone on patients who had been crippled with rheumatoid arthritis was indeed dramatic: some who had been unable to walk because of badly inflamed knee joints were able to run shortly after an injection of the drug. Unfortunately, continual treatment is necessary to maintain the beneficial anti-inflammatory effect, which is attributed to cortisone's glucocorticoid activity, and serious side-effects, principally due to overlapping mineralocorticoid activity, soon become apparent. Cortisone and other mineralocorticoids produce retention of sodium in cells and the corresponding inflow of water leads to oedema (fluid retention causing swelling) and raised blood pressure. The alteration of calcium and protein metabolism produces brittle bones (osteoporosis) and a characteristic moon-shaped face develops in the patients.

Considerable efforts were then directed in a search for steroids which showed anti-inflammatory (glucocorticoid) activity without mineralocorticoid activity. Complete separation has not been achieved but a moderately successful struc-

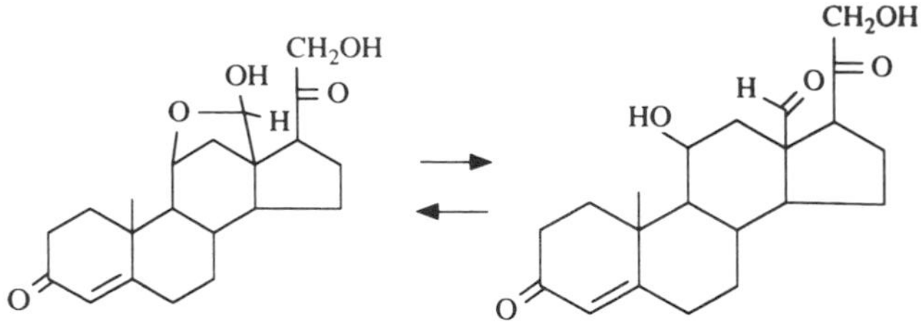

52 Aldosterone

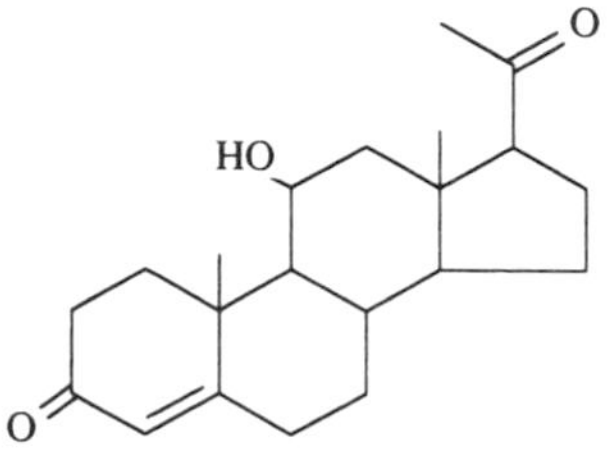

53 Corticosterone

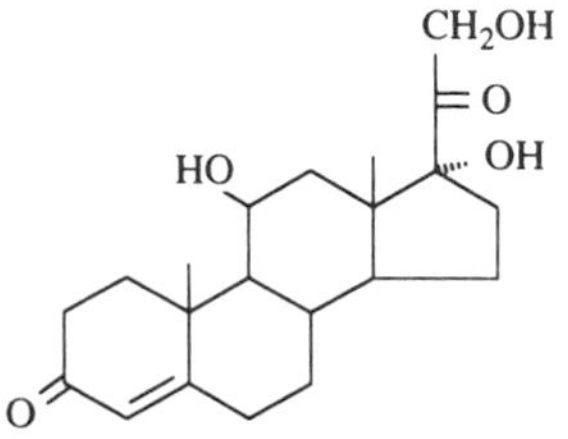

54 Hydrocortisone (cortisol)

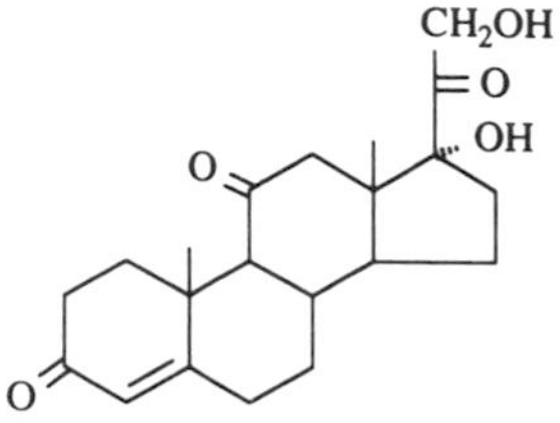

55 Cortisone

tural framework (**56**) based on hydrocortisone has been identified. Prednisolone, triamcinolone, betamethasone, and dexamethasone are synthetic glucocorticoids of this type, widely used against rheumatoid arthritis, severe asthma, and other inflammatory conditions. For use in creams and ointments triamcinolone acetonide (**57**) has ten times the activity of triamcinolone itself primarily because it is less polar. High topical activity with little systemic activity is similarly achieved by lipophilic derivatives of other glucocorticoids such as the 17-valerate of betamethasone. There are many synthetic glucocorticoid drugs fitting this structural pattern (**56**), some having additional 6α-substituents, but the widely used drug prednisone (**58**) is based on cortisone instead. Treatment with the endogenous hydrocortisone and cortisone is generally reserved for replacement therapy, such as in Addison's disease, where the natural steroids are at low levels. Synthetic mineralocorticoids like fludrocortisone (**59**) have also been developed for use in replacement therapy.

6.3.5 Bile Salts, Cholesterol, and Vitamin D

Most vertebrates including birds, fish, reptiles, and man secrete bile into the gut in order to help in the digestion and absorption of fats and other lipids in the small intestine. The active agents are bile salts which are composed of a non-polar steroid nucleus with a polar side-chain resulting in detergent properties. In the presence of these steroidal detergents lipids are emulsified, promoting their enzymic hydrolysis and facilitating absorption into the lymphatics. Chemical hydrolysis of the bile salts leads to bile acids whose structures (*e.g.* **60**) are normally that of C_{24} steroids having *cis* AB ring junctions and a carboxylic acid group at the end of the side-chain. In the major bile salt, known confusingly as glycocholate (**61**), cholic acid is linked as its amide to glycine, NH_2CH_2COOH;

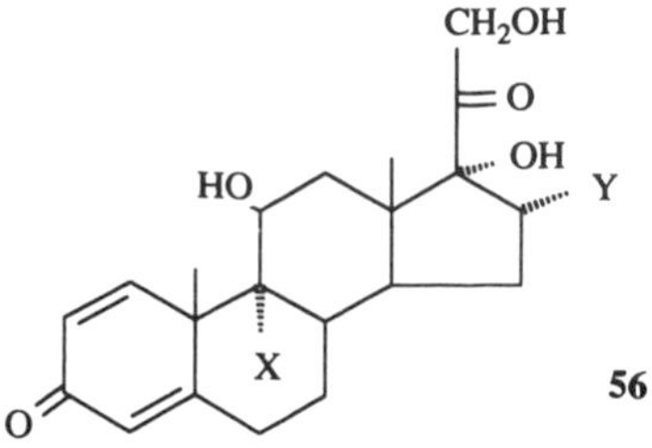

X	Y	
H	H	Prednisolone
F	OH	Triamcinolone
F	CH_3	Dexamethasone
F	β-CH_3	Betamethasone

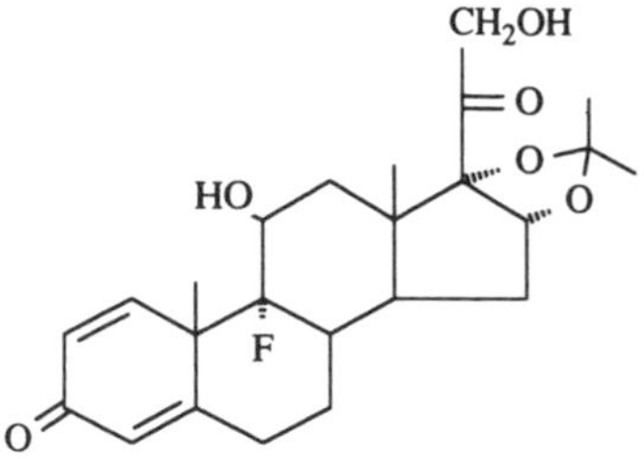

56

57 Triamcinolone acetonide

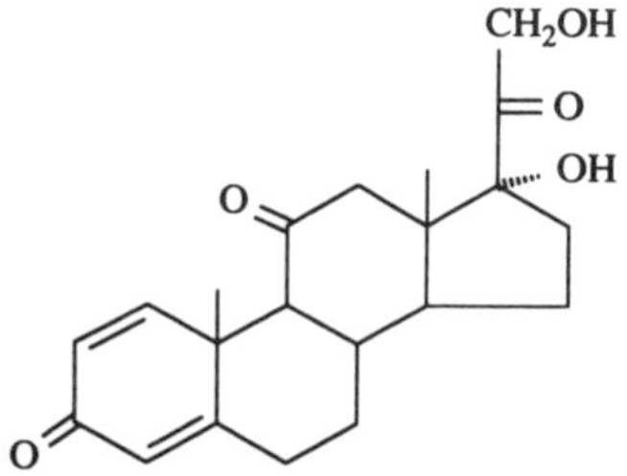

58 Prednisone

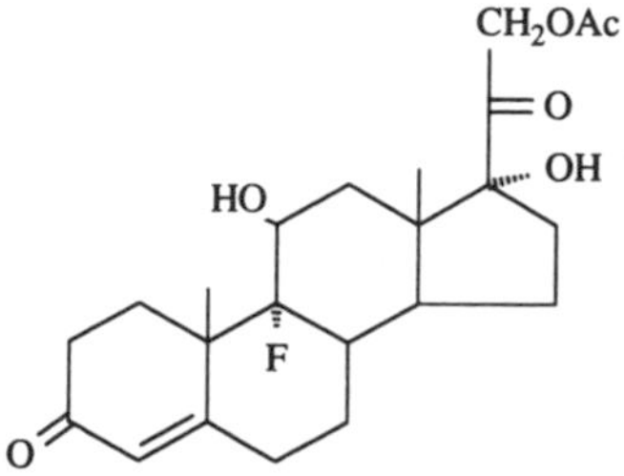

59 Fludrocortisone

other bile salts are similarly comprised of a bile acid amide of glycine or taurine $NH_2CH_2CH_2SO_3H$. In addition to the function of bile salts as detergents their excretion is the human body's principal means of eliminating excess cholesterol from which they are biosynthesized.

Much ingenious chemistry was applied to the elucidation of cholesterol's structure (see Fieser and Fieser, 'Steroids', Reinhold Publishing Corp. 1959; or

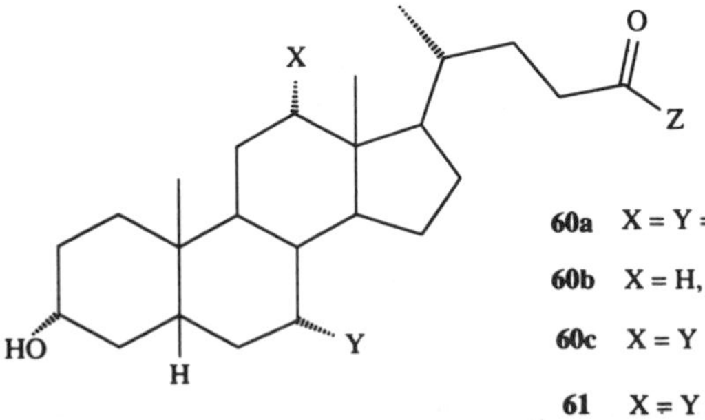

60a	X = Y = Z = OH	Cholic acid
60b	X = H, Y = Z = OH	Desoxycholic acid
60c	X = Y = H, Z = OH	Lithocholic acid
61	X = Y = OH, Z = $NHCH_2CO_2^-$	Glycocholate

Templeton, 'Chemistry of Terpenoids and Steroids', Butterworth & Co, 1969, for details) and then similar efforts were expended on establishing its biosynthetic route. Each of the carbons in cholesterol derives from acetic acid which is converted sequentially into *R*-mevalonic acid, isopentenyl pyrophosphate, squalene, lanosterol, and finally cholesterol (Scheme 6.4). Cholesterol or a close analogue is essential in cell membranes of terrestrial animals and it also acts as the biosynthetic precursor of the hormonal steroids.

Vitamin(s) D refers to the family of compounds able to prevent or treat rickets, a bone-softening disease caused by poor control of calcium and phosphorus metabolism, which used to be common in children exposed to little sunlight. Most natural foods have a low content of this vitamin but fish liver oils are a rich source and were used to treat rickets. Today the disease is rare since milk, margarine, and other foods in the USA and Europe are artificially fortified with the vitamin but bone-softening due to vitamin D deficiency is still found in strictly Muslim Arab women who ensure that none of their skin is exposed to sunlight. In sunlight human skin produces vitamin D_3 (**64**) by photochemical reaction of 7-dehydrocholesterol; vitamin D_2 (**63**) may also be produced from

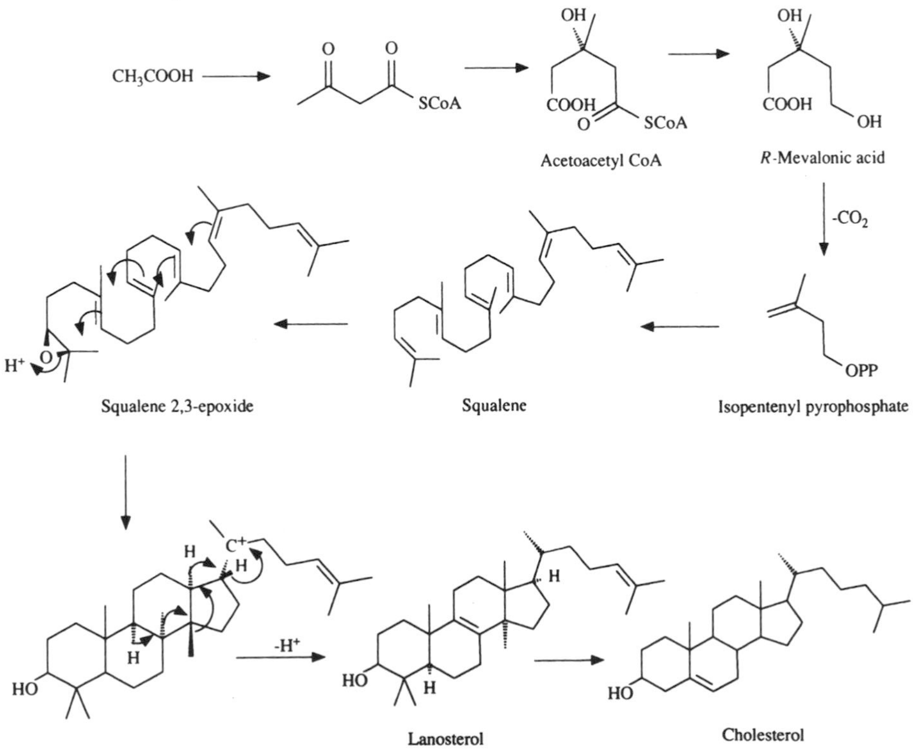

Scheme 6.4 *Biosynthesis of cholesterol*

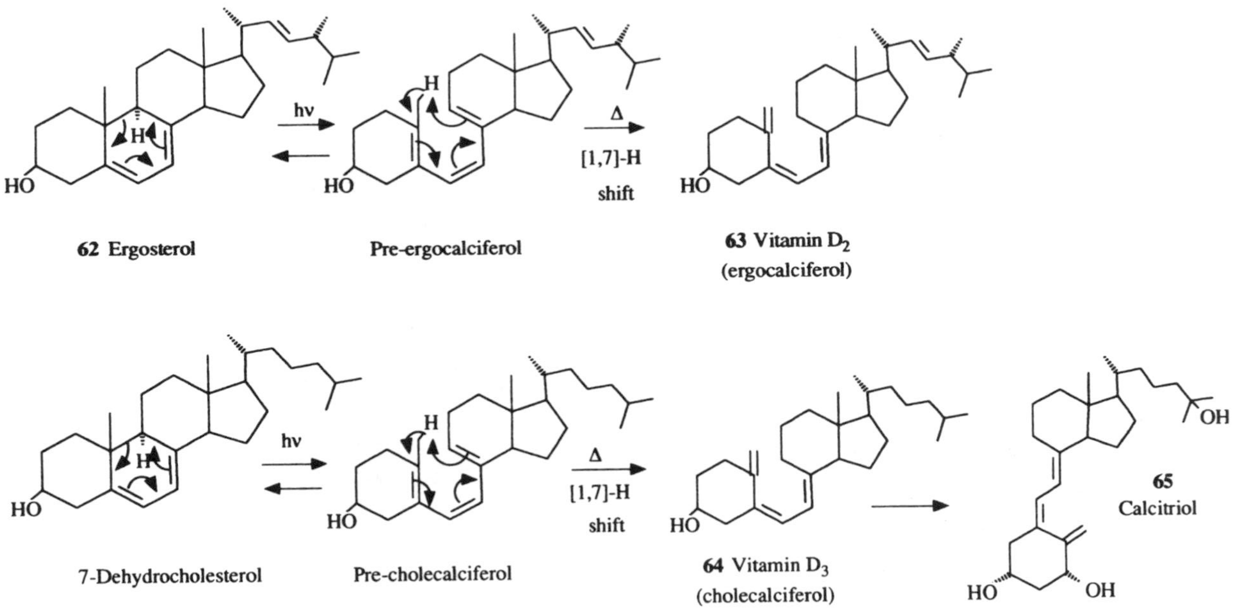

62 Ergosterol

hν

Pre-ergocalciferol

Δ

[1,7]-H shift

63 Vitamin D_2 (ergocalciferol)

7-Dehydrocholesterol

hν

Pre-cholecalciferol

Δ

[1,7]-H shift

64 Vitamin D_3 (cholecalciferol)

65 Calcitriol

Scheme 6.5 *Biosynthesis of D vitamins*

ergosterol (**62**), a steroid abundant in yeast. Photochemical activation promotes the initial ring-opening reactions (Scheme 6.5) of ergosterol and 7-dehydrocholesterol to give pre-ergocalciferol and pre-cholecalciferol by population of their lowest unoccupied molecular orbitals which have the correct symmetry for the conrotatory six-electron electrocyclic reactions, symmetry-forbidden from their ground states. The alternative disrotatory ring-openings which are symmetry-allowed from the ground state under thermal conditions would generate *trans* double bonds in either ring A or C and so are impossible. The next step is a thermal [1,7]-H shift which is symmetry-allowed and geometrically feasible if antarafacial. The active form of the vitamin, calcitriol (**65**), is produced by metabolic dihydroxylation of vitamin D_3 and has been shown to stimulate intestinal phosphate and intestinal calcium transport and, paradoxically, to cause a rise in plasma calcium at the expense of bone.

6.3.6 Plant Steroids and Toad Poisons

Plant steroids are classified into sterols, cardiac glycosides (toad poisons are similar), saponins, and alkaloids. Sterols are 3-hydroxy-steroids containing an aliphatic side-chain; we have already encountered cholesterol and ergosterol. Sterols are widely distributed, occurring both free and in combination not only in plants but also in most other organisms including bacteria, fungi, algae, marine invertebrates such as sponges and sea anemones, and higher animals. The major function of plant sterols is thought to be, as in animals, to act as a building block in membranes. Spinasterol (**66**) and sitosterol (**67**) have typical structures for dominant plant sterols but sterols occurring in minor proportions are frequently found to bear one to three methyl groups at positions 4, 4′, and 14 and may have additional double bonds (Δ^7, Δ^8, Δ^{24}, or Δ^{28}).

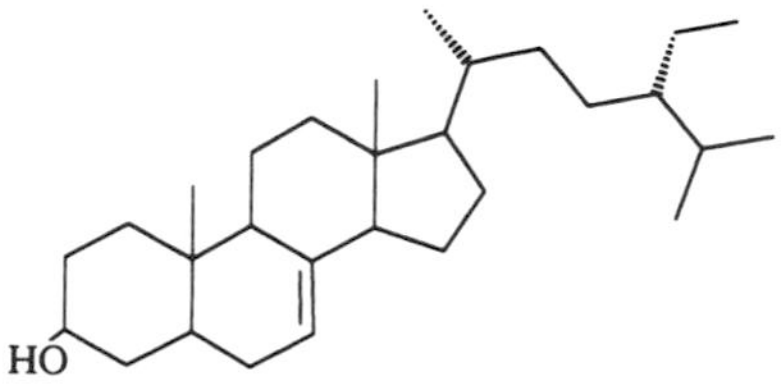

66 Spinasterol

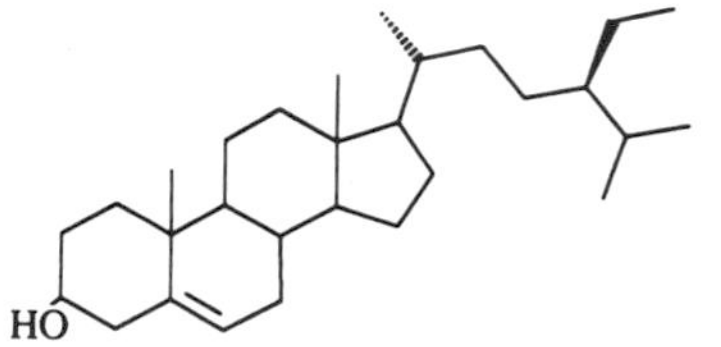

67 Sitosterol

There are two types of plant steroidal glycosides, the cardiac glycosides which have a powerful action on heart muscle, and saponins, which (as the name implies) give foams with water. Medicinally useful plants of the *digitalis* family such as the foxglove (an extract of its seeds and leaves, 'digitalis', was first used in heart therapy by William Withering in 1785) contain digitoxin (**68**). Similarly active steroids are found in many other plants and also in the skin secretions of toads and have been used both medicinally and in arrow poisons. The steroidal aglycone (or genin) is responsible for cardiotoxicity and generally has *cis*-AB and *cis*-CD ring junctions, 3β- and 14β-hydroxyl groups, and a 17β-(α,β-unsaturated

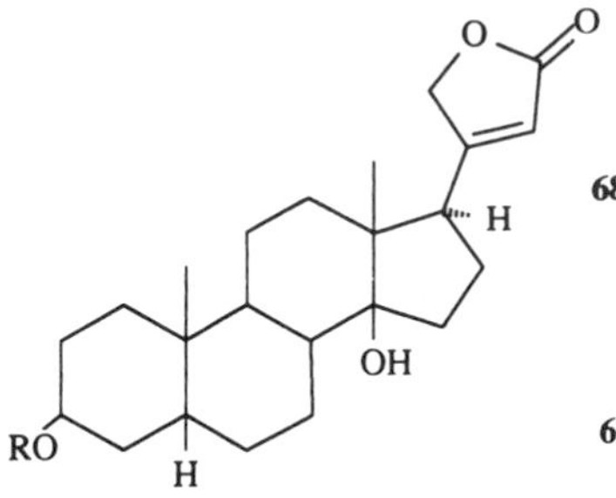

68 R = 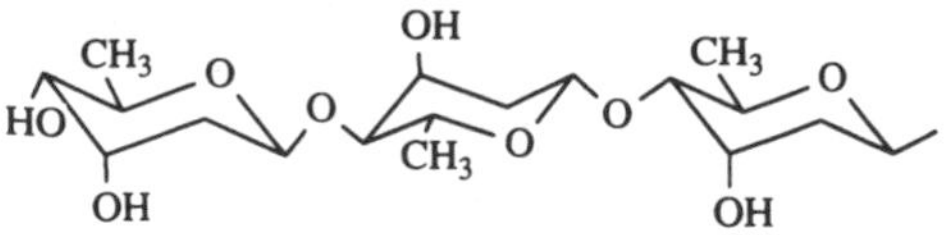 Digitoxin

69 R = H Digitoxigenin

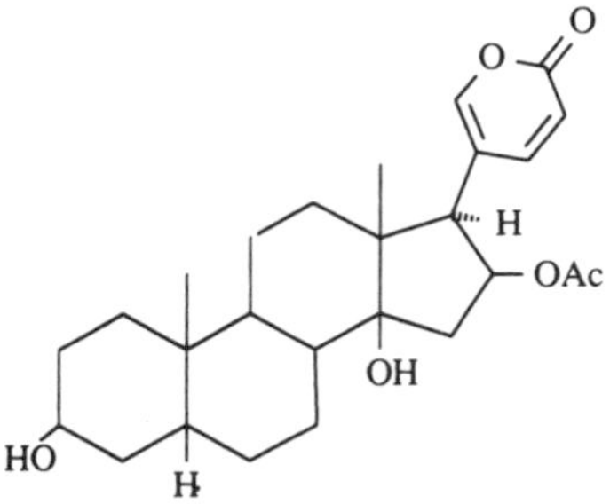

70 Bufotalin

five- or six-membered lactone) as exemplified by digitoxigenin (**69**) (from hydrolysis of digitoxin) and bufotalin (**70**) (from toad skin secretions). About twenty different sugars (both L- and D-configurations) have been found in cardiac glycosides and except for D-glucose they are 6-deoxy- or 2,6-dideoxy-hexoses which may be methylated on O3 and are uncommon or unknown elsewhere in nature. The unusual oligosaccharides are essential for the full activity of the cardiac glycosides, giving them their specificity for heart muscle, but full structures of only a few have been determined.

Saponin aglycones or sapogenins are C_{27} steroids with a spiroketal side-chain as exemplified by sarsasapogenin (**71**). The CD ring-fusion is *trans* and the only common structural variation to ring D and the side-chain is that C27 may be either axial or equatorial. They may be found in substantial amounts in some species, suiting them as starting materials for steroid partial synthesis (Section 6.5.1), but their biological functions are largely unknown.

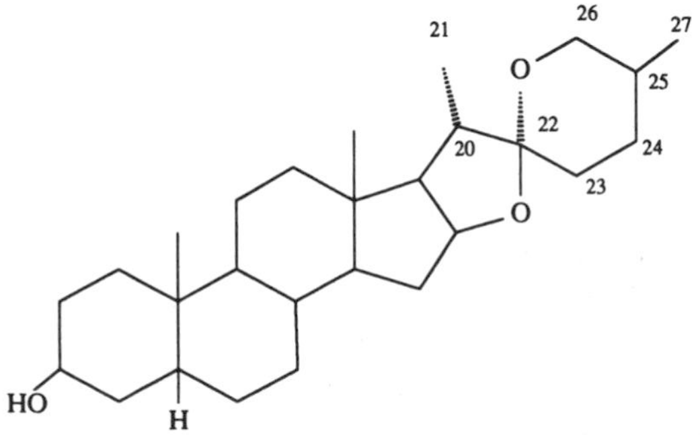

71 Sarsasapogenin

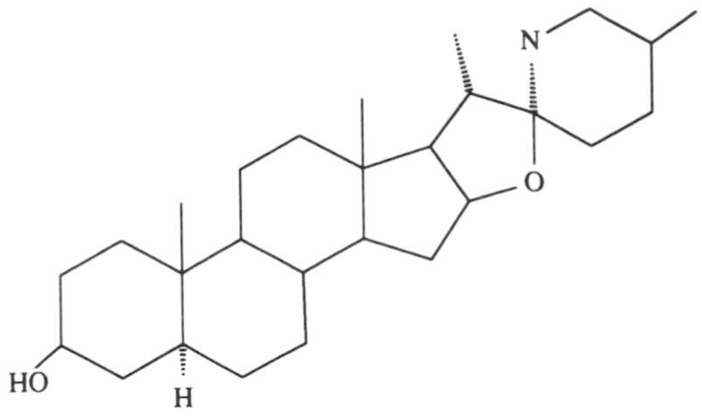

72 Tomatidine 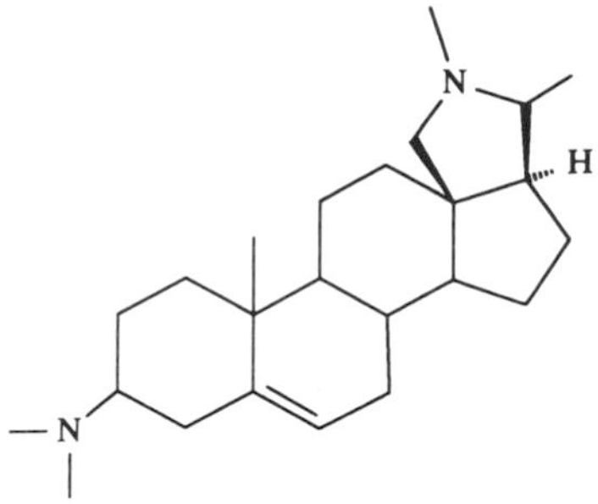 **73** Conessine

Steroidal alkaloids are naturally occurring nitrogen-containing steroids widely distributed amongst plants. Some, like tomatidine (**72**) the principal aglycone of the tomato plant, have antifungal and antibacterial activities while others (such as conessine, **73**) are antiamoebic.

6.3.7 Marine and Insect Steroids

Marine animals and plants contain a wide variety of sterols and their study is an active area of current research. Most marine sterols (*e.g.* **74**) are structurally very similar to cholesterol (or analogues having other unsaturation at positions-5 or -7) except in their side-chains, but some show more fundamental differences such as the A-nor-steroids (**75**) found in sponges. The majority of marine animals contain between ten and thirty different sterols, and some, like sponges and bivalve molluscs, do not have cholesterol as the predominant (or sometimes even detectable) sterol. The functions and biosynthetic pathways of marine sterols are under investigation, but in species with little cholesterol it seems inevitable that these sterols take its place in the plasma membrane.

Moulting, which is essential to insect development, is controlled by steroidal hormones, the most important of which is ecdysone (**76**). 20-Hydroxyecdysone has been identified as an alternative hormone and there are others with side-chains containing five- or six-membered lactones.

6.4 CHEMICAL PROPERTIES OF STEROIDS

Steroids are generally stable, crystalline compounds with similar chemical properties to those of simpler analogues except that their reactions tend to be highly stereo- and regio-selective. Much of our current understanding of stereochemistry is in fact founded on stereoselectivities initially observed in steroid reactions.

6.4.1 Regioselectivity

The reactivities of steroidal functional groups depend on their positions. The commonly encountered positions of steroidal carbonyl groups are 3, 11, 17 and 20 and their reactivities towards nucleophilic attack (*e.g.* hydride reduction,

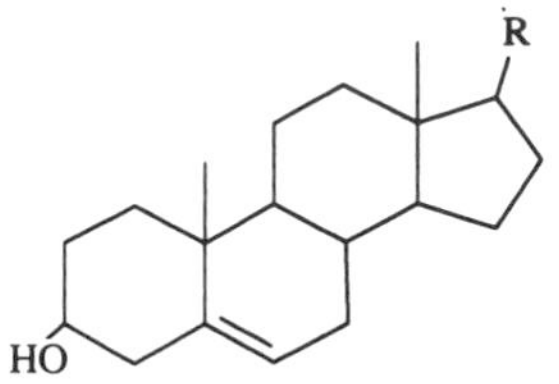

R

Gorgosterol

Strogylosterol

Cyclasterol

74 Marine steroids

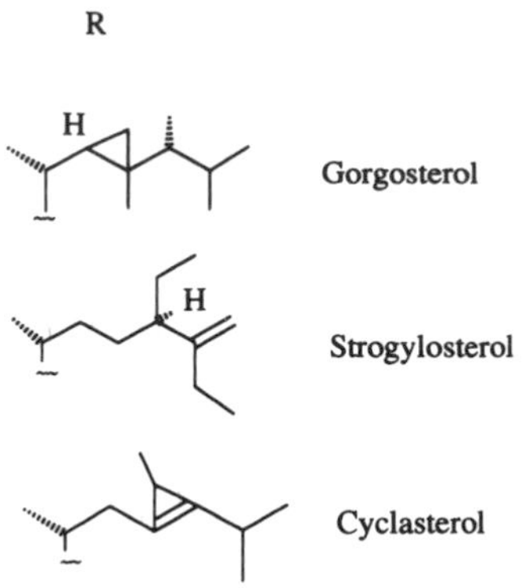

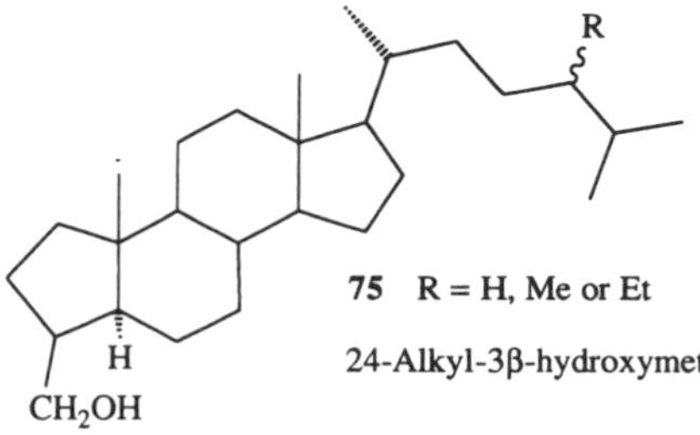

75 R = H, Me or Et

24-Alkyl-3β-hydroxymethyl-A-norcholestanes

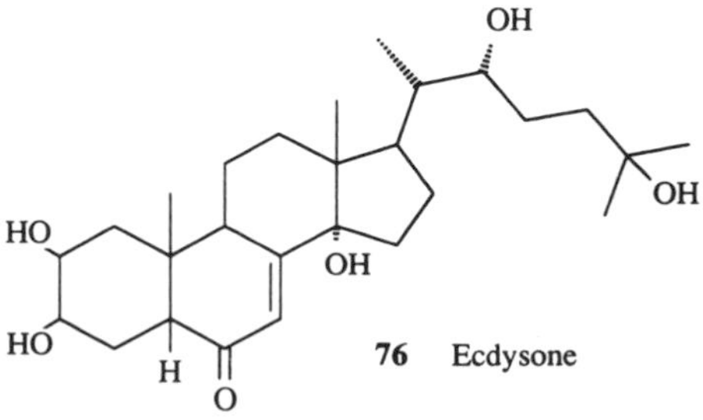

76 Ecdysone

ketal formation, and Grignard addition) are in the order 3 > 17 ⩾ 20 > 11, which is also the order of increased steric crowding. Products resulting from regioselective reaction just at the more reactive carbonyl groups may be obtained by addition of the correct number of molecular equivalents of reagent. For example 5α-androstane-3,17-dione (**77**, Scheme 6.6) may be selectively reduced to 3-hydroxy-5α-androstan-17-one. α,β-Unsaturation decreases reactivity towards reduction (but not ketal formation) and so sodium borohydride selectively reduces androst-4-ene-3,17-dione (**78**) at the 17-carbonyl giving testosterone (**28**). In each case the hydride adds to the less hindered lower face giving β-hydroxyl groups.

Selective ketal formation at less hindered positions is important for the blocking of chosen steroidal carbonyl groups. Derivatives having ketals at more hindered positions are also available by the partial hydrolysis of polyketals, which is selective for more exposed positions. Reaction (Scheme 6.7) of cortisone with ethylene glycol and acid thus protects the two less hindered carbonyl groups

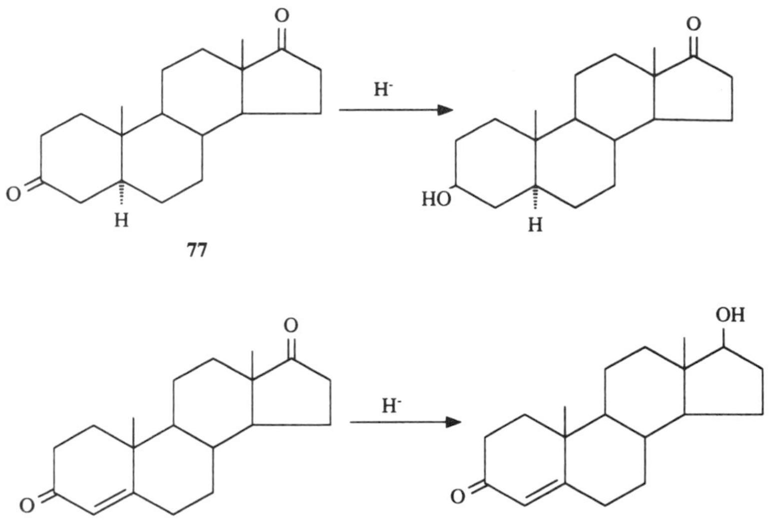

77

78

28 Testosterone

Scheme 6.6 *Regio- and stereo-selective reductions*

and dilute acid treatment of this diketal (**79**) gives a product (**80**) having a single ketal at position-20. Cortisone is a Δ^4-steroid and its carbon–carbon double bond is stabilized by conjugation with the 3-keto group. On modification of the carbonyl to a ketal this stabilization is removed and under the acidic conditions isomerization to a $\Delta^{5(6)}$-steroid occurs. Subsequent regeneration of the carbonyl gives back energetic advantage to a Δ^4-steroid. The shift of the double bond to the 5-position on ketalization of a Δ^4-3-ketone has also been used to protect it against hydrogenation since a double bond at position-5 (in common with those at 7, 8, or 14) is amongst the least reactive.

The reactivities of hydroxyl and most other functional groups are similarly affected by steric hindrance: those at the exposed position-3 tend to be the most reactive provided the reactions have transition states with greater steric demands than their starting materials. For groups at sp^3 carbons the orientation is also important; in general, axial groups are more hindered than equatorial ones and β-groups more hindered than α-ones. Addition and substitution reactions are usually slower in congested positions: for example, hydrocortisone (**54**) requires forcing conditions to acetylate its highly hindered axial 11β-position, and tertiary hydroxyl groups (such as in 5-hydroxy- and 17-hydroxy-pregnanes) are also difficult to acetylate. Hydrolysis of esters is also slow in the more hindered positions but certain reactions, such as chromic acid oxidation, are faster for sterically hindered positions because their rate-limiting steps have transition states with lower steric demands than have their starting materials. The triol (**81**, Scheme 6.8) may thus be oxidized at the 11β-hydroxyl without affecting the other hydroxyl groups.

Esters of 3β-hydroxy-Δ^5-steroids hydrolyse very much faster than their 3α-

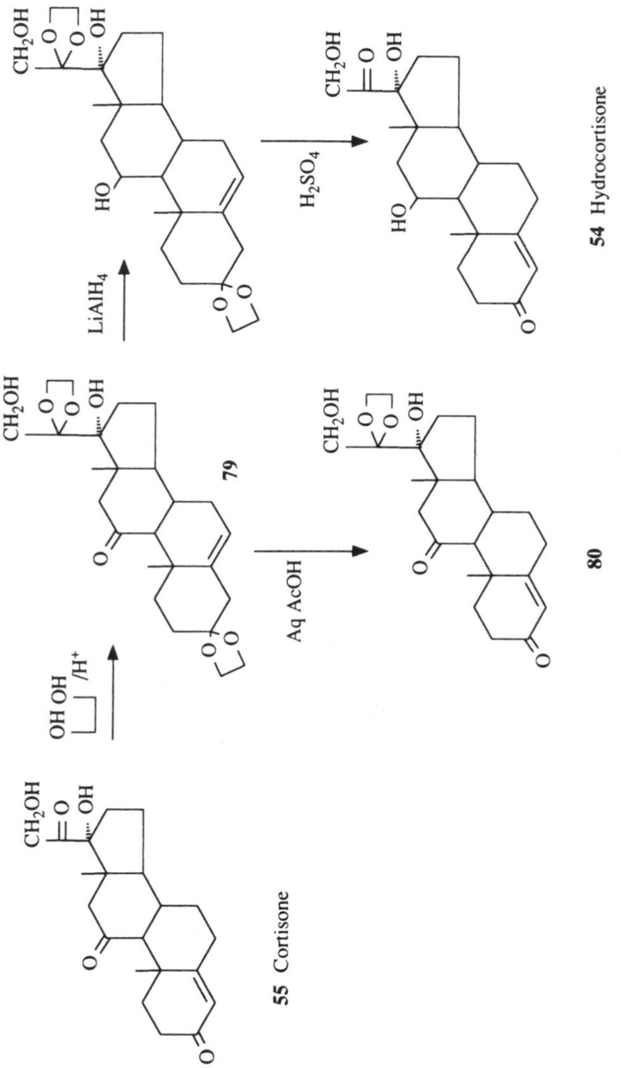

Scheme 6.7 *Selective ketal formation and reduction of cortisone*

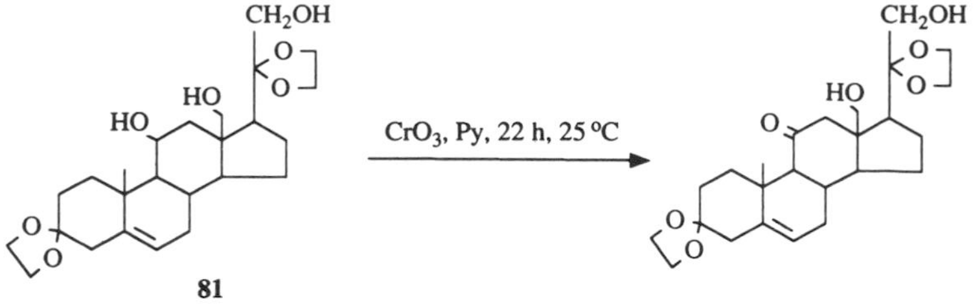

Scheme 6.8 *Selective oxidation at C11*

epimers and produce the 3β-hydroxy-Δ^5-steroids with retention of configuration. The explanation is that the 5,6-double bond participates in the displacement of the ester to give an intermediate non-classical carbocation (**83**, Scheme 6.9). Convincing evidence for this mechanism is provided in the acetolysis of cholesteryl tosylate (**82**) which gives a good yield of the '*i*-cholesterol acetate' (**85**) having a cyclopropane ring, with cholesteryl acetate (**84**). It is essential that the attacking π-orbital is behind the departing ester group (compare S_N2 displacements) and similar acetolysis of the 2α-analogue which can not achieve this correct disposition of the reacting orbitals (at least not in the chair conformation) yields not an *i*-steroid but mainly the diene (**86**, Scheme 6.9) by an *E*1 elimination.

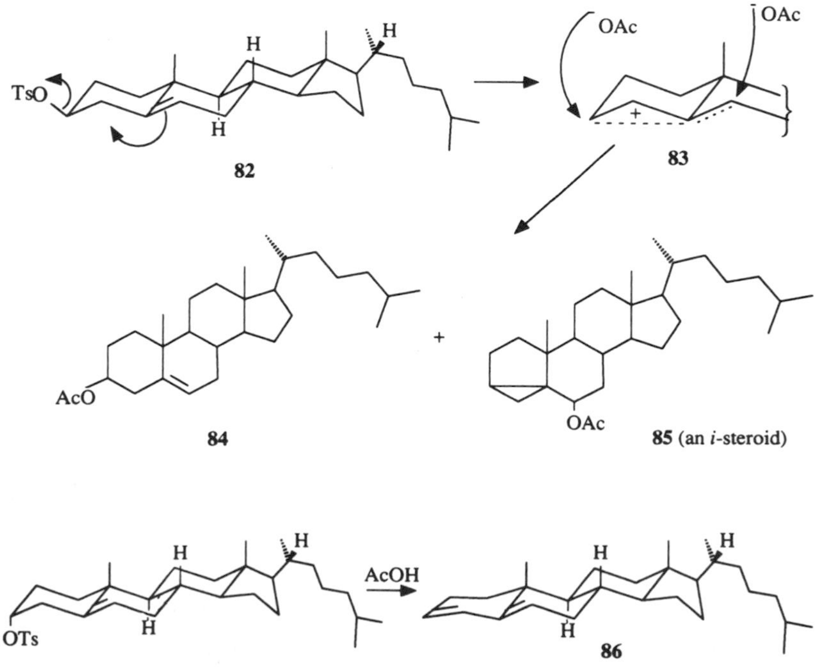

Scheme 6.9 *Acetolysis of cholesteryl tosylate and its 3α-epimer*

6.4.2 Stereoselectivity in Steroids

Many steroid reactions may yield two or more products differing in the orientation (α or β) of the groups participating in the reaction, and the predominant isomer may alter with the reaction conditions. If the reaction is reversible the thermodynamically preferred (*i.e.* more stable) isomer results, but for essentially irreversible reactions the most rapidly formed product is favoured and is that leading from the lower-energy transition state (normally resulting from reagent attacking the less hindered side).

6.4.2.1. Kinetically Controlled Reactions. Kinetically controlled reactions must be essentially irreversible and are typically fast and high-yielding reactions, like hydride reductions. In kinetically controlled reactions one expects (and generally obtains) products resulting from attack on the less hindered side of the starting material (termed steric approach control). The expectation of nucleophilic attack on the less hindered lower face of steroidal carbonyl groups is generally realized in crowded situations. Thus the hydride reduction of the diketal (**79**, Scheme 6.7) of cortisone shows steric approach control to give, almost exclusively, the 11β-hydroxy derivative even though this product is much more hindered than its 11α-epimer having an equatorial hydroxyl group. Removal of both ketal protecting groups yields hydrocortisone. Reduction of ketones at less hindered positions may also show steric approach control as noted in the hydride reductions (Section 6.4.1) of androstane-3,17-diones which give β-hydroxyl derivatives. However, hydride reductions of carbonyl groups having little steric congestion, particularly 3-keto groups such as in cholestan-3-one, tend to yield the equatorial alcohol especially using reagents requiring only narrow access, such as sodium borohydride and lithium aluminium hydride. Such production of the thermodynamically preferred isomer in kinetically controlled reactions has been termed 'product development control' and for borohydride reductions is thought to be due to the transition state being product-like and so having a preference for the incipient OH group to be equatorial.

Halogenation and other electrophilic additions to C═C are usually kinetically controlled and the factor deciding product stereochemistry is maximization of overlap between the reagent's reacting orbital and the C═C π-orbital. Since the greatest electron density in the π-orbital is perpendicular to the C—C σ-bonds, attack is generally from an axial direction. Bromination of cholest-2-ene (**87**, Scheme 6.10) thus gives the *trans*-diaxial dibromide.

It is well known that catalytic hydrogenation proceeds via *cis*-addition to the

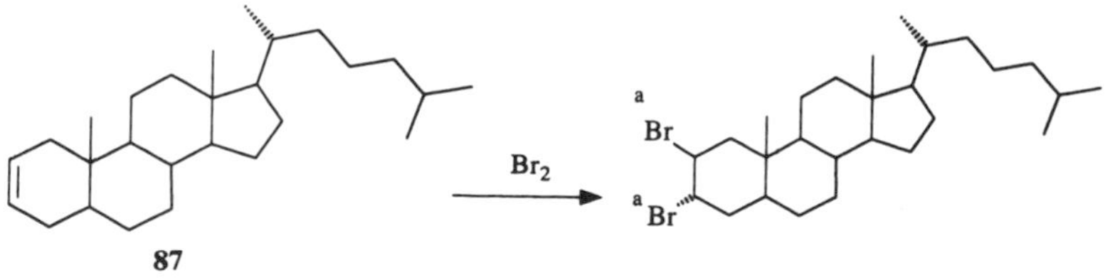

87

Br_2

a Br

a Br

Scheme 6.10 *Bromination giving a diaxial product*

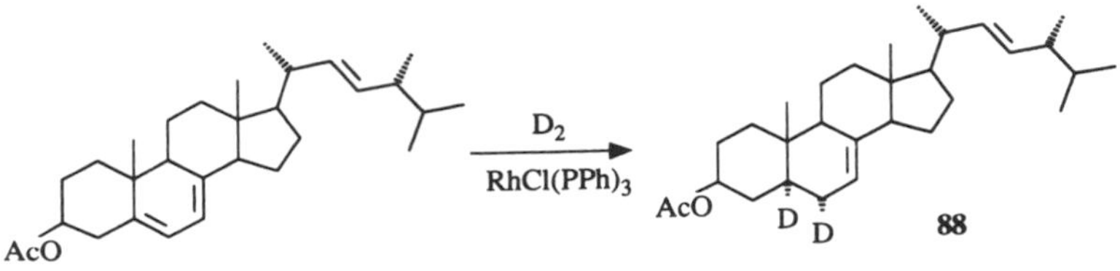

D2
RhCl(PPh)3

AcO
Ergosterol acetate
AcO D D **88**

Scheme 6.11 *Stereoselective hydrogenation of ergosterol acetate*

less hindered side of a double bond, but less widely appreciated that the catalyst also promotes isomerization often causing the observed products to be other than those of simple *cis*-addition. Hydrogenation of Δ^5-steroids, for example, may give either AB-*cis*- or AB-*trans*-steroids. To obtain a *cis*-addition product such as **88** (Scheme 6.11) mild catalysts should be used; homogeneous catalysts, such as Wilkinson's catalyst, $(Ph_3P)_3RhCl$, which are soluble in the reaction medium, are especially discriminating but all stereochemical predictions of hydrogenations must be tentative. Further examples are given in the discussion on total synthesis of steroids (Section 6.4.2.2).

Useful predictions on the stereochemical outcomes of reactions may be made by simple consideration of steric and stereoelectronic (*i.e.* the disposition of charge and orbitals in space) factors as given above but more subtle stereoelectronic effects or conformational changes may result in an unexpected product by altering energetic preferences amongst competing transition states.

6.4.2.2 Thermodynamically Controlled Reactions. Reactions which proceed reversibly (or whose products may equilibrate in some other way) eventually attain an equilibrium mixture containing a predominance of the most stable product regardless of the relative energies of competing transition states. Keto–enol tautomerism, alkene isomerization, acid-catalysed acetalization, and the Meerwein–Pondorf reduction of ketones using aluminium alkoxides are examples of importance in steroid chemistry.

Equatorial groups are sterically favoured over axial ones but it is often difficult to predict which isomer will be thermodynamically favoured since polar groups near heteroatoms may require an axial orientation to optimize electronic interactions (*cf.* anomeric effect). Thus equilibrations of 1-bromo-2-keto-steroids by treatment with hydrogen bromide in acetic acid generate the products having axial bromines, *i.e.* 1β-bromo-2-keto-5α-steroids isomerize to their 1α-isomers whilst 1α-bromo-2-keto-5β-steroids isomerize to their 1β-isomers.

Dissolving metal reductions, such as reductions of α,β-unsaturated ketones to ketones using lithium in liquid ammonia, generally yield the thermodynamically preferred products but the mechanisms of stereocontrol are complex.

6.4.2.3 Thermodynamic versus *Kinetic Control.* Bromination of the 12-keto-steroid **89**, Scheme 6.12) can show either kinetic or thermodynamic control. Ketones are halogenated at their α-positions via their enol forms and so the first step is enol formation. Addition of bromine is from an axial direction under kinetic control leading to the 11β-bromo derivative in which the axial bromine and the angular

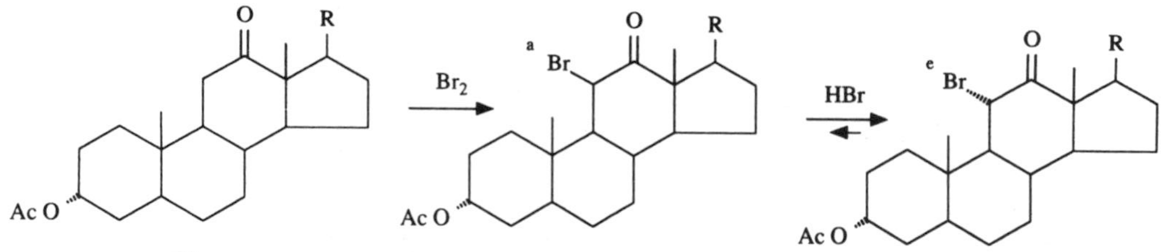

89

Scheme 6.12 *Bromination of a 12-keto-steroid*

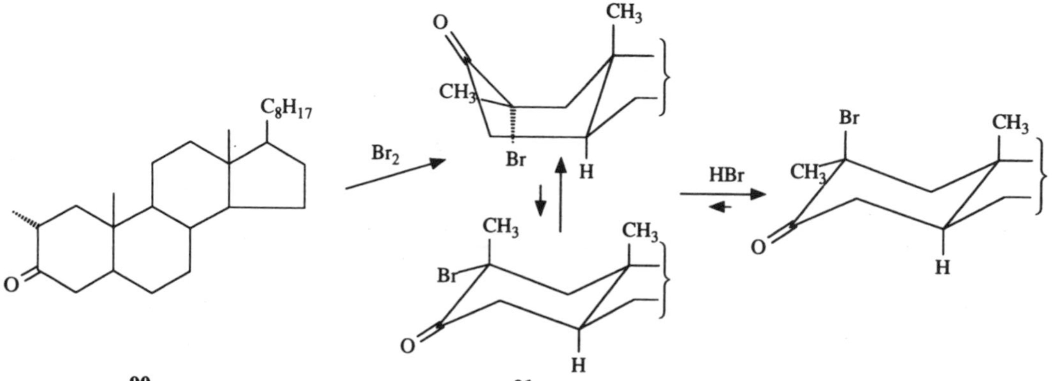

90 91 92

Scheme 6.13 *Bromination of a 3-keto-steroid* via *a boat form*

methyl groups experience a severe steric clash. This is the product obtained from reaction of bromine in base, but in the absence of base hydrogen bromide by-product promotes isomerization (presumably by reversal of the reaction) to give the thermodynamically controlled 11α-bromo product.

Kinetically controlled bromination normally gives the axial bromide but there are several exceptions especially for bromination in ring A. The probable explanation is that ring A flips to a boat form so that the bromine may approach the double bond from an axial direction without experiencing severe hindrance; subsequent flipping to a chair puts the bromine in an equatorial position. Some evidence for this is provided in the kinetically controlled bromination (Scheme 6.13) of the 2α-methyl-3-keto-steroid (**90**) which gives the product (**91**) showing a conformational preference for the boat form. In the presence of hydrogen bromide this isomerizes to the thermodynamically preferred isomer (**92**) which has the bromine in its stereoelectronically preferred axial orientation and has decreased steric hindrance since a bromine atom is smaller than a methyl group.

6.5 SYNTHESIS OF STEROIDS

Initial investigation of the properties of steroids was conducted on samples isolated from biological sources. Oestrone is one of the easiest sex hormones to isolate and it has been obtained from the urine of many animals, paradoxically in most abundance (up to 17 mg l^{-1}) from stallion urine. Once there was demand for medicinal use other means of preparing large amounts of oestrone and the other steroidal hormones were sought. Two strategies, both requiring major chemical advances, were followed: the conversion of a readily available steroid into desired hormones (partial synthesis), and the total synthesis of steroidal hormones from simple chemicals.

6.5.1 Partial Synthesis

For about thirty years the best and most widely used raw material for steroid synthesis was the sapogenin diosgenin (**93**) obtained from the roots of varieties of *Dioscorea*, but with attempts by the major producer, Mexico, to control steroid production and increase local profits, pharmaceutical companies have developed alternative methods based on cholesterol and plant sterols.

6.5.1.1 Diosgenin to Progesterone. In the 1930s progesterone, when available, cost *ca.* $1000 per gram and Russell Marker, who was working at Pennsylvania State College, saw a need for greater supplies at much lower cost. He discovered that certain Mexican yams contained high proportions of a sapogenin known as diosgenin and developed a process for its conversion into progesterone. In 1944 he collaborated in the formation of a new company 'Syntex' to exploit this process but the eccentric Marker fell into dispute over profits and responded by setting up a rival operation. Although a greatly skilled chemist, harassment, including the killing of one of his workers and the shooting of another, forced him to sell out after only one year of operation and he retired from chemistry in 1949.

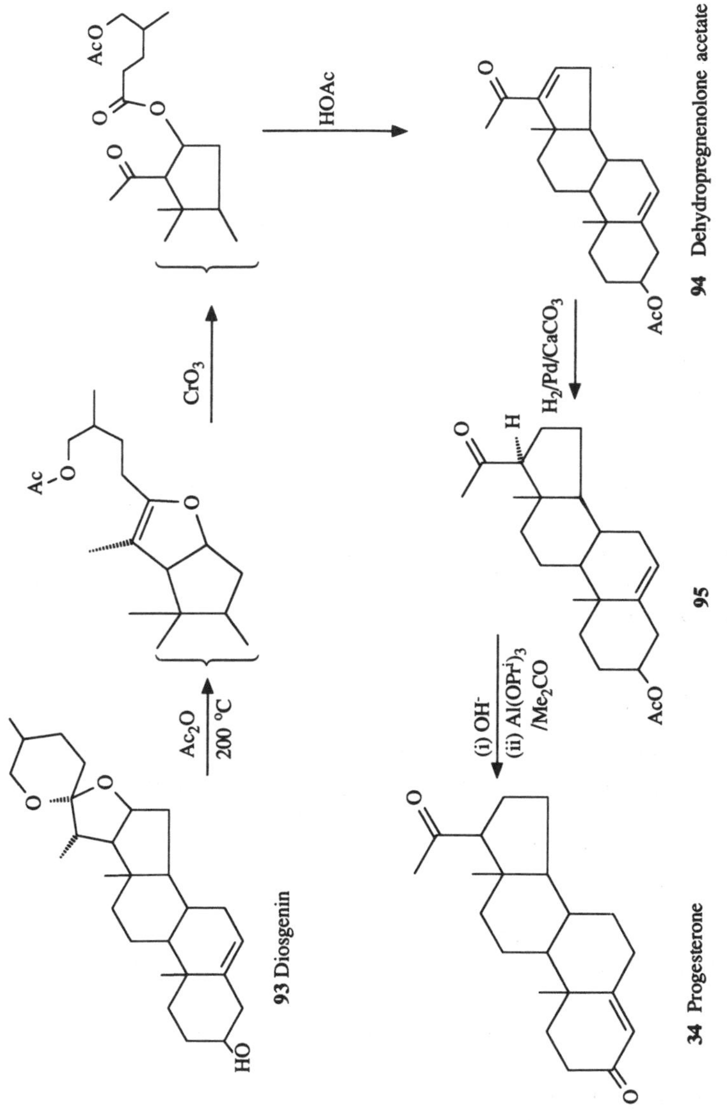

93 Diosgenin

Ac_2O / 200 °C

CrO_3

HOAc

94 Dehydropregnenolone acetate

$H_2/Pd/CaCO_3$

95

(i) OH^-
(ii) $Al(OPr^i)_3$ / Me_2CO

34 Progesterone

Scheme 6.14 *Partial synthesis of progesterone from diosgenin*

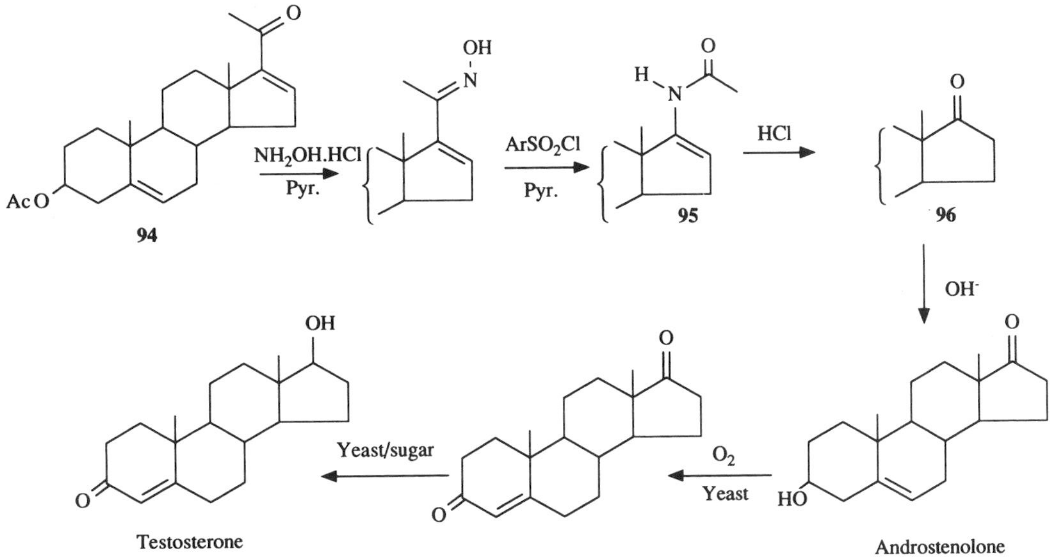

AcO
94
$NH_2OH.HCl$
Pyr.
OH
N
$ArSO_2Cl$
Pyr.
H N
O
95
HCl
O
96
OH^-
O
HO
Androstenolone
O_2
Yeast
O
O
Yeast/sugar
OH
O
Testosterone

Scheme 6.15 *Partial synthesis of testosterone from dehydropregnenolone acetate*

Diosgenin has a spiroketal fused to ring D which must be removed (by the Marker degradation), leaving two carbon atoms for a progestane, or cleaved completely (generally by a Beckmann rearrangement of the progestane) for androstanes and oestranes. The Marker degradation involves treatment (Scheme 6.14) with acetic anhydride at 200 °C followed by regioselective chromic acid cleavage to reveal the required C17 acetyl group and finally acetolysis to eliminate the superfluous ester at C16, giving the important intermediate 16-dehydropregnenolone acetate (**94**). Regio- and stereo-selective hydrogenation adds hydrogen to the less hindered α-side of C17 giving pregnenolone acetate (**95**), and hydrolysis followed by Oppenauer oxidation is all that is required to give progesterone.

6.5.1.2 Diosgenin to Androstenolone Acetate and Testosterone. Removal of the carbons 20 and 21 from 16-dehydropregnenolone acetate (**94**, Scheme 6.15) (obtained by the Marker degradation of diosgenin, Section 6.5.1.1) is accomplished using a Beckmann rearrangement. First the oxime is produced and, on treatment with a sulphonyl chloride in pyridine, the more stable *trans*-isomer rearranges by migration of C17 to nitrogen yielding the amide (**95**). Since this amide is α,β-unsaturated it is also an emamine and so readily hydrolyses producing androstenolone acetate (**96**). Further hydrolysis gives androstenolone itself.

Conversion of androstenolone into testosterone requires reduction of the 17-carbonyl to OH and oxidation of the 3-OH to a carbonyl and has been performed in many ways both chemical and microbiological. High-yielding chemical routes use protecting groups for regiospecificity but a shorter and higher-yielding process uses yeast. The oxidation at C3 is accomplished (Scheme 6.15) by shaking an aqueous suspension of androstenolone with yeast in oxygen and then C17 is reduced by yeast fermenting on sugar in the absence of oxygen.

6.5.1.3 Androstenolone to Oestrone. The fundamental change necessary to give oestrone is dehydrogenation of ring A to an aromatic benzene ring with loss of the C10 methyl group. Starting from androstenolone a carbonyl (for α-bromination) is required at C3. First androstenolone is (Scheme 6.16) reduced at the superfluous Δ^5-double bond and then oxidized at the 3-hydroxy group to give 5α-androstane-3,17-dione (**97**) with a little of the 5β-isomer. Two double bonds are then introduced into ring A by high-yielding bromination/dehydrobromination steps. First kinetically controlled dibromination gives 2,2-dibromocholestane-3,17-dione but this rearranges in the presence of hydrogen bromide to the 2,4-dibromo-isomer (**98**). Dehydrobromination is accomplished using collidine (trimethylpyridine): after boiling with collidine for 30 seconds 2-bromocholest-4-ene-3,17-dione is formed and after 30 min the desired diene (**99**) is obtained.

The remaining aromatization step is the most difficult. Early aromatizations involved dissolving the diene in mineral oil and passing it through a tube of glass helices at 600 °C. Such free-radical thermolytic aromatization has been widely applied in oestrane synthesis but generally gives poor yields: such thermolysis of the dienone (**99a**) to oestrone gives optimum yields of *ca.* 20%. More recently

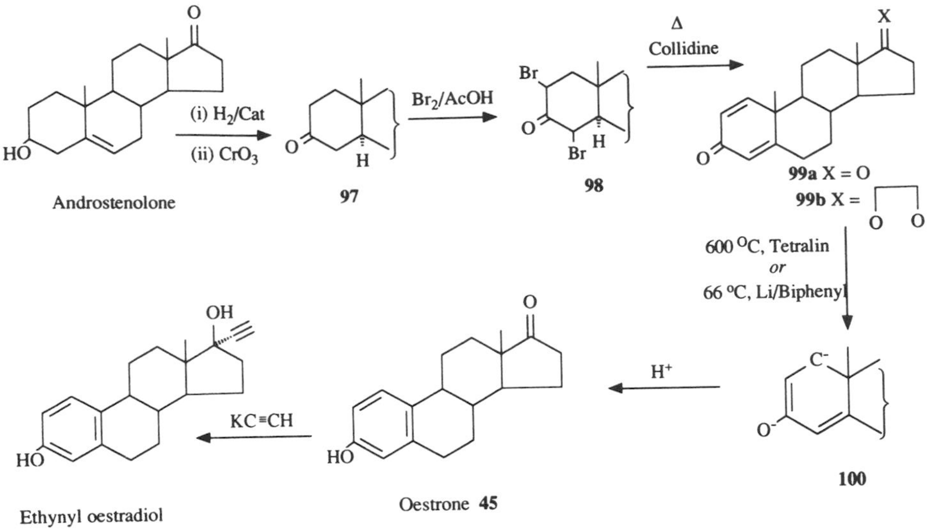

Scheme 6.16 *Partial synthesis of oestrogens*

reductive thermolysis of the 17-ethylene ketal (**99b**) using lithium metal in refluxing biphenyl/THF at 35 °C containing diphenylmethane (to react with any dimethyl-lithium by-product) has given oestrone in 75% yield after hydrolysis of the ketal. The basic mechanism of all dissolving-metal reductions is that the metal dissolves with liberation of electrons which serve as the reducing agent. Here it is believed that addition of electrons to the dienone gives a dianion (**100**) which stabilizes itself by loss of the angular methyl group as a carbanion in a *trans*-diaxial elimination involving the electron pair at C1 and concomitant formation of a phenoxide ion. The orally active progestogen ethynyloestradiol is easily and stereospecifically prepared from oestrone (**45**, Scheme 6.16) by reaction with potassium acetylide in liquid ammonia.

6.5.1.4 Androstenolone Acetate to Norethisterone. Norethisterone (norethindrone in US, 17β-hydroxy-19-norpregn-4-en-20-yn-3-one) is important for its use as an oral contraceptive. Although closely related to testosterone, manufacturing processes generally use androstenolone acetate either directly or first converted into oestrone as described above (Section 6.5.1.3). Oestrone is also available from cholesterol or from total synthesis.

Starting from oestrone (Scheme 6.17) the benzene ring is reduced to a cyclohexene using the Birch reduction; to allow this the 3-hydroxyl and 17-keto groups are first protected as a methyl ether and a ketal respectively. The reduction employs lithium in liquid ammonia containing a little ethanol which donates electrons to the aromatic ring giving a radical-anion which picks up a proton from the ethanol. Addition of another electron gives an anion which is protonated, yielding the diene (**101**). Hydrolysis removes the 17-ketal protective group and converts the enol ether into the β,γ-unsaturated ketone which then isomerizes to the α,β-unsaturated ketone (**102**). α,β-Unsaturated ketones readily

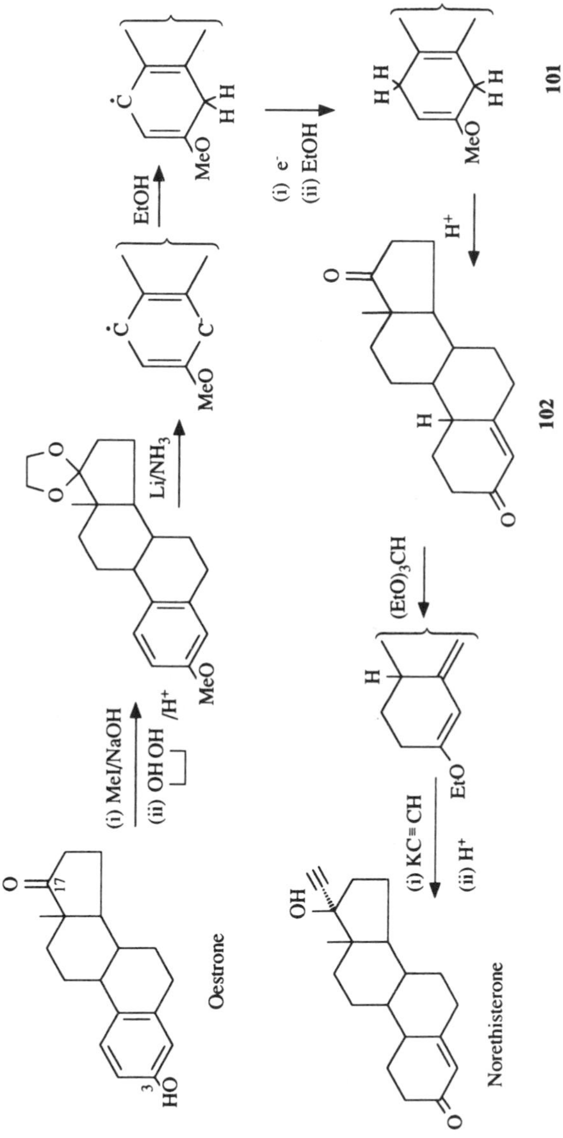

Scheme 6.17 *Partial synthesis of norethisterone from oestrone*

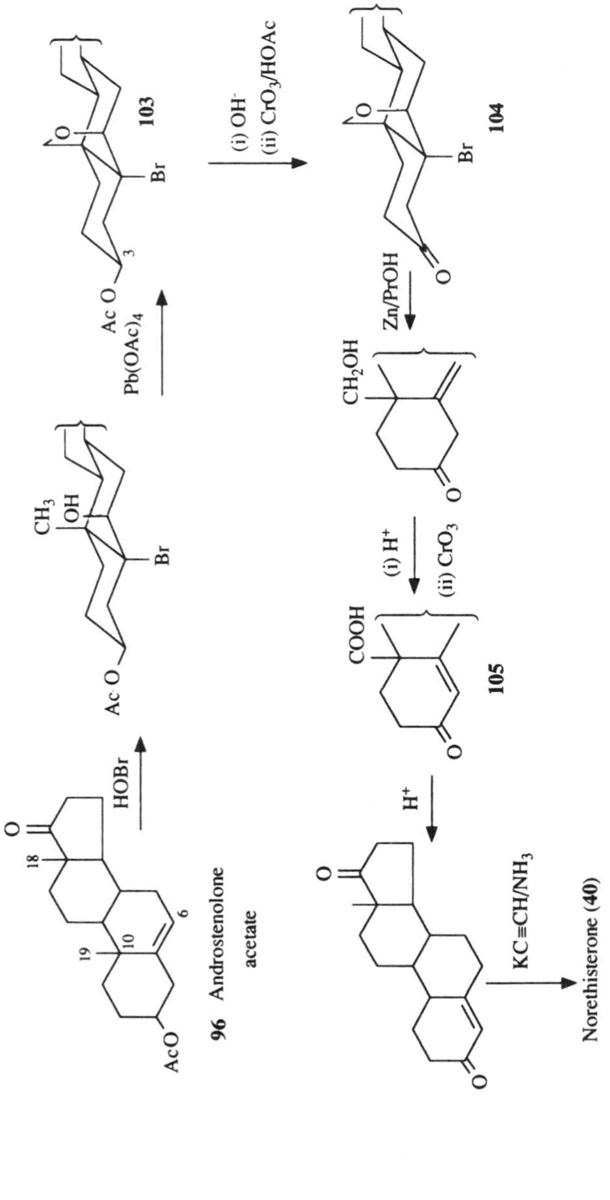

96 Androstenolone acetate

HOBr

Pb(OAc)$_4$

103

(i) OH^-
(ii) CrO_3/HOAc

104

Zn/PrOH

(i) H^+
(ii) CrO_3

105

H^+

KC≡CH/NH_3

Norethisterone (40)

Scheme 6.18 *Partial synthesis of norethisterone from androstenolone acetate*

form enol ethers thus enabling selective protection of the 3-keto group by treatment with triethyl orthoformate. Finally reaction with potassium acetylide introduces the ethynyl group exclusively from the less hindered β-side and acidic hydrolysis regenerates the 3-carbonyl group to yield norethisterone.

The direct route to 19-nor-steroids from androstenolone acetate makes use of 'remote functionalization' of the angular methyl group at C10 by reaction with a perfectly positioned, 1,3-diaxially related, 6β-oxygen radical. Androstenolone acetate is reacted (Scheme 6.18) with bromine water causing *trans*-diaxial addition of hypobromous acid to the double bond. Lead tetra-acetate then gives the required oxygen radical which reacts with the angular methyl group to produce the cyclic ether (**103**). Several variants on this strategy are used in 19-nor-steroid synthesis and functionalization at C18 in the synthesis (Section 6.5.1.6) of aldosterone is also similar. Reductive ring-opening of the C3 modified ether (**104**) gives a 19-hydroxy steroid which is oxidized to the carboxylic acid (**105**) and then decarboxylated to the diketone precursor (**102**, Scheme 6.18) of 19-norethisterone.

6.5.1.5 Partial Synthesis of Cortisone. The first (1948) manufacturing route to cortisone started from deoxycholic acid obtained from bile. Although the process (outlined in Scheme 6.19) was very successful it involved about thirty steps and

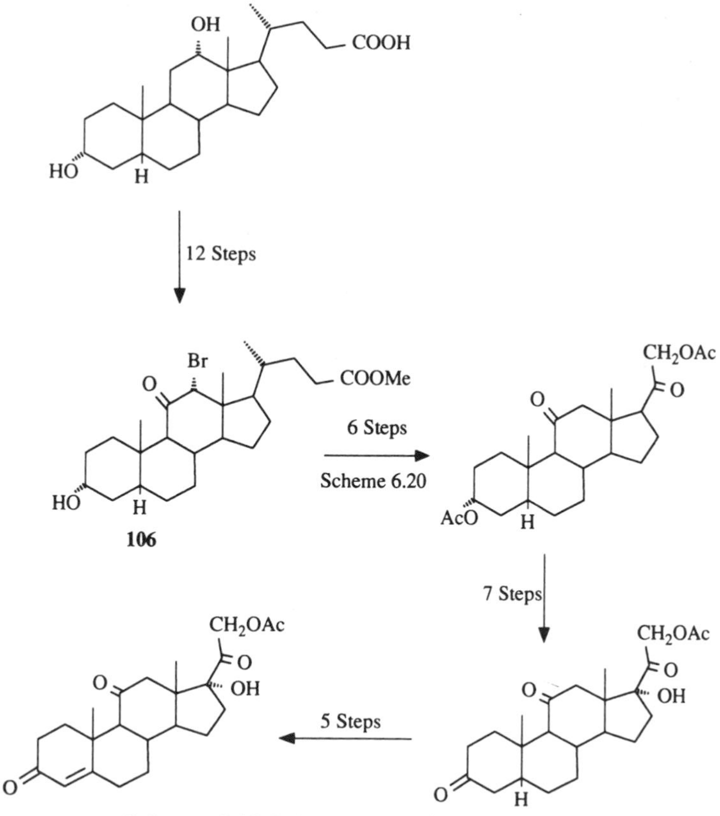

Scheme 6.19 *Bile acid process for cortisone synthesis*

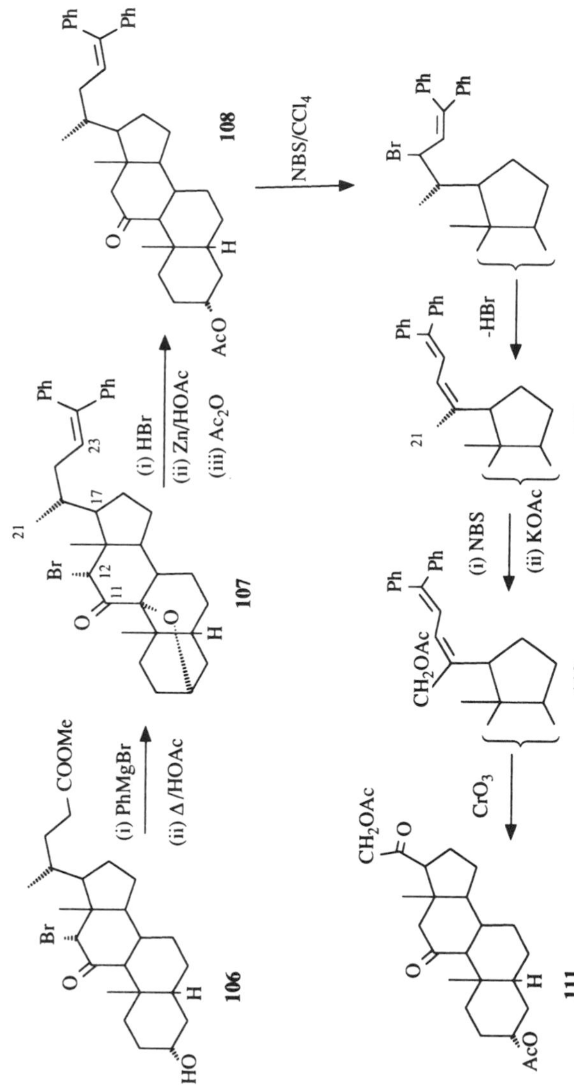

Scheme 6.20 *The Meystre–Miescher degradation*

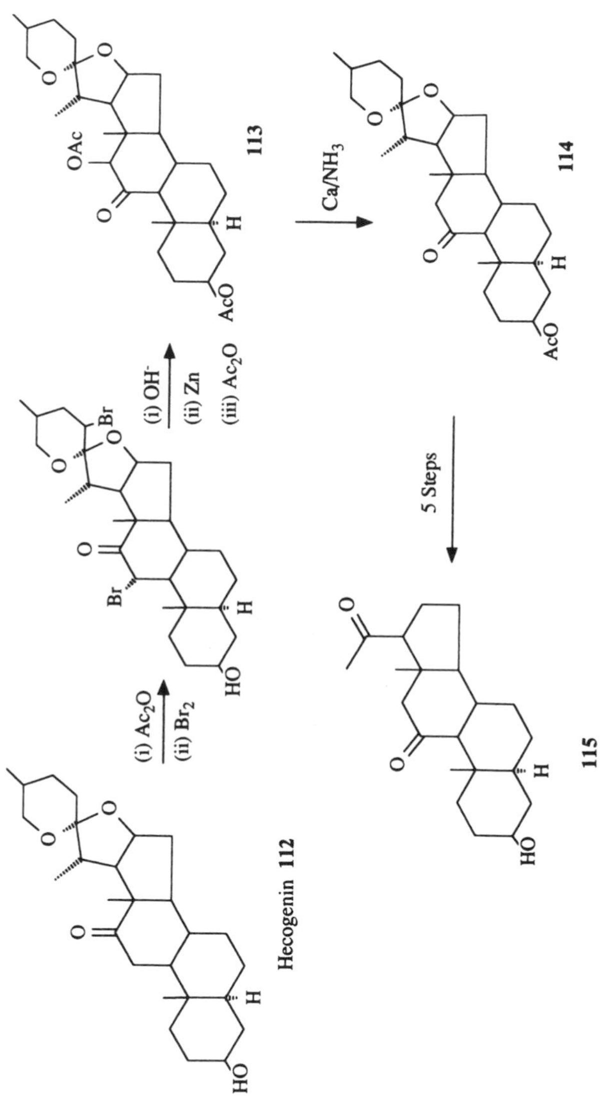

Hecogenin **112**

(i) Ac_2O
(ii) Br_2

(i) OH^-
(ii) Zn
(iii) Ac_2O

113

Ca/NH_3

114

5 Steps

115

Scheme 6.21 *Partial synthesis of cortisone from hecogenin*

will not be described fully here. Three major changes are involved: the 12-OH group is moved to the 11-position, the Meystre–Miescher degradation transforms the side-chain into $COCH_2OH$, and a 17α-OH is added. The next two paragraphs include descriptions of interconversions of the first and third types and the ingenious Meystre–Miescher degradation is shown in Scheme 6.20. Commencing with a methyl ester (**106**), Grignard addition of PhMgBr gives a diphenylcarbinol which dehydrates giving a Δ^{23}-sterane (**107**). After removal of the 12-bromine, reaction of the Δ^{23}-sterane (**108**) with *N*-bromosuccinimide adds a bromine at the allylic position and dehydrobromination then yields the diene (**109**). A second allylic bromination selectively introduces Br to C21 which is displaced to give the acetate (**110**). Finally chromic acid cleavage of the double bonds leaves a product (**111**) with the required side-chain.

A more recent process converts the plant sapogenin hecogenin (**112**), obtained from the leaves of *Agave sisalana* as a side-product in sisal production, into the cortisone precursor (**115**) by an 11-step route (Scheme 6.21). It is first acetylated and then bromine is introduced α to the carbonyl group at C12. Treatment with base causes rearrangement to a 12-hydroxy-11-ketone (*cf.* base-catalysed isomerization of mannose to glucose, Section 2.3.1) which is acetylated and debrominated at C23 yielding **113**. Removal of the acetoxy group using calcium in liquid ammonia gives the 11-keto-sapogenin (**114**) which may be converted into a pregnane derivative (**115**) similarly to diosgenin (Section 6.5.1.1).

Microbiological processes are now the most important for introduction of the 11-oxygen function. Progesterone, for example, may be converted (Scheme 6.22) into 11α-hydroxyprogesterone (**116**) in excellent yield by moulds of the *Rhizopus* family. After saturation of the 5–6 double bond in **116** a 17-OH group may be introduced (Scheme 6.22) by conversion of the acetyl side-chain into an enol acetate (**117**), followed by epoxidation and hydrolysis to give **118**. Bromination selectively attacks C21 and mild alkaline hydrolysis then liberates the 3,17,21-triol (**119**). Selective oxidation at O3 is accomplished using *N*-bromoacetamide and finally the 5–6 double bond is re-introduced by selective α-bromination followed by dehydrobromination.

Another important microbiological method, facilitating the conversion of diosgenin into cortisone, is the oxidation of cortexolone (**124**) to hydrocortisone (**54**) by a variety of micro-organisms including *Curvularia lunata*. Cortexolone may be produced from dehydropregnenolone acetate, which is readily available (Section 6.5.1.1) from diosgenin. One of the best routes (Scheme 6.23) starts by treatment of dehydropregnenolone acetate (**94**) with alkaline sodium hydroxide to give the epoxide followed by hydrogen bromide to generate the bromohydrin (**120**). Removal of the bromine by catalytic hydrogenation occurs without saturation at $\Delta^{5(6)}$ provided that a weak base (NH_4OAc) is present to remove the HBr formed. Selective formylation gives an O3-protected product (**121**) which is treated with bromine, saturating the double bond and adding to C21, α to the carbonyl. Treatment with sodium iodide regenerates the 5–6 double bond and substitutes an iodine for the bromine at C21; acetolysis of the iodide yields the diester (**122**). The 17-OH is also protected as an acetate and then Oppenauer oxidation affecting only the readily hydrolysed formate ester gives Δ^4-cortexolone

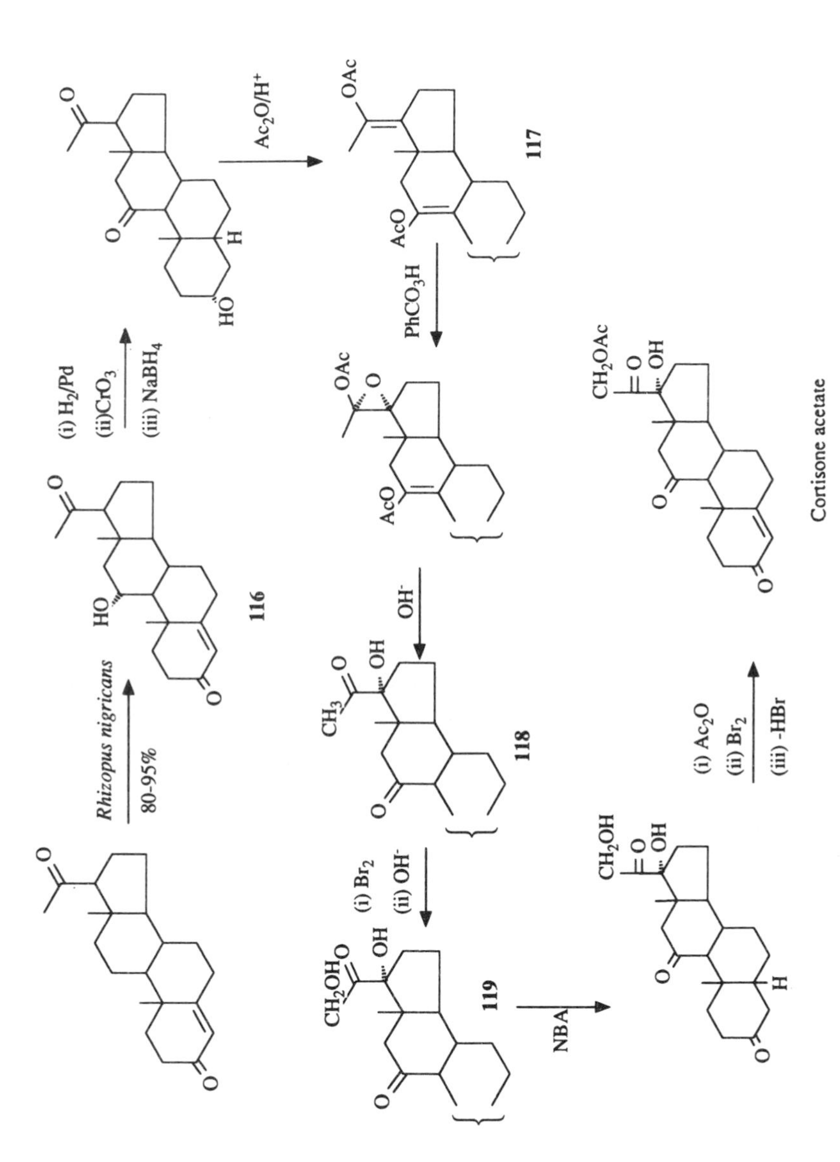

Scheme 6.22 *Partial synthesis of cortisone from progesterone*

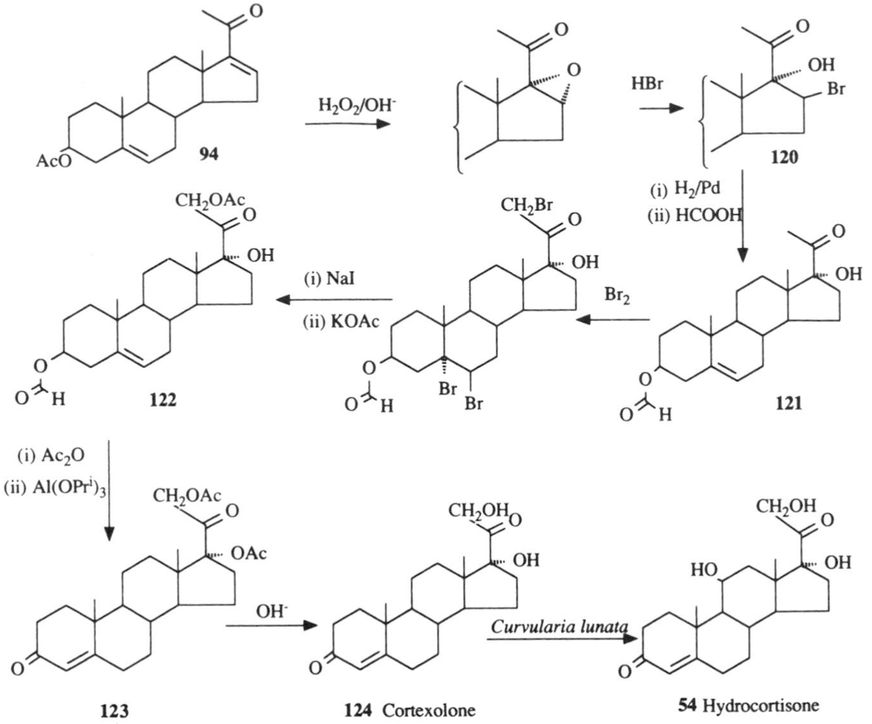

Scheme 6.23 *Partial synthesis of hydrocortisone*

diacetate (**123**). The overall yield is 35% and many steps can be performed without isolation of intermediates.

6.5.1.6 Partial Synthesis of Aldosterone. Many partial syntheses of aldosterone appeared in the early 1960s, and the neatest (Scheme 6.24) uses a Barton reaction in which the nitrite ester of corticosterone acetate (**125**) is irradiated with UV light. On photolysis the nitrite ester liberates NO to give an oxygen radical 1,3-diaxially related to the 18-methyl group requiring an oxygen function. Transfer of hydrogen to the 11-oxygen and NO to the 19-carbon gives the nitroso compound (**126**) which isomerizes to the oxime. In the presence of nitrous acid this is hydrolysed to aldosterone 21-acetate.

6.5.1.7 Cholesterol and Plant Sterols in Partial Synthesis. The double bonds in the side-chains of stigmasterol (**127**, Scheme 6.25), present to *ca.* 20% in soybean oil, and ergosterol (**62**, Section 6.3.5) facilitate side-chain cleavage using ozonolysis and these sterols may be converted into progesterone in four and six steps respectively. Chemical methods of removing the C17 side-chains in other abundant sterols generally lack selectivity and are useless for partial synthesis, but some other sterols are viable starting materials for steroidal hormones since microbiological methods for their conversion into 17-keto-steroids have been developed. Such microbiological conversions tend to be expensive, and are only

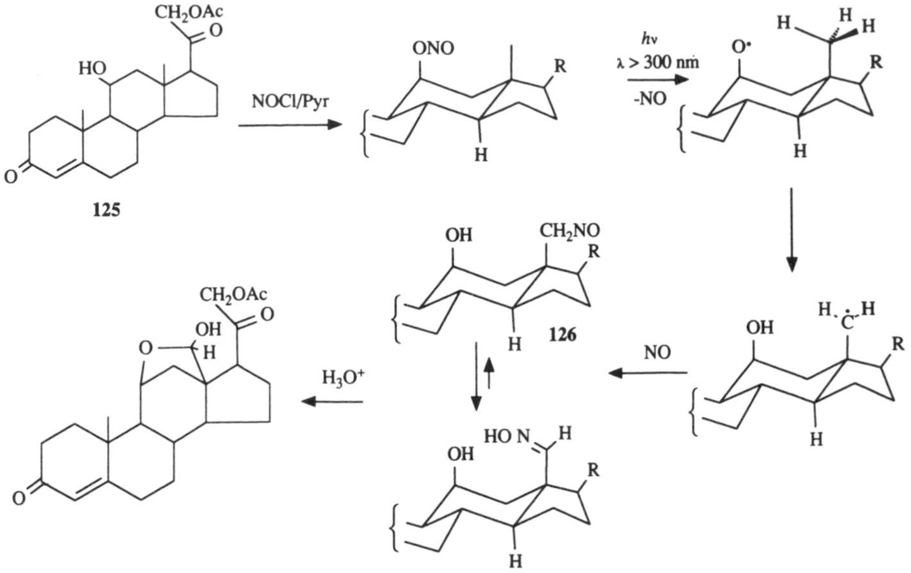

Scheme 6.24 *Barton reaction in partial synthesis of aldosterone acetate*

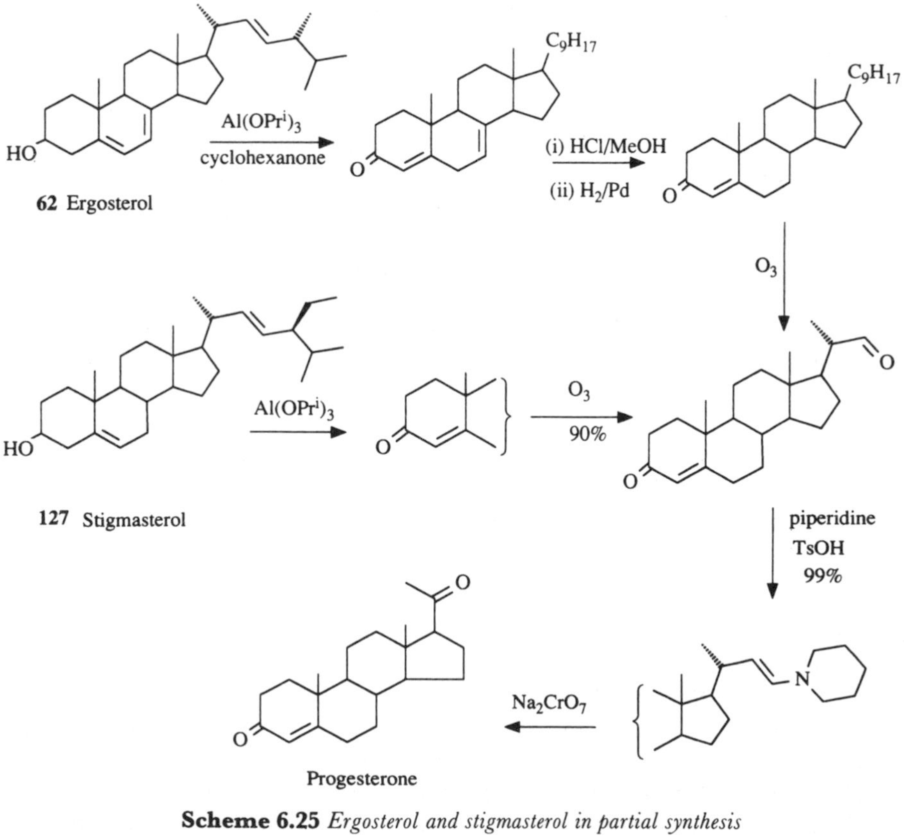

Scheme 6.25 *Ergosterol and stigmasterol in partial synthesis*

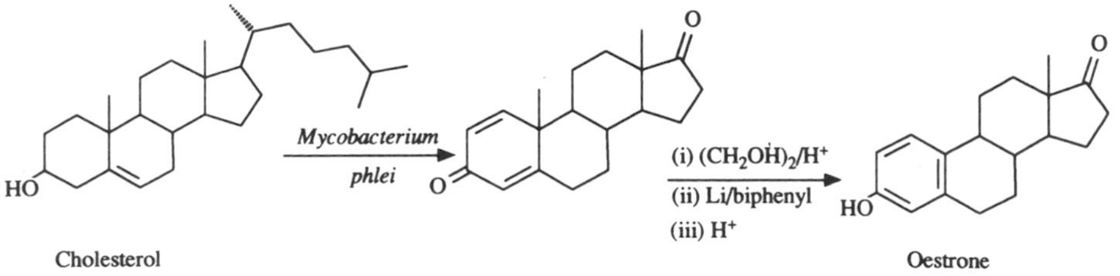

Cholesterol —(Mycobacterium phlei)→ androsta-1,4-diene-3,17-dione —(i) $(CH_2OH)_2/H^+$ (ii) Li/biphenyl (iii) H^+→ Oestrone

Scheme 6.26 *Microbiological process for oestrone partial synthesis*

used commercially where they replace several conventional chemical steps. The conversion (Scheme 6.26) of the major animal sterol, cholesterol, into androsta-1,4-diene-3,17-dione using cultures of *Mycobacterium phlei* readily meets this criterion, making oestrone available in just three more steps (desribed in Section 6.5.1.3). No comparable microbiological process appears to have been published using the major sterol in most vegetable oils, sitosterol (24*R*-ethylcholesterol).

6.5.2 Total Synthesis of Steroids

The total synthesis of complex steroids such as cholesterol presented a formidable intellectual challenge. Nevertheless the simplest naturally occurring steroid, equilenin (**128**), was first synthesized in 1939 only four years after its structure determination, the more complex oestrone was synthesized in 1948, and most other important natural steroids succumbed to chemical synthesis in the 1950s. The challenge alone was a powerful incentive but the inaccessibility of the potent antirheumatic drug cortisone by partial synthesis provided additional impetus for its total synthesis. Total synthesis still has more than intellectual value today, providing the practical route to steroids having an angular ethyl group at C13 such as the potent orally active progestogen levonorgestrel.

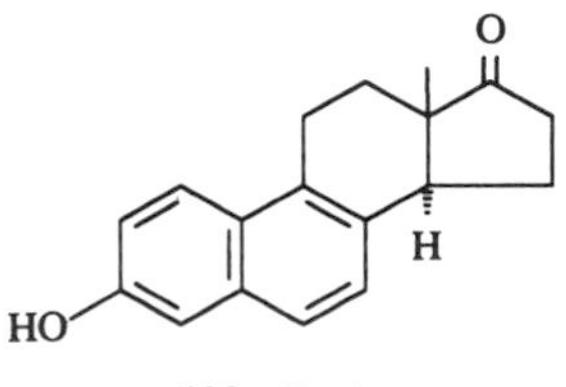

128 Equilenin

The many different ways of synthesizing the tetracyclic ring structure have been classified according to the order of ring construction such as C→BC→ABC→ABCD used in the total synthesis of cortisone (see Scheme 6.29). Such classification has little chemical significance and stress is more appropriately placed on methods of obtaining a single desired stereoisomer. The comparatively simple steroid oestrone has four chiral centres and thus 16 (2^4) conceivable stereoisomers; cortisone having six chiral centres and 64 stereoisomers is more typical and many, like cholesterol with up to 256 stereoisomers, have daunting stereochemical complexity. We have already seen that high stereoselectivity is possible during modification of functional groups on a steroid and so the fundamental challenge in steroid synthesis is stereoselectivity (including enantioselectivity) in the construction of the carbon skeleton—in essence ring junction formation.

6.5.2.1 Obtaining a trans BC *Ring Junction.* Natural steroids have the thermodynamically more stable *trans* BC ring junction which is generally easy to obtain using thermodynamically controlled reactions. Furthermore, provided that there is a carbonyl group adjacent to the ring junction, *cis*-isomers may be converted into *trans*-isomers via their enolates by treatment with base. Such base-catalysed isomerization is normally accomplished concomitantly with another step as in the regioselective Oppenauer oxidation (**129**) used in a synthetic approach to adrenal steroids. Isomerization of *cis*-fused to *trans*-fused rings was frequently used in early syntheses but today it is regarded as somewhat risky (other chiral centres may be affected) and needless since a route generating each chiral centre correctly and unambiguously can generally be devised now stereoselectivity is better understood.

$Al(OPr^i)_3$
Cyclohexanone

(129)

6.5.2.2 Obtaining a trans CD *Ring Junction.* In contrast to the decalins *cis*-hydrindane (**130a**) is more stable than *trans*-hydrindane (**131a**). An alkyl group at the ring junction causes increased relative destabilization of the *trans*-hydrindane (**131b**) since the alkyl group is closer to its 1,3-diaxially related substituents than in the *cis*-hydrindane (**130b**). Consequently the *trans* CD ring junction present in most natural steroids is difficult to synthesize. One approach is to join ring D as a six-membered ring originally and then to ring-contract to the required five-membered ring as illustrated in Scheme 6.27. This strategy is not widely used since introduction of the angular methyl group (C18) is seldom stereospecific and the cyclization step generally gives low yields.

Another approach (Scheme 6.28), widely investigated, is to reduce an unsaturated hydrindane derivative under kinetically controlled conditions. Many

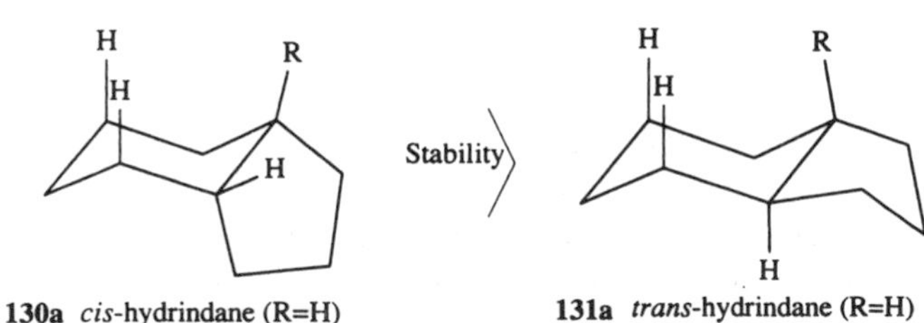

130a *cis*-hydrindane (R=H)
130b *cis*-alkylhydrindanes

131a *trans*-hydrindane (R=H)
131b *trans*-alkylhydrindanes

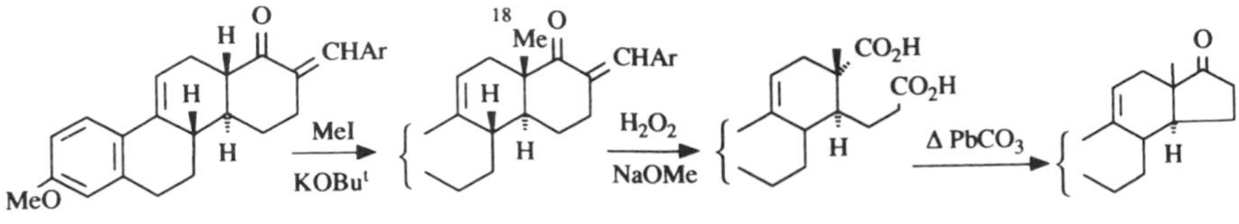

Scheme 6.27 *A ring contraction in steroid synthesis*

difficulties are encountered and *cis*-hydrindanes are frequently the unwelcome products as in reaction **132**. Only a few successful examples have been discovered and one of the most widely applicable is the catalytic reduction of the unsaturated hydrindane carboxylic acid (**133**). Some syntheses delay hydrogenation until the steroid skeleton has been constructed thus facilitating addition of hydrogen from the less hindered side as in **134** where hydrogenation gives the *trans*-fused CD rings followed by dissolving-metal reduction to generate the thermodynamically preferred *anti-trans*-fused BC junction (*anti* refers to the 8- and 14-hydrogens).

Perhaps the most obvious way of forming the *trans*-fused CD ring junction is to use a 1,2-*cis*-disubstituted cyclohexane and then form ring D by cyclization between the substituents. This approach was adopted in Sarett's synthesis of

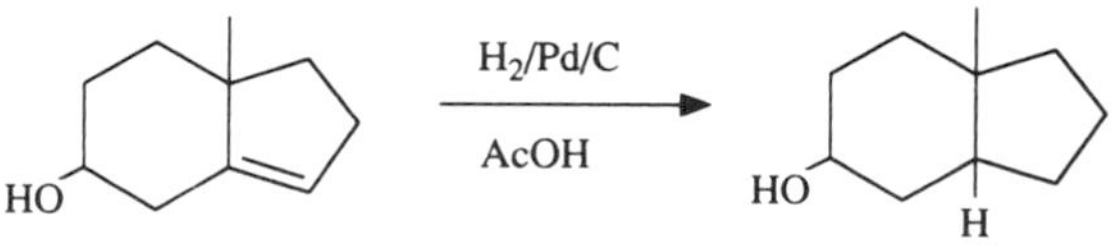

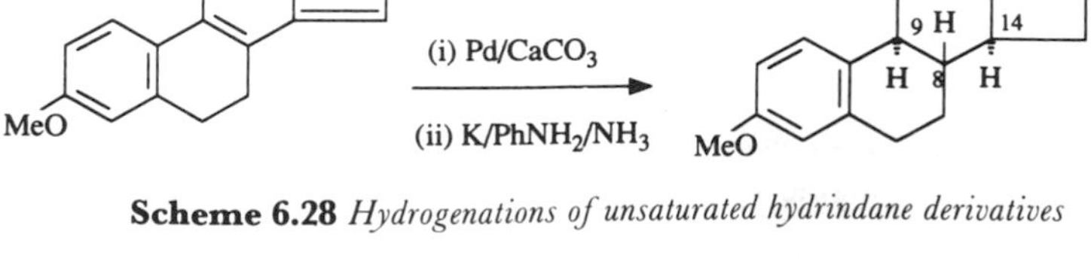

Scheme 6.28 *Hydrogenations of unsaturated hydrindane derivatives*

cortisone (Scheme 6.29), which is worthy of detailed consideration. Initially rings B and C were (incorrectly) *cis*-fused by Diels–Alder reaction of benzoquinone and 3-ethoxypenta-1,3-diene. Hydrogenation saturated the 12,13-double bond and stereospecific hydride reduction from the opposite side to the methyl gave a single diol. Acid treatment released the ketone (**135**) from its enol ether to take part in a Michael addition to methyl vinyl ketone (a Robinson annulation) forming ring A. The carbonyl group of the tricyclic compound (**136**) was protected and then the 14-OH converted into a ketone by Oppenauer oxidation which also gave the required *trans*-fused junction by base-catalysed isomerization at C8. A methyl group and then a methallyl group were attached to C13 by thermodynamically controlled base-catalysed α-alkylations of the ketone (**137**) resulting in the larger group adopting the equatorial position each time. To avoid the elimination of the 11-OH it was oxidized allowing introduction of the last two skeletal carbons by regioselective Grignard addition of ethoxyethynylmagnesium bromide to the 14-carbonyl, giving the tricyclic acid (**138**). Hydrolysis followed by dehydration yielded a double bond to C-14 which on dissolving-metal reduction and tosylation yielded the compound (**139**) having the CH_2CO_2H group *trans* to the methallyl group to which it will be joined for

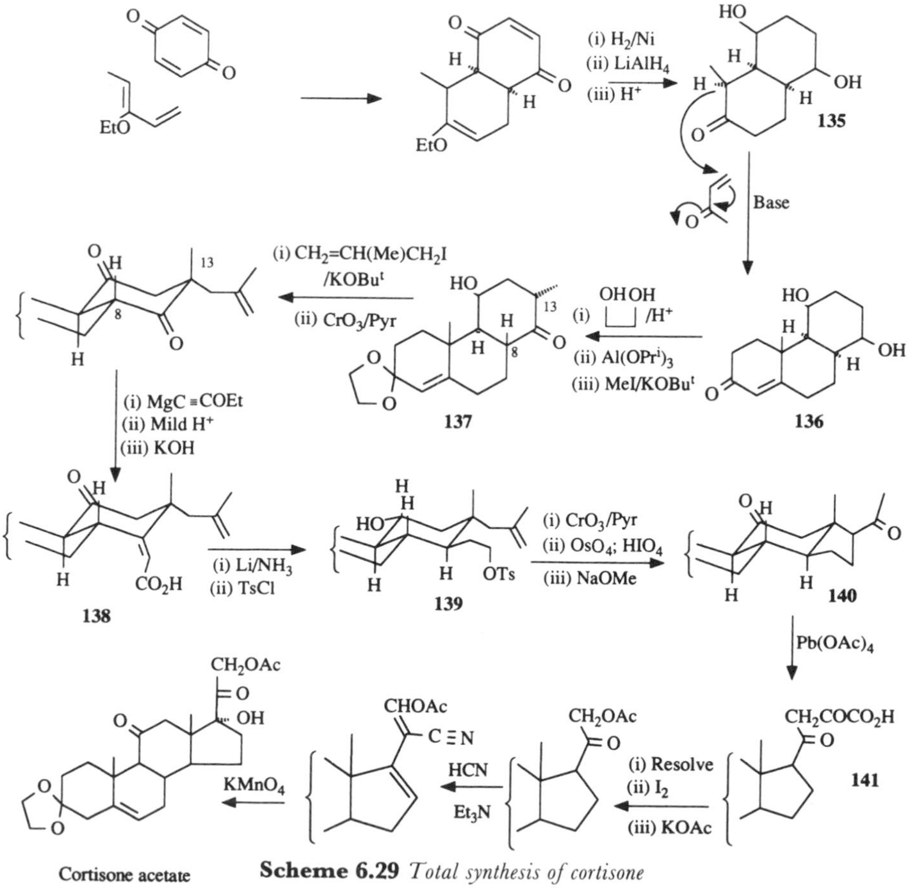

Scheme 6.29 *Total synthesis of cortisone*

ring D. That the CH_2CO_2H is equatorial seems not to be responsible for the stereocontrol, rather the observed product is that having the CH_2CO_2H *trans* to the 11-OH regenerated in an equatorial orientation during the reduction. All the chiral centres are now correctly formed and the ring-joining reaction giving the steroid (**140**) and the remaining functional group transformations merely have to preserve this stereochemistry. By modern standards this 25-step route to cortisone is cumbersome and inelegant but it was the first stereospecific synthesis to be reported and represented the pinnacle of organic synthesis in 1952.

A modern variant of the above strategy is to start with a 1,2-*trans*-disubstituted cyclopentane for ring D and to form ring C from the substituents. Particularly attractive is the synthesis (Scheme 6.30) of 19-nor-steroids by Diels–Alder cyclization of *o*-quinodimethanes such as **144** and **146** with high stereoselectivity for the desired *trans-anti-trans* geometry. The *o*-quinodimethanes are unstable intermediates generated from rearrangement of benzocyclobutenes (*e.g.* **145**) or *o*-methylacetophenones (*e.g.* **143**). The most widely used and highly versatile means of synthesizing the required 1,2-*trans*-disubstituted cyclopentanes is addition of organocuprate reagents to α,β-unsaturated cyclopentanones as illustrated by the preparation of the cyclopentanone **143** in Scheme 6.30. We will discuss further reactions of this type during our consideration of prostaglandin synthesis (Section 7.4.3).

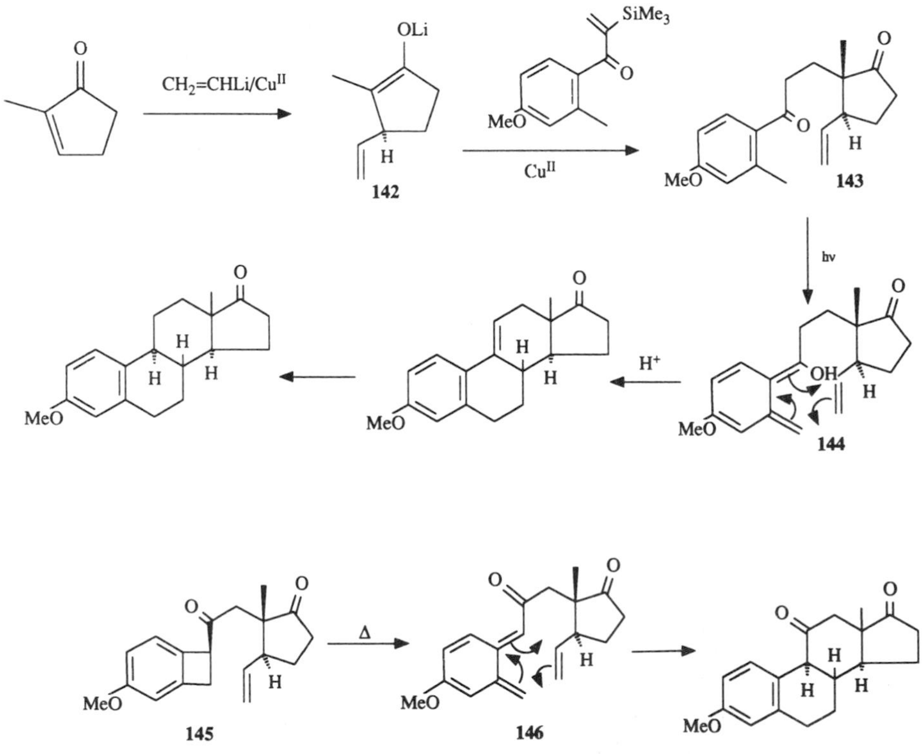

Scheme 6.30 *Steroid synthesis using* o-*quinodimethanes*

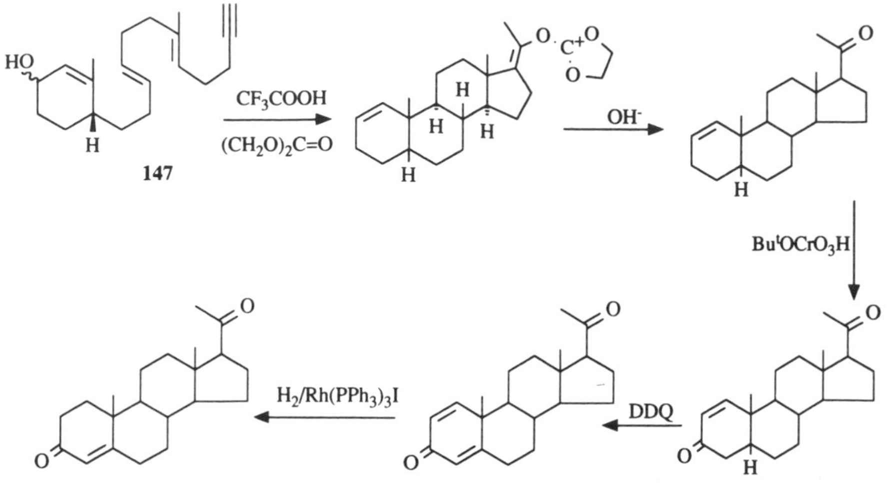

DDQ is the oxidizing agent dichlorodicyanobenzoquinone

Scheme 6.31 *Biomimetic synthesis of progesterone*

Formation of two rings simultaneously is thus a very successful approach to steroid synthesis but recently has been out-done by cycloadditions generating three rings at once. Thus progesterone has been prepared (Scheme 6.31) by cyclization of the cyclohexylpolyene (**147**). The approach has been extended to the generation of 11-keto-steroids and, with improvements in synthesis of the polyene starting materials, could be of commercial importance for corticosteroid synthesis. Such cyclizations show remarkable stereoselectivity and have been termed biomimetic since they resemble the biosynthesis of lanosterol (Scheme 6.4) by cyclization of squalene epoxide.

6.5.2.3 Enantioselectivity in Steroid Synthesis. The oldest method of obtaining one enantiomer from a racemic mixture is resolution. In the above synthesis of cortisone (Section 6.5.2.2.) a steroidal acid (**141**) was generated specifically for resolution with the optically active base strychnine. Such resolution is often technically difficult and at best yields only 50% of the desired enantiomer. The other long established method of enantiospecific synthesis is to use a naturally occurring optically active starting material but this tactic has seen little application in steroid total synthesis.

Throughout steroid synthesis there has been much interest in using microbiological methods (*i.e.* fungi, yeasts, or bacteria) to generate chiral products. Many microbiological reactions have been used in steroid total synthesis but since their value as enantioselective reactions has been largely supplanted by the chemical asymmetric syntheses described below we will only consider one representative example. Racemic hexahydroindenedione (**148**, Scheme 6.32) is readily prepared by the Michael addition of methyl vinyl ketone to 2-methylcyclopentane-1,3-dione and may be reduced by various hydrogenases present in many organisms. Using *Curvalvaria falcata* both (8*S*) and (8*R*) enantiomers are reduced to the (1*S*) alcohols, giving a mixture of diastereoisomers separable by

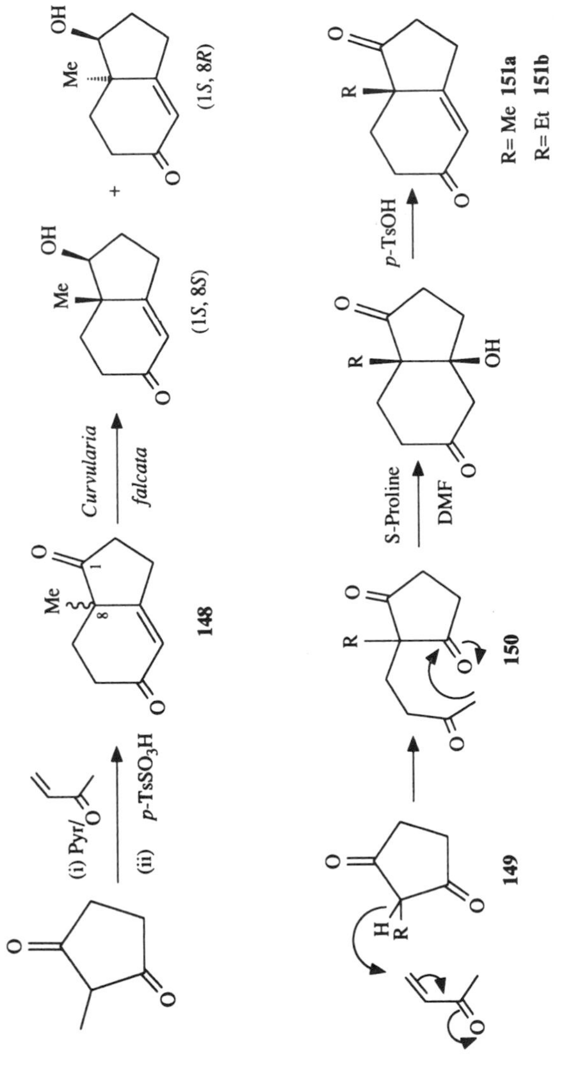

(i) Pyr/
(ii) p-$TsSO_3H$
Me
O
1
8
148
Curvularia
falcata
Me OH
(1*S*, 8*S*)
+
Me OH
(1*S*, 8*R*)

H
R
149
R O
O
150
S-Proline
DMF
R O
OH
p-TsOH
R O
R= Me **151a**
R= Et **151b**

Scheme 6.32 *Asymmetric synthesis of chiral synthons for steroids*

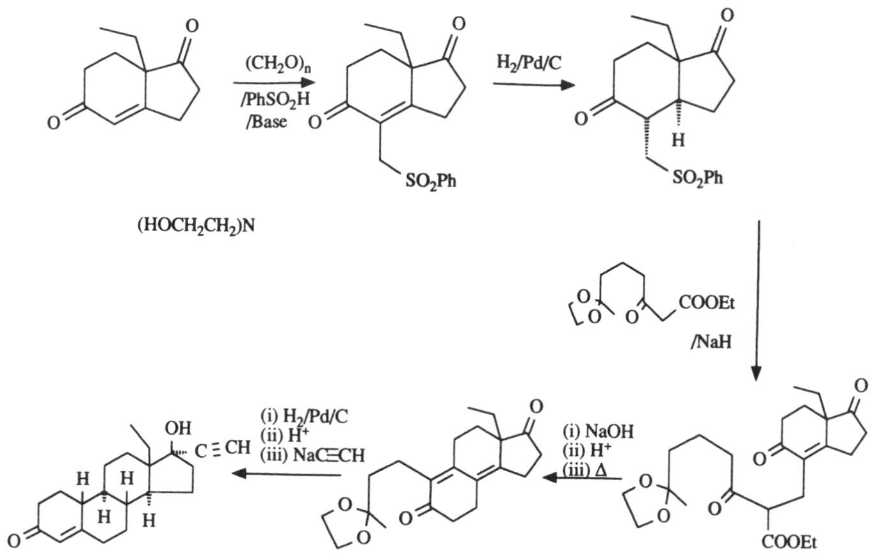

Scheme 6.33 *Total synthesis of levonorgestrel*

physical methods. The *S*-enantiomer is reduced much more quickly than the *R*- and so partial reaction yields a mixture of *R*-dione with the desired (1*S*,8*S*)-hydroxyketone. The enantiomerically pure bicyclic derivatives so prepared have been used as CD ring starting materials.

The (*S*)-hydrindenedione (**151a**, Scheme 6.32), which is very suitable as a CD building block, has been prepared in up to 97% optical yield (*i.e.* contaminated by only 1.5% of the *R*-enantiomer) and over 90% chemical yield by a Robinson annulation of 2-alkylcyclopentane-1,3-dione (**149**) in the presence of *S*-proline. The starting material (**149**) is symmetrical as is the initial product (**150**) of Michael reaction with methyl vinyl ketone; the asymmetry-inducing step is the aldol reaction but precisely how the proline interacts to give a nearly enantiospecific reaction remains a mystery. The product (**151a**) has seen wide use as a chiral starting material in steroid synthesis and the similarly prepared ethyl analogue (**151b**) is particularly important as a chiral synthon for levonorgestrel as illustrated in Scheme 6.33.

The cyclopentyl enolate (**142**, Scheme 6.34) used in the *o*-quinodimethane steroid synthesis (Scheme 6.30) has been prepared by ring expansion of the chiral cyclopropane (**153**) which may be prepared by asymmetric alkylation of di(8-phenylmenthyl) malonate (**152**). It does not matter that the ring expansion and decarboxylation steps give an indeterminant stereochemistry at position-2 in **154** and **155** since subsequent generation of the enolate (**142**) leaves only one chiral centre (C3) to exert stereocontrol over the subsequent reactions.

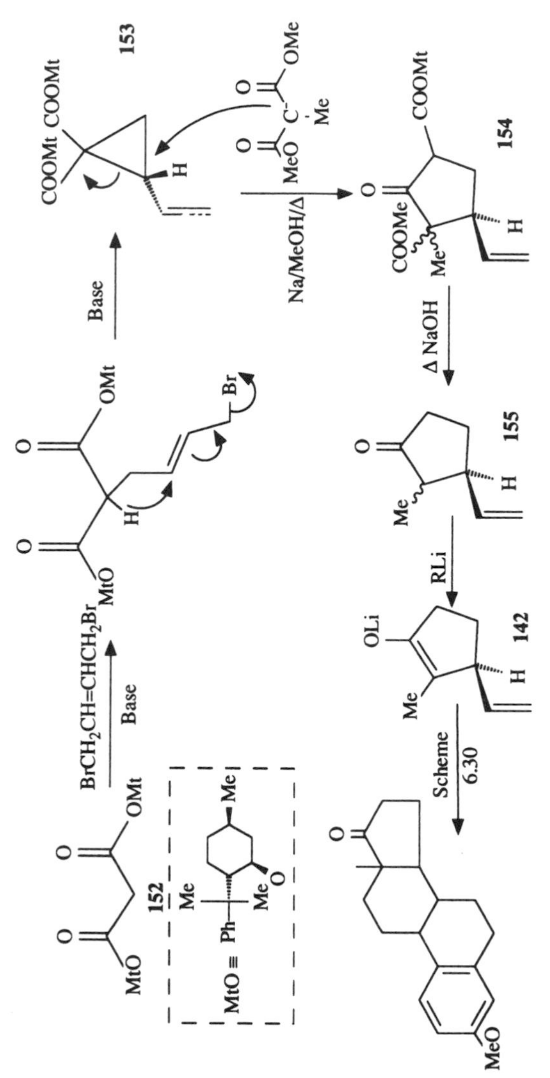

Scheme 6.34 *Asymmetric synthesis of 3-methyloestrone*

CHAPTER 7

Prostaglandins

Prostaglandins, discovered in 1930 amongst secretions from the human prostate gland, are found in minute amounts in most animal cells and exert powerful and diverse biological effects on virtually all animals including humans. They may either stimulate or inhibit smooth (*i.e.* involuntarily controlled) muscle and serve in the regulation of gastric secretion, blood pressure, respiratory muscles, inflammation, and various nervous, reproductive, and metabolic processes. Many clinical uses of prostaglandins are therefore feasible but it has proved difficult to isolate the various types of biological activity and, as yet, the only routine medicinal uses are to terminate unwanted pregnancy, to induce labour, and for treatment of ulcers. In the 1970s research into this group of compounds was stimulated by the discoveries of prostacyclin (prostaglandin I_2), which prevents platelet aggregation, thromboxane A_2, which causes blood platelets to aggregate, and leukotrienes, potent bronchoconstrictors released in lungs during asthmatic attacks. It was also discovered that the anti-inflammatory effects of aspirin and other non-steroidal drugs correlate with their efficacy in inhibition of prostaglandin biosynthesis. This led to many novel medical uses of non-steroidal anti-inflammatory drugs.

Prostaglandins are fairly small but stereochemically complex molecules. Their structures were determined in the early 1960s and chemists quickly realized that these were molecules presenting tantalizing challenges for synthesis. Total synthesis is the only means of preparing sufficient quantities of prostaglandins for extensive biological testing and many novel synthetic methods were developed which are of great value to organic synthesis.

7.1 STRUCTURE AND NOMENCLATURE

Prostaglandins (PG) are cyclic derivatives of 20-carbon carboxylic acids and with thromboxanes and leukotrienes are known as eicosanoids [after eicosanoic acid, $CH_3(CH_2)_{18}COOH$], or, with synthetic prostaglandin analogues, as prostanoids. Prostanoic acid (**1**) has a cyclopentane ring with seven- and eight-carbon side-chains characteristic of prostaglandins and is used as the parent skeleton for numbering. Fully systematic names for prostaglandins are based on prostanoic acid but semi-systematic names based on the nine common patterns of substitution, each denoted by a suffix (A to I, Scheme 7.1), are more usual. The C9 configuration is indicated by a subscript and is α (down) in naturally occurring prostaglandins such as $PGF_{2\alpha}$ but may be β (up) in synthetic prosta-

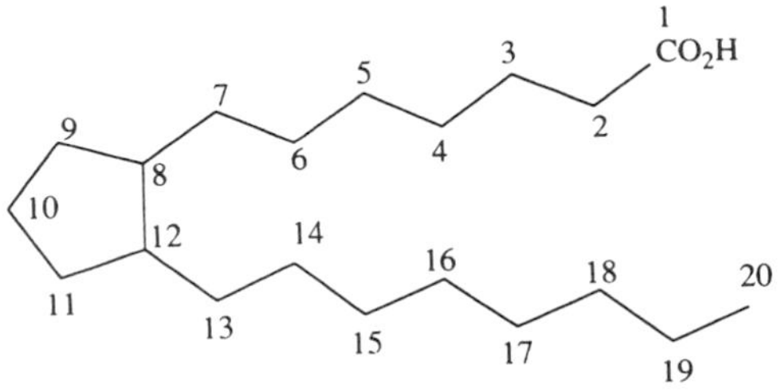

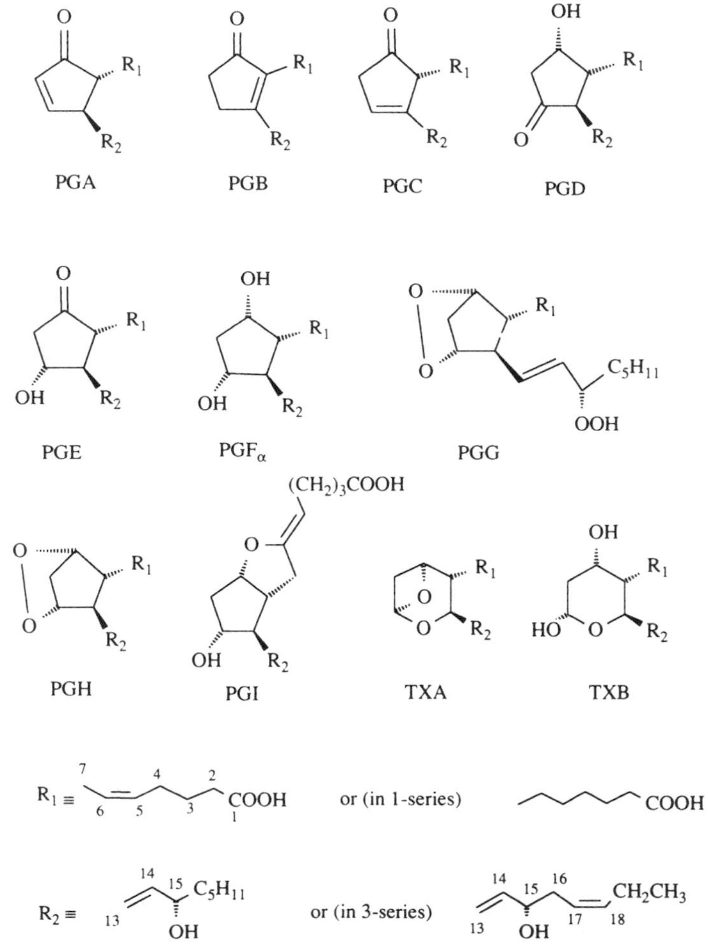

Scheme 7.1 *Structures of prostaglandins and thromboxanes*

glandins. Thromboxanes (TX) are named similarly (Scheme 7.1) and by ignoring the extra oxygen atoms in the skeleton the numbering is directly comparable to that of prostaglandins. Thromboxane TXA_2 is believed to have potent biological activity but is too unstable for isolation, decomposing to the weakly active TXB_2.

Excepting prostaglandins G and I, variation in the side-chains is considered separately. The side-chain attached to C8 is the α-chain, that on C12 the ω-chain, and subscripts are used to denote the total number of alkene bonds in the side-chains. The most plentiful and biologicaly active prostaglandins are those of the 2-series (*i.e.* having two double bonds in the side-chains) and they have 5,6-*Z* and 13,14-*E* carbon–carbon double bonds. Replacement of one side-chain with an alternative (Scheme 7.1) results in a prostaglandin of the 1- or 3-series.

The naming of leukotrienes (LT) such as the unstable biosynthetic intermediate leukotriene A_4 (LTA_4, Scheme 7.2) follows a similar pattern. Other leukotrienes have the same carbon skeleton without the oxirane but with other more complex functional groups attached. LTC_4, for example, has the tripeptide glutathione, joined through sulphur at C6.

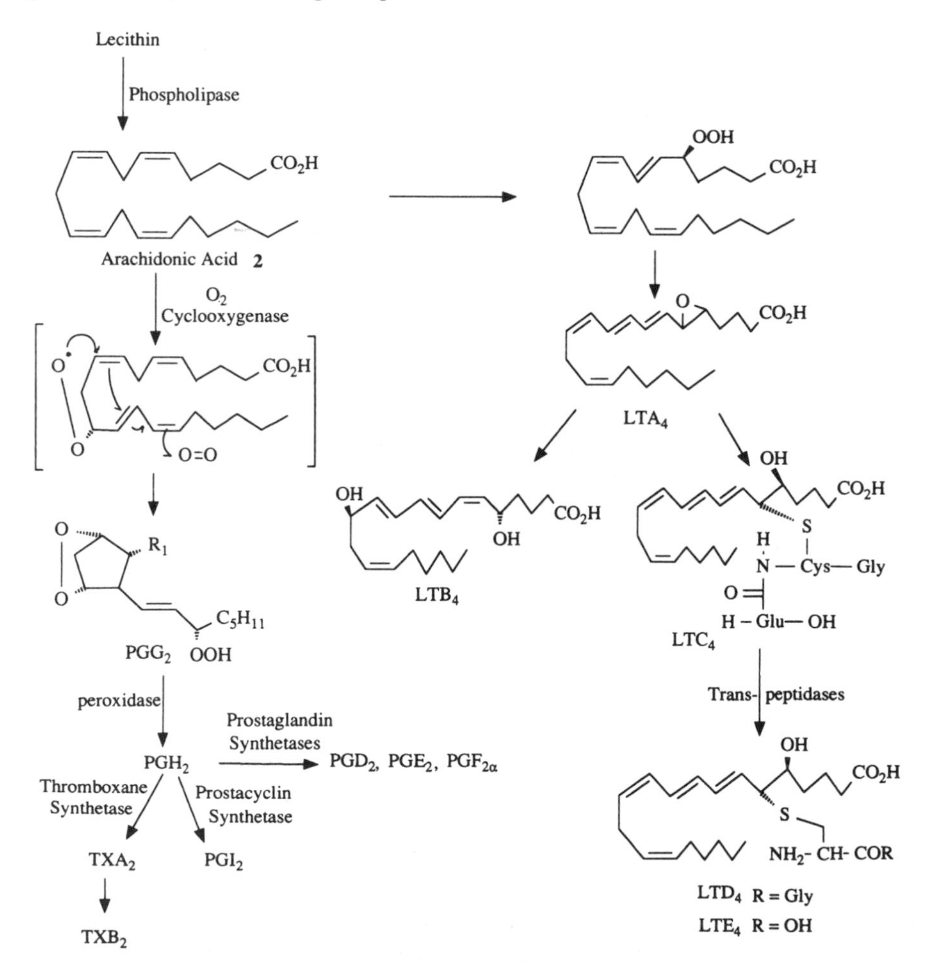

Scheme 7.2 *Structures and biosynthesis of eicosanoids*

7.2 BIOSYNTHESIS

Prostanoids are synthesized from three essential eicosanoid acids, which have three, four, or five double bonds and give prostaglandins having two fewer double bonds in their side-chains. Arachidonic acid (**2**), Scheme 7.2, which is the precursor for prostaglandins of the most important 2-series, is released from the phospholipid lecithin, reacts with oxygen, and cyclizes in a cyclo-oxygenase-catalysed reaction to give the prostaglandin endoperoxide PGG_2. This endoperoxide is converted into a second endoperoxide, PGH_2, in a peroxidase-catalysed reaction. Various synthetases then catalyse conversions of PGH_2 into other prostaglandins and TXA_2. Arachidonic acid may also be converted by another cyclo-oxygenase into a 5-hydroperoxy derivative from which leukotrienes are biosynthesized.

Many drugs have been found to exert their influence through interference with the biosynthesis of prostaglandins. The anti-inflammatory action of corticosteroids correlates with their ability to prevent release of arachidonic acid from lecithin. The corticosteroid promotes synthesis of a peptide which inhibits the enzyme phospholipase A_2. Aspirin has been found to acetylate the cyclo-oxygenase enzyme irreversibly and much of its pharmacological activity is due to inhibition of this enzyme and the consequent prevention of prostaglandin endoperoxide (PGG and PGH) biosynthesis. Other anti-inflammatory drugs (such as indomethacin and ibuprofen) and antipyretic agents (such as paracetamol) also owe their activity to inhibition of endoperoxide biosynthesis although by different mechanisms.

7.3 INTERCONVERSIONS OF PROSTAGLANDINS

Unlike the steroids there are no prostaglandins available in bulk which may be used as precursors for partial synthesis; the best natural source is a Gorgonian coral from the Caribbean, known as the sea whip, which gives prostaglandins as *ca.* 2% of its dry weight. Nevertheless, interconversions are valuable to widen the applicability of total syntheses; to provide synthetic analogues which are more stable, longer acting, or more specific than natural prostaglandins; and to exemplify methods of stereoselective synthesis.

7.3.1 Changes to the Ring Substituents

7.3.1.1 Transformation of PGE into Other Prostaglandins. As indicated in Scheme 7.3, PGE is readily converted into other prostaglandins. It is a β-hydroxyketone and so eliminates water, giving PGA on contact with acids or bases or just by heating. In strong base the initial product, PGA, isomerizes first to PGC then to PGB, and by appropriate choice of reagent and care during isolation either may be obtained.

Reduction of PGE_2 with sodium borohydride gives a mixture of the C9 epimers $PGF_{2\alpha}$ and $PGF_{2\beta}$, but reduction with a very bulky borohydride such as the borophenalene derivative (**3**) delivers the hydride to the less hindered side of

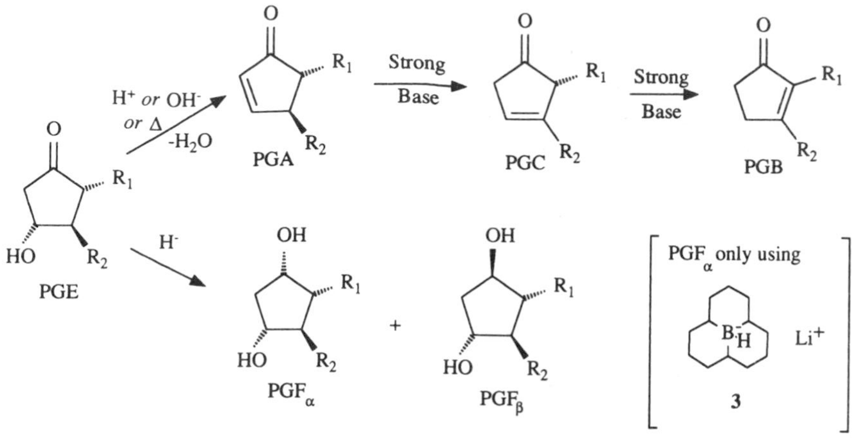

Scheme 7.3 *Reactions of prostaglandin E*

the carbonyl group, opposite to the α-side-chain, giving $PGF_{2\alpha}$ stereospecifically.

7.3.1.2 Transformation of PGF into Other Prostaglandins. The transformation of $PGF_{2\alpha}$ into PGE_2 requires oxidation of the hydroxyl at C9 without reaction of other hydroxyls at C11 and C15. Using the methyl ester of $PGF_{2\alpha}$ (**4**, Scheme 7.4) the C9 hydroxyl is the most hindered OH group since it is *cis* to the α-side-chain and a silylating reagent such as trimethylsilyl chloride in base leaves it free while the other two hydroxyls are protected. An alternative protecting group, popular for use with prostaglandins, is tetrahydropyranyl (THP) which

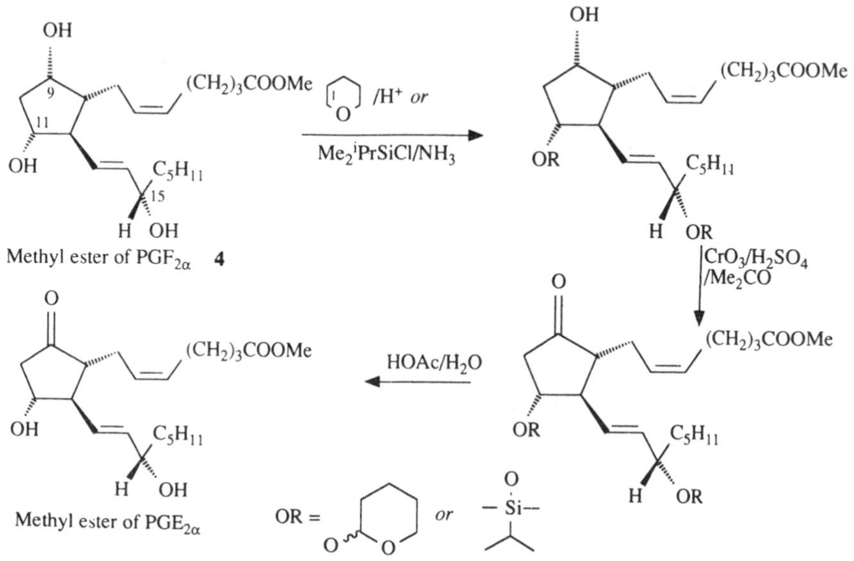

Scheme 7.4 *Conversion of $PGF_{2\alpha}$ derivatives into $PGE_{2\alpha}$ derivatives*

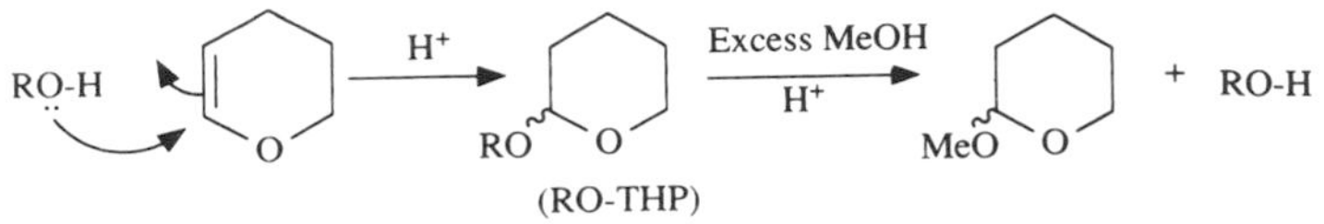

RO-H + dihydropyran $\xrightarrow{H^+}$ RO-THP $\xrightarrow[H^+]{\text{Excess MeOH}}$ MeO-THP + RO-H

(RO-THP)

Scheme 7.5 *Formation and removal of tetrahydropyranyl ethers*

may be introduced by acid-catalysed reaction (Scheme 7.5) of alcohols with dihydropyran. An extra chiral centre is introduced with a THP group, often with little selectivity, which makes THP derivatives difficult to purify. This difficulty is outweighed by the ease of preparation and of deprotection of THP derivatives under mildly acid conditions, compatible with preservation of reactive prostaglandins. Examples of the use of THP derivatives in prostaglandin chemistry may be found in Schemes 7.5, 7.10, 7.13, 7.15, and 7.18.

Vigorous conditions must be avoided in the oxidation of the 11,15-disubstituted $PGF_{2\alpha}$ derivatives (Scheme 7.4) to PGE_2 derivatives to prevent attack on the alkenyl groups, or removal of blocking groups. Chromic acid in acetone or in pyridine, known as Jones' or Collins' reagent respectively, is employed, being milder than chromic acid in other solvents. Finally the blocking groups are removed using aqueous acetic acid to yield PGE_2 methyl ester.

The unstable endoperoxides PGH_2 and PGG_2 have been prepared from $PGF_{2\alpha}$ methyl ester, obtained by reaction of $PGF_{2\alpha}$ with diazomethane. First the two ring hydroxyls are transiently blocked by reaction with butane-1-boronic acid (Scheme 7.6) and the protected prostaglandin is silylated at position-15 which also removes the boronate protecting group. Reaction with the oxazolium salt (**5**) activates the hydroxyls to displacement and tetraethylammonium bromide gives the 9β,11β-dibromide (**6**).

The silyl group may then be hydrolysed and the released OH group transformed into Cl by reaction with the oxazolium salt (**5**) and a tetra-alkylammonium chloride. Enzymatic hydrolysis of the ester is followed by reaction with hydrogen peroxide and silver trifluoroacetate which cause displacement of each halide to give PGG_2. Alternatively the dibromide ester (**7**) may be hydrolysed to the free acid and then reacted with peroxide to yield PGH_2.

Prostacyclin (PGI_2) is an important target for medicinal chemists since its ability to inhibit platelet aggregation (blood clotting) suggests that it could be used to prevent heart attacks or strokes. Unfortunately it is too unstable for this purpose but analogues with similar biological properties but greater stability have been found; the calcium salt of its 9-methyl derivative ('ciprostene calcium') is one of the first prostaglandins in use as a platelet antiaggregatory drug. Prostacyclin has been synthesized in two steps (Scheme 7.7) from $PGF_{2\alpha}$ by regioselective iodination of the 5,6-double bond followed by treatment with base. The iodonium ion first formed reacts with the hydroxyl on C9 to give a cyclic ether. The base diazabicyclononene (DBN) is used to eliminate HI, giving PGI_2, since it is strongly basic but non-nucleophilic.

TXB_2 has been synthesized (Scheme 7.8) from the methyl ester of 9,15-diacetoxy-$PGF_{2\alpha}$. The key step is cleavage of the 11,12 C—C bond with lead

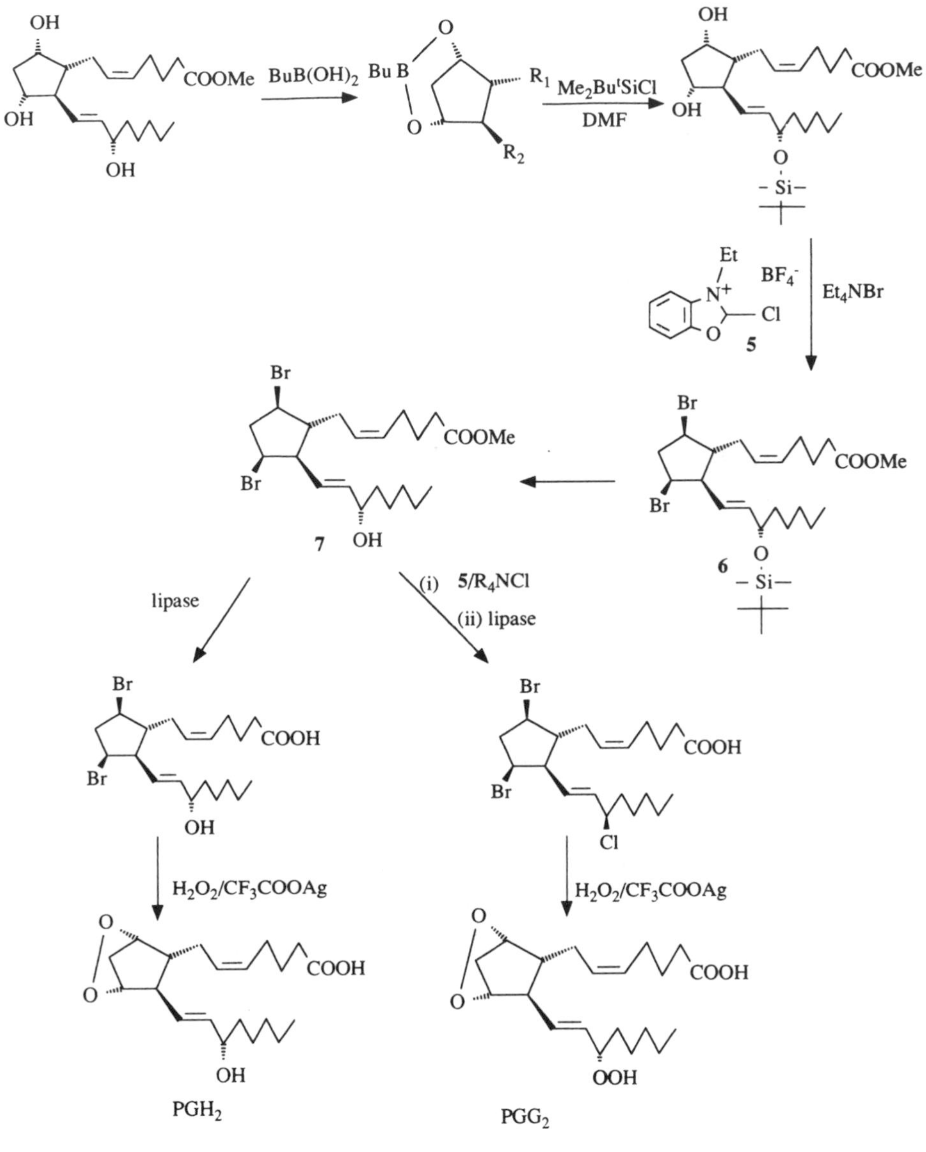

Scheme 7.6 *Transformation of $PGF_{2\alpha}$ into prostaglandin endoperoxides*

tetra-acetate (LTA). This bond is not part of a 1,2-diol (LTA attacks diols, see Section 2.3.9.4) but is activated by the 13,14-double bond. The aldehyde is then protected as a dimethyl acetal in order to allow hydrolysis of the ester groups without aldehyde condensation reactions. Mild acid hydrolysis of the acetal liberates the aldehyde which reacts intramolecularly with the hydroxyl group to give the cyclic hemiacetal, TXB_2.

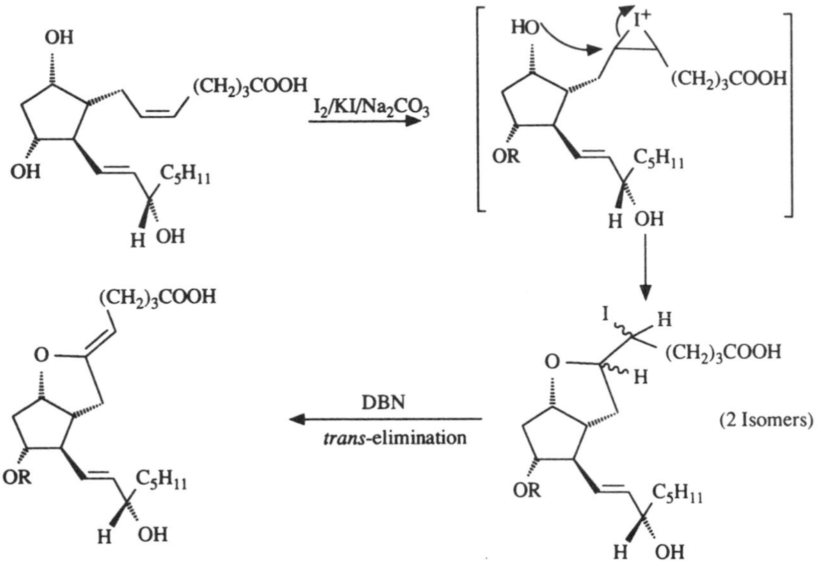

OH
$(CH_2)_3COOH$
OH
C_5H_{11}
H OH
$I_2/KI/Na_2CO_3$
HO
I^+
$(CH_2)_3COOH$
OR
C_5H_{11}
H OH
I
H
$(CH_2)_3COOH$
O
H
(2 Isomers)
OR
C_5H_{11}
H OH
DBN
trans-elimination
$(CH_2)_3COOH$
O
OR
C_5H_{11}
H OH

Scheme 7.7 *Conversion of $PGF_{2\alpha}$ into PGI_2*

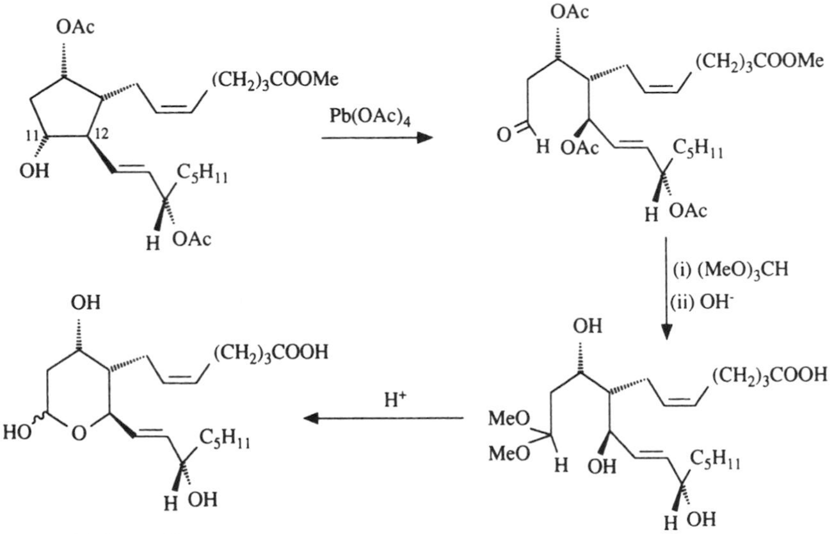

OAc
$(CH_2)_3COOMe$
11
12
OH
C_5H_{11}
H OAc
$Pb(OAc)_4$
OAc
$(CH_2)_3COOMe$
O
H OAc
C_5H_{11}
H OAc
(i) $(MeO)_3CH$
(ii) OH^-
OH
$(CH_2)_3COOH$
MeO
MeO
H OH
C_5H_{11}
H OH
H^+
OH
$(CH_2)_3COOH$
HO
O
C_5H_{11}
H OH
TXB_2 (25 % overall)

Scheme 7.8 *Conversion of a $PGF_{2\alpha}$ derivative into TXB_2*

7.3.1.3 Transformation of PGA into PGE. A high-yielding conversion of PGA_2 into PGE_2 has been achieved by use of distant stereochemical control (Scheme 7.9). The ω-side-chain on the upper face of prostaglandins is insufficiently bulky to induce highly stereoselective additions to the 10,11-double bond of PGA_2 but introduction of a tri-*p*-xylylsilyl group at C15 makes the side-chain large enough to block the upper face of the cyclopentene ring. Epoxidation then proceeds (Scheme 7.9) with 94% stereoselectivity giving the α-epoxide which may be

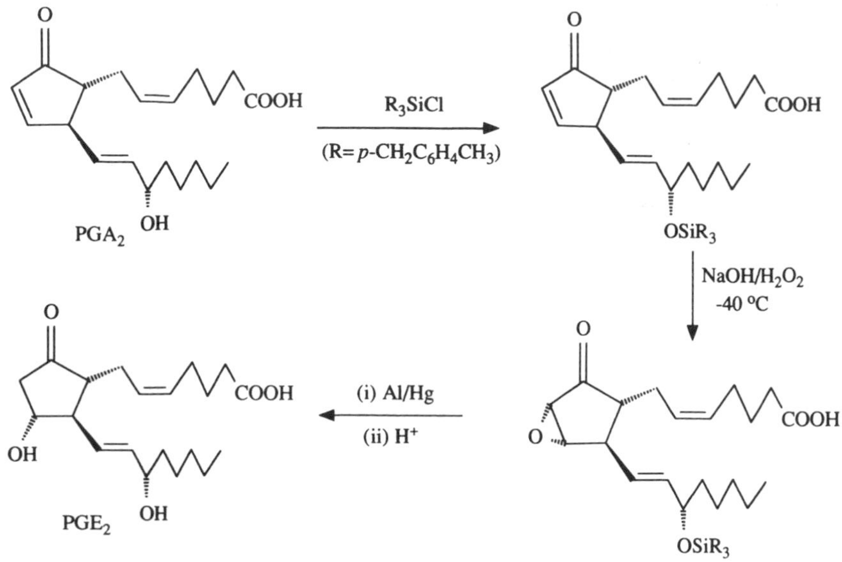

Scheme 7.9 *Conversion of PGA_2 into PGE_2*

reduced regiospecifically to give an α-OH group on C11 as required. Removal of the tri-*p*-xylylsilyl group (known as a controller group) by treatment with acid releases PGE_2.

7.3.2 Changes to the Side-chains

All naturally occurring prostaglandins have a short duration of action in biological tissues and much work has been devoted to the preparation of side-chain modified analogues that might show longer activity and better tissue selectivity. Synthesis of grossly modified prostaglandins from preformed prostaglandins is fairly unusual since it is usually easier to obtain such analogues directly by total synthesis. However, reduction of the α-side-chain of 2-series prostaglandins to give 1-series prostaglandins is important and so are reactions at C15.

Selective hydrogenation of the α-side-chain may be achieved without reduction of the 13,14-double bond using Wilkinson's catalyst (tristriphenylphosphine-rhodium(II) chloride). Hydrogenation over palladium catalysts saturates both side-chains but the α-side-chain is selectively reduced when steric hindrance for reagents approaching the 13,14-alkene is increased by the introduction of a large blocking group at C15 in the ω-side-chain. Thus reaction of $PGF_{2\alpha}$ with acidic dihydropyran or with basic chlorodimethylisopropylsilane gives 11,15-disubstituted $PGF_{2\alpha}$ which, using mild conditions (hydrogenation over palladium on charcoal at −5 °C), may be reduced regiospecifically (Scheme 7.10) at the 5,6-double bond to give $PGF_{1\alpha}$ after removal of the blocking groups. The use of smaller blocking groups such as in the palladium-catalysed hydrogenation of 11,15-bis(trimethylsilyl)-prostaglandin $F_{2\alpha}$ results in some accompanying reduction of the ω-side-chain. Hydrogenations of PGE_2 to PGE_1 proceed similarly.

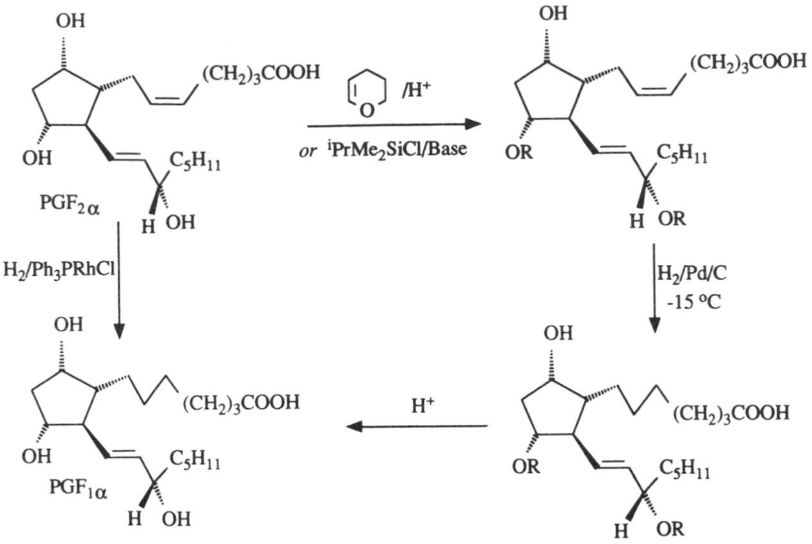

Scheme 7.10 *Conversion of $PGF_{2\alpha}$ into $PGF_{1\alpha}$*

Oxidation of the C15 hydroxyl group to a ketone is the first step in prostaglandin metabolism and so analogues having extra groups near C15 which might interfere with this deactivation have been important synthetic targets. Chemically the same oxidation may be accomplished in high yield using the benzoquinone DDQ (**8**), which is selective for this position. The ketone resulting from oxidation of PGF_2 has been reacted with Grignard reagents (after protection of the hydroxyl groups as trimethylsilyl ethers and esters) to give pairs of diastereomeric 15-alkyl-$PGF_{2\alpha}$s (Scheme 7.11). Metabolic oxidation is blocked in these derivatives and most are more potent and longer lived than PGF_2 itself; 15(*R*)-15-methyl-PGE_2 (obtainable by Collins oxidation of O11-protected 15-methyl-PGF_2) is of importance as an anti-ulcer compound. The most successful analogue so far is the anti-ulcer drug 'misoprostol' in which PGE_1 is modified by inclusion of a 16-hydroxyl and a 16-methyl group and removal of the C15 hydroxyl group. Movement of the hydroxyl group from C15 to C16 reduced side effects and conferred oral activity while the methyl group increased the duration of action.

Reduction of the carbonyl group at C15 normally gives a near 1:1 mixture of epimers since the asymmetric cyclopentyl ring is too distant to influence the reaction. Introduction of the bulky *p*-phenylphenylurethane group at C11 (**9**), coupled with the use of the highly hindered borohydride derivative (**10**, derived from limonene and thexylborane), leads to high stereoselectivity in such reductions (*e.g.* Scheme 7.12). The PhC_6H_4NHCO group is termed a controller group (*p*-phenylbenzoate, PhC_6H_4COO, also works but gives a lower selectivity) and was designed to maximize favourable π–π interaction between the enone (C13=C14—C15=O) and C_6H_4 units, thus shielding the under side of the ω side-chain. Similar selectivity is observed for racemic and optically active borohydrides but the procedure has found little application since mixtures of C15

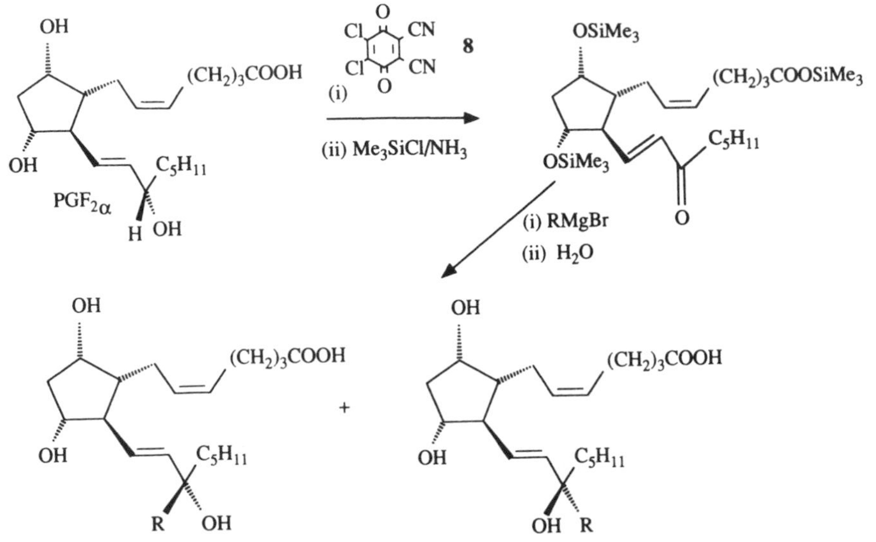

Scheme 7.11 *Oxidation and alkylation at C15 of* $PGF_{2\alpha}$

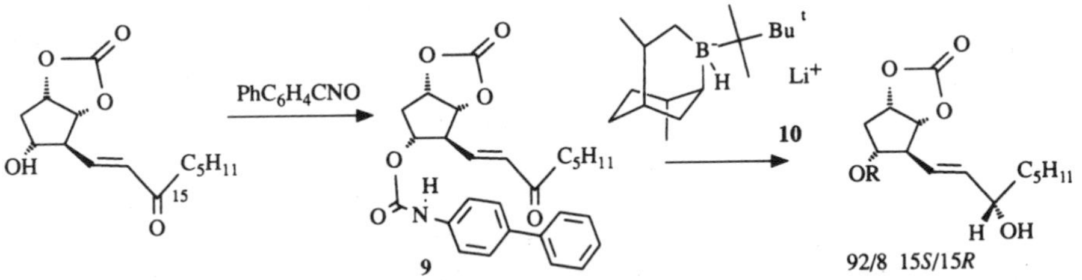

Scheme 7.12 *Use of a controller group for stereoselective reduction at C15*

epimers from simple hydride reductions are normally easily separated and the unwanted epimer may be oxidized back to ketone and recycled.

A simpler method for stereoselective reduction of 15-keto prostaglandins is to use the chiral aluminium hydride reagent **11** which generally gives the 15*S*-alcohols contaminated with less than 1% of the 15*R*-epimer regardless of the nature (or absence) of protection at position-11.

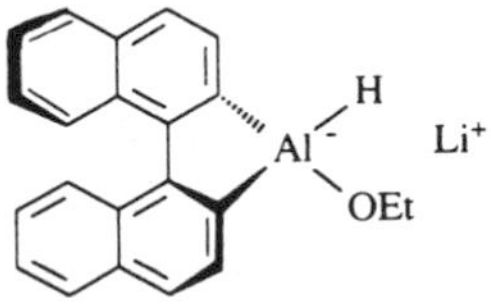

11

7.4 TOTAL SYNTHESIS OF PROSTAGLANDINS

The structure of PGE_1 was established in 1963 by *X*-ray analysis and since then hundreds of prostaglandin syntheses have been published. The main difficulty in synthetic routes is to generate up to five chiral centres with the correct stereochemistry but generation and retention of reactive functionalities such as the β-ketol system of PGE must also be tackled. Cyclization of acyclic precursors used to be a popular approach but recently the preferred strategies have been cleavage of bicyclic molecules of defined stereochemistry and stereocontrolled conjugate addition to cyclopentenone derivatives.

7.4.1 Cyclization of Acyclic Precursors

Aldol, Michael, and Dieckmann reactions have each been used to construct a cyclopentane ring in prostaglandin synthesis. Stereoselectivity depends mainly on the thermodynamic stability of 1,2-*trans*-cyclopentane derivatives over 1,2-*cis* isomers. If the desired *trans*-geometry is not achieved in the initial reaction, subsequent isomerization may be successful.

The first synthesis (Scheme 7.13) of PGE_1 in its natural form is a good example of this approach. Intramolecular cyclization of the readily available nitroketal (**12**) in the presence of the Lewis acid stannic chloride gives only two diastereoisomers with positions 8, 11, and 12 (PG numbering) having the correct relative stereochemistry. The carbonyl group in the side-chain is reduced with little stereoselectivity using zinc borohydride so giving four diastereoisomers. The $CHNO_2$ group is sufficiently acidic so that treatment with weak base causes equilibration and gives the *trans,trans,trans*-geometry around the ring. This leaves just two C15 epimers (and their enantiomers) and so permits chromatographic separation of the 'natural' epimer from the unnatural one, which is oxidized and

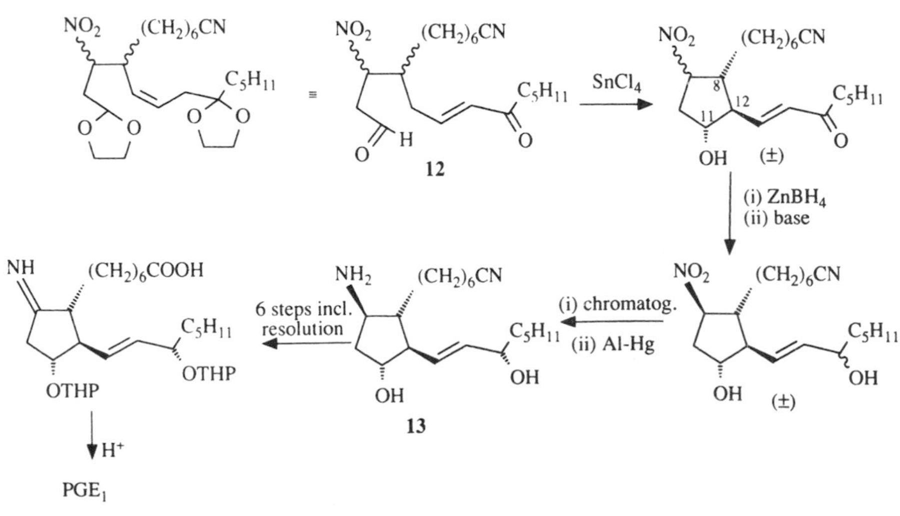

Scheme 7.13 *First total synthesis of PGE_1 with its natural stereochemistry*

recycled. The amine (**13**) obtained on reduction of the pure nitro derivative with aluminium amalgam may be resolved with an organic acid and is readily converted into PGE_1. First a formyl group is attached to the nitrogen, then THP protecting groups are attached at C11 and C15, and the cyano group is hydrolysed to its carboxylic acid. Reaction with *N*-bromosuccinimide gives the *N*-bromo derivative which is dehydrobrominated to an imine (C=NH). Careful hydrolysis of the imine at pH 2 gives PGE_1.

7.4.1.1 Mild conditions for β-Ketol Formation. A problem in the conversion of prostaglandin derivatives such as **13** into PGE derivatives is that further reaction to give PGA may occur. An alternative method is reaction with 3,5-di-t-butyl-1,2-benzoquinone as shown in Scheme 7.14. Only the less sterically hindered carbonyl reacts giving an imine (**14**) which tautomerizes to the imine having an aromatic ring. This imine may be hydrolysed to a PGE derivative under extremely mild conditions.

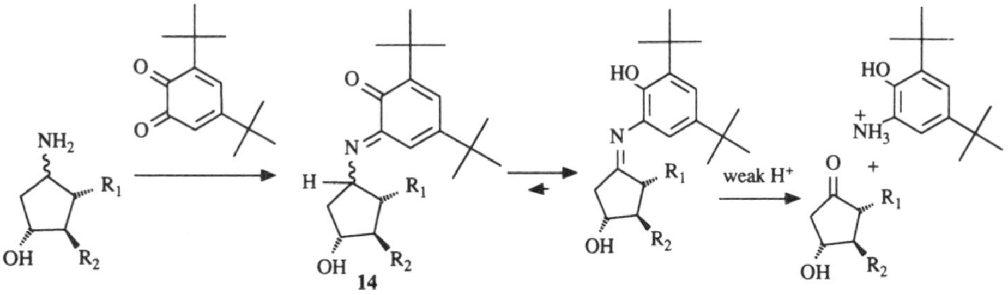

Scheme 7.14 *Mild route to PGE derivatives*

7.4.2 Total Synthesis of Prostaglandins Using Bicyclic Precursors

The basis of this strategy is that five-membered rings have strong thermodynamic preferences for joining with three-, four-, and five-, six-membered rings in a *cis* fashion. Such bicyclic molecules with a rigid, defined stereochemistry may be prepared in high yield and may be cleaved to give diastereomerically pure di- or tri-substituted cyclopentyl derivatives which have proved to be versatile precursors for prostanoids.

The most famous route (Scheme 7.15) of this type is Professor Corey's of Harvard University. A Diels–Alder reaction between the cyclopentadiene (**15**) and 1-chloroacryloyl chloride below 0 °C gives the bicyclic derivative having three chiral centres with the desired relative stereochemistry. The Cl and COCl groups were chosen for their ease of conversion into a carbonyl group by reaction with azide, thermal Schmidt rearrangement and hydrolysis:

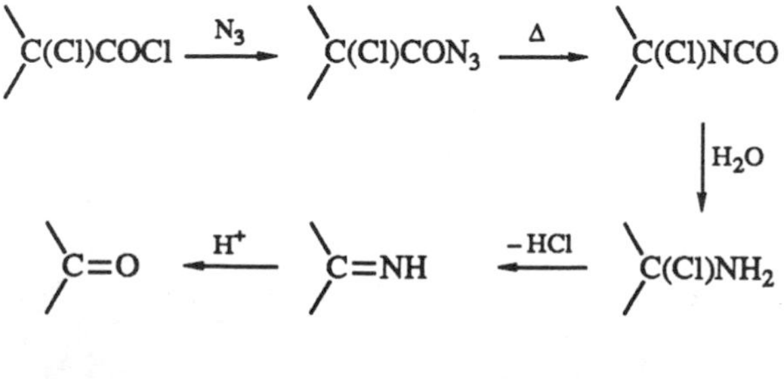

$$\text{C(Cl)COCl} \xrightarrow{N_3} \text{C(Cl)CON}_3 \xrightarrow{\Delta} \text{C(Cl)NCO} \xrightarrow{H_2O} \text{C(Cl)NH}_2 \xrightarrow{-HCl} \text{C=NH} \xrightarrow{H^+} \text{C=O}$$

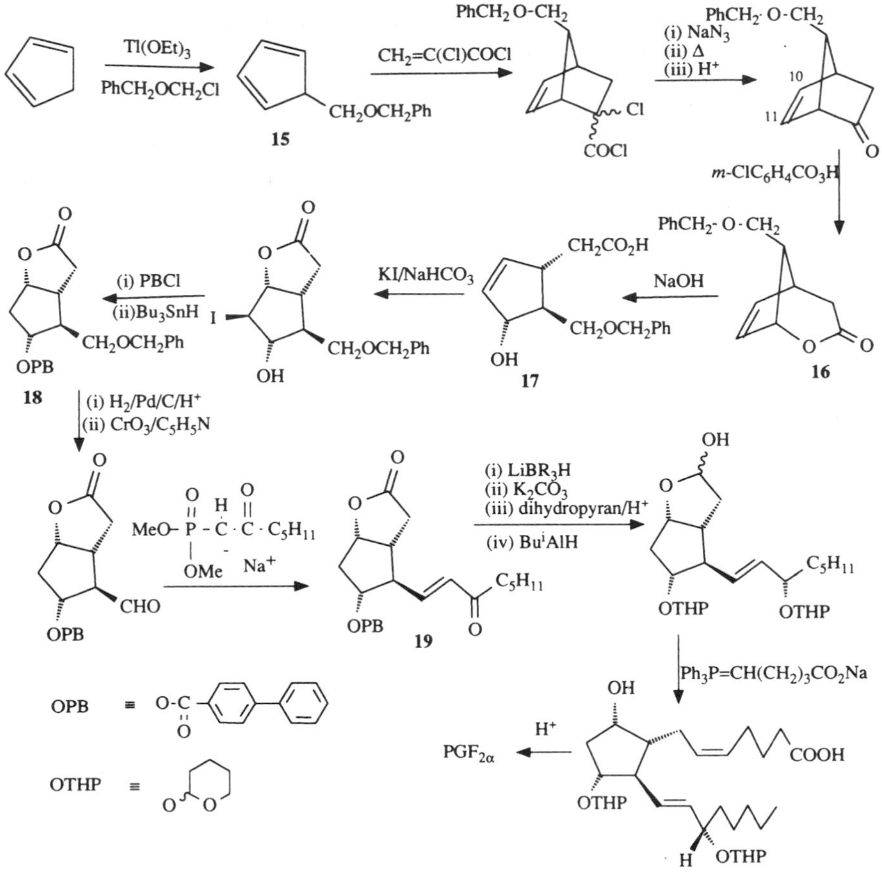

Scheme 7.15 *Total synthesis of $PGF_{2\alpha}$ by the Corey route*

The sequence from the cyclopentadiene (**15**) to the ketone proceeds in 85% yield.

Baeyer–Villiger oxidation with 3-chloroperbenzoic acid is then used to insert an oxygen next to the carbonyl group. The alkyl group which migrates in this rearrangement is the one most able to stabilize a positive charge and since one of the alkyl groups next to the carbonyl is secondary and has an adjacent double bond while the other is primary and not allylic, a single product (**16**) results. The 10,11-double bond is too hindered to be epoxidized under the mild conditions employed.

Hydrolysis of the lactone (**16**) gives a cyclopentene (**17**) with substituents correctly disposed for PGE synthesis but for PGF_{α} an additional α-OH is required. The cyclopentene is reacted with iodine to give an iodonium ion on the less hindered β-side, which is displaced by the carboxylate to give an iodolactone. Each stereocentre is now in place but the iodine substituent is superfluous and is removed by reaction with tri-n-butyltin hydride after protection of the hydroxyl with a bulky group such as a *p*-phenylbenzoate (OPB). The product (**18**) and similar lactones have been dubbed 'the Corey lactone' and they are such useful precursors for prostaglandins that many other synthetic routes have been reported.

Hydrogenolysis of the Corey lactone (**18**) removes the benzyl group and the primary alcohol revealed is oxidized to an aldehyde using chromium trioxide in pyridine (Collins' reagent). The ω-side-chain is then attached by reaction with a phosphonate; this modification of the Wittig reaction gives a *trans*-double bond as desired. The C15 carbonyl is reduced (see Section 7.3.2) using lithium tri-s-butylborohydride to give mainly the product with the 15(*S*)-configuration and the *p*-phenylbenzoate group is removed using mild aqueous base.

The hydroxyl groups are protected with tetrahydropyranyl groups and then the lactone (**19**) is reduced with di-isobutylaluminium hydride to produce a hemiacetal (in equilibrium with an aldehyde) which serves as point of attachment for the α-side-chain. A Wittig reaction using a reactive phosphorane in the absence of inorganic salts produces the *cis*-double bond required. The product is a di-THP derivative of $PGF_{2\alpha}$ and may be converted into a variety of prostaglandins as already discussed (Section 7.3.1.2).

One recent alternative to the Corey route employs a photochemically induced ring expansion of a cyclobutanone (Scheme 7.16), so avoiding the selective but hazardous reducing agent di-isobutylaluminium hydride for production of the aldehyde required for attachment of the α-side-chain.

O OH

O

$h\nu/H_2O$ 70%

R R

OH OH

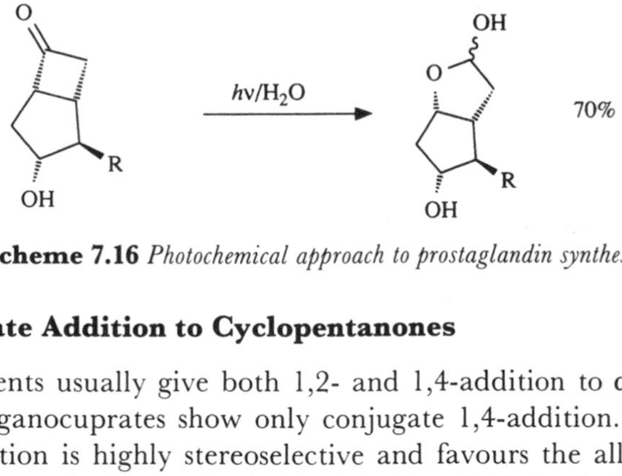

Scheme 7.16 *Photochemical approach to prostaglandin synthesis*

7.4.3 Conjugate Addition to Cyclopentanones

Grignard reagents usually give both 1,2- and 1,4-addition to α,β-unsaturated ketones but organocuprates show only conjugate 1,4-addition. In cyclopentenones the reaction is highly stereoselective and favours the all-*trans* geometry desirable for prostaglandin synthesis. Exploitation of this has led to some elegant syntheses of prostaglandins in which the relative configurations of C8, C11, and C12 are generated in a single reaction.

A good example is the attachment of the ω-side-chain to the cyclopentenone (**20**, Scheme 7.17) by addition of the appropriate lithium organocuprate which gives a PGE_1 derivative directly. A mixture of isomers results unless both the cyclopentenone and the organocuprate are enantiomerically pure and so most of the effort in such syntheses lies in the preparation of the organocuprate and the cyclopentenone derivatives.

A variation used in a short total synthesis of PGE_1 (Scheme 7.18) is attachment

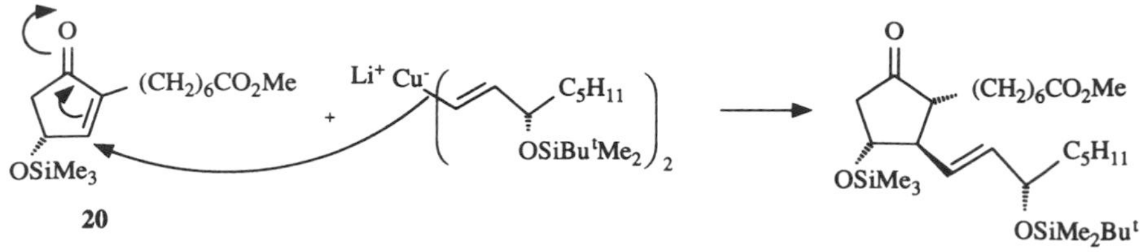

Scheme 7.17 *Conjugate addition of a lithium organocuprate to synthesize a prostaglandin derivative*

of the ω-side-chain by a conjugate addition to a protected 4-hydroxycyclopent-2-enone (**21**). This gives an enolate intermediate (**22**) which is trapped with an aldehyde, thus joining the α-side-chain. The resulting alcohol is dehydrated with methanesulphonyl chloride and base to an alkene which is reduced stereospecifically using zinc dust in acetic acid. Aqueous acetic acid removes the THP groups and finally PGE_1 is released by hydrolysis of the methyl ester using porcine liver esterase.

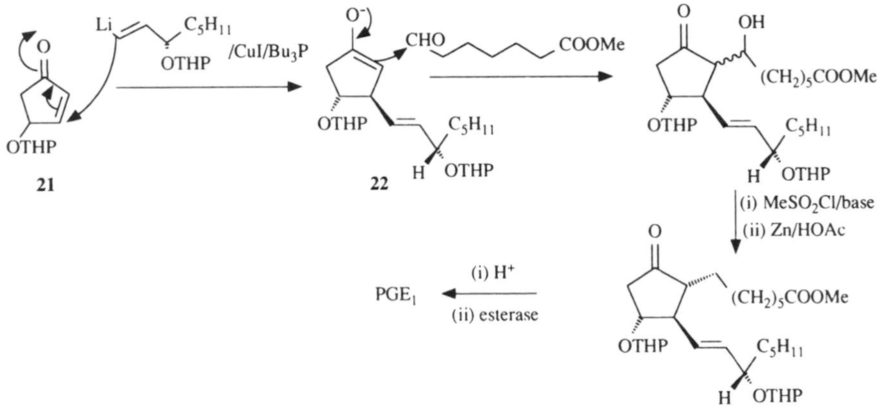

Scheme 7.18 *Short total synthesis of PGE_1*

7.4.4 Enantioselectivity in Prostaglandin Synthesis

If no chiral materials are used in the synthesis of a prostaglandin the product will be racemic. This is not acceptable for new medicinal products since enantiomers have different biological activities. For example, the biological activity of the enantiomer of PGE_1 (termed *ent*-PGE_1) is only 0.1% of that of the natural hormone in the stimulation of smooth muscle contraction. A wide range of methods for producing optically pure synthetic prostaglandins has been explored.

7.4.4.1 Resolution of Prostanoids. Resolution is the oldest method of obtaining optically pure material from a racemate but it is inefficient since the maximum yield is 50%. To reduce waste resolution should be as early as possible in a synthetic route; in the Corey synthesis the hydroxy-acid (**17**) was resolved with L-arginine.

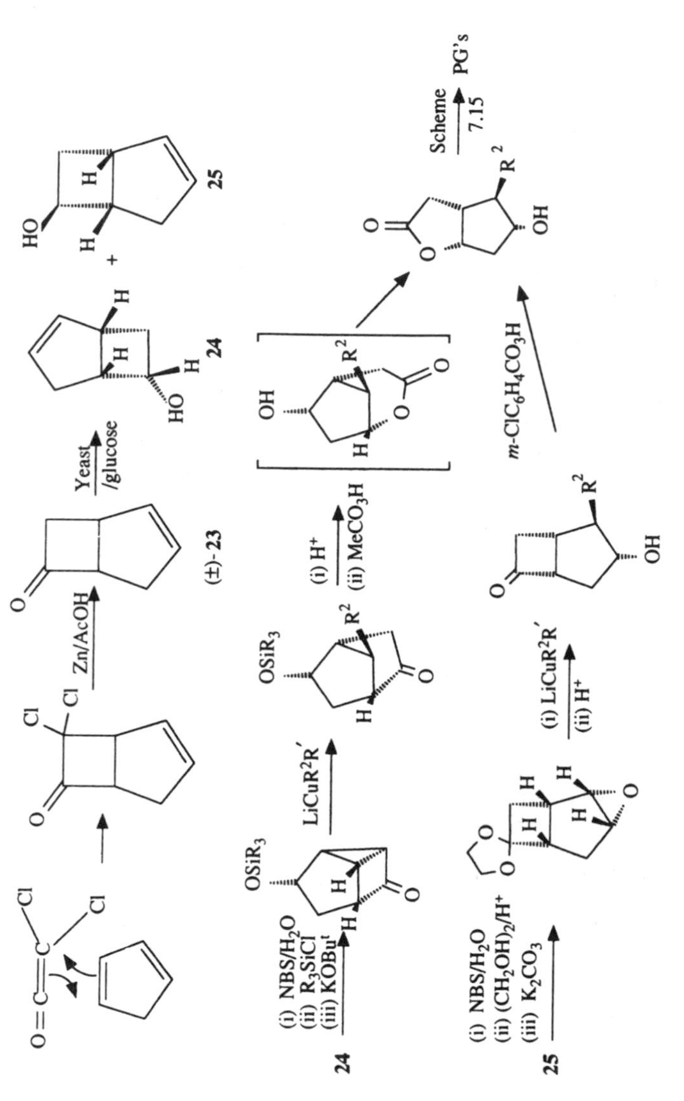

Scheme 7.19 *Glaxo route to prostaglandins*

7.4.4.2 Microbiological Resolution. The use of yeasts to distinguish between enantiomers also normally results in 50% wastage but the synthetic scheme developed by Glaxo (outlined in Scheme 7.19) achieves much greater efficiency. Reaction of cyclopentadiene with dichloroketene followed by dechlorination with zinc in acetic acid gives the racemic bicyclic ketone (**23**). This ketone is reduced by a dehydrogenase enzyme in an actively fermenting yeast giving a pair of diastereoisomers (**24** and **25**) each with the *S*-configuration at the new chiral centre. These may be separated by distillation, and subjecting each to a different eight-step synthetic sequence produces the natural $PGF_{2\alpha}$ from both isomers.

7.4.4.3 Use of Chiral Starting Materials. Optically active starting materials, such as D-glucose and D-glyceraldehyde, have given prostaglandins in lengthy syntheses but *S*-malic acid may be converted (Scheme 7.20) into an optically active cyclopentenone, suitable for prostaglandin synthesis, in just four steps. Two isomers are produced in the cyclization step, but they are separable by crystallization and the more useful one is the major product.

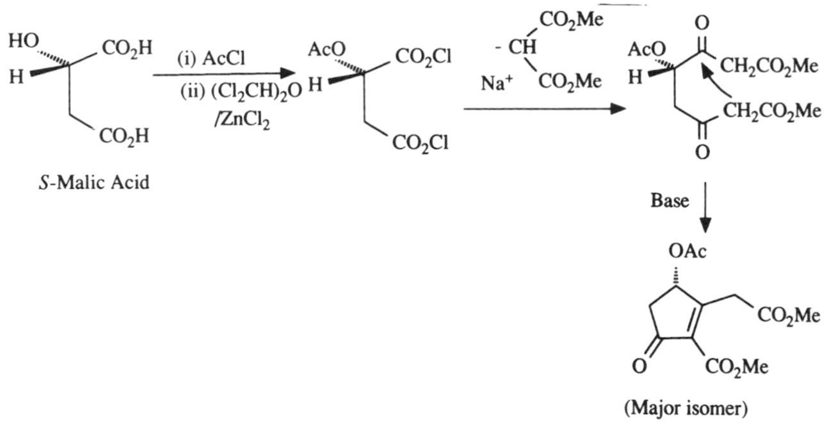

HO, H, CO_2H, CO_2H — *S*-Malic Acid → (i) AcCl (ii) $(Cl_2CH)_2O$ /$ZnCl_2$ → AcO, H, CO_2Cl, CO_2Cl → $^{-}CH(CO_2Me)_2$ Na^+ → AcO, H, CH_2CO_2Me, CH_2CO_2Me → Base → OAc, CO_2Me, CO_2Me, O (Major isomer)

Scheme 7.20 *Synthesis of a chiral intermediate for prostaglandin synthesis*

7.4.4.4 Asymmetric Synthesis. The most intellectually satisfying way to produce a chiral product from symmetric starting material is to use a chiral reagent or chiral reaction conditions to produce new asymmetric centres with the desired chirality, *i.e.* an asymmetric synthesis. This approach allows use of all the product and is beginning to find wide use in prostaglandin synthesis as understanding of stereochemical control improves. One reaction that gives a useful synthon for prostaglandins is the hydroboration of the cyclopentadiene ester (**26**, Scheme 7.21) using di-isopinocampheylborane. The asymmetric borane is simply prepared from pinene and the reaction gives better than 96% optical yield.

The chiral aluminium hydride (**11**, see also Section 7.3.2) has been used to give chiral cyclopentenones (*e.g.* **27**) in good yield. The same reagent has been employed for preparation of chiral ω-side-chains (**28**) in high enantiomeric

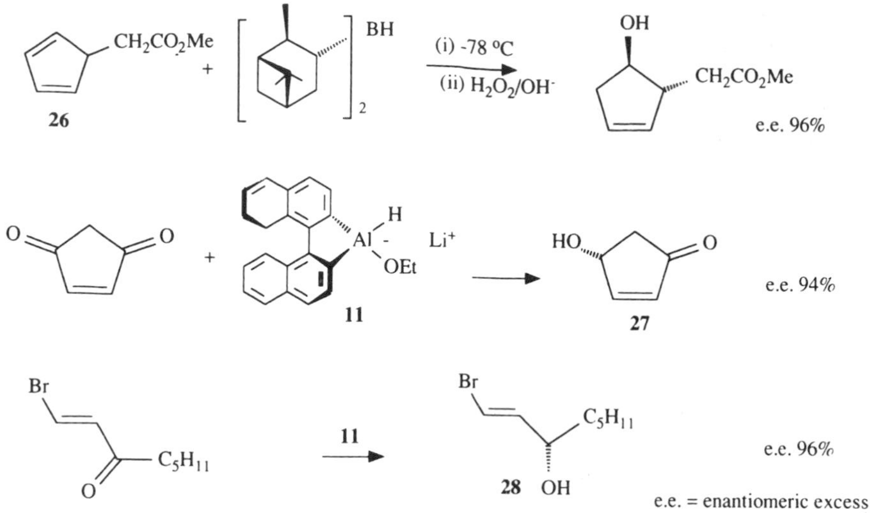

e.e. = enantiomeric excess

Scheme 7.21 *Asymmetric syntheses applicable to prostaglandin synthesis*

excess and high chemical yield. The availability of such chiral prostaglandin precursors has increased the importance of the conjugate addition approach to prostaglandin synthesis.

CHAPTER 8

β-Lactam Antibiotics

In 1928 Alexander Fleming observed that a plate on which *Staphylococci* bacteria were being cultured at St. Mary's Hospital, London had been contaminated with a mould and that in the immediate vicinity of this mould (*Penicillium notatum**) was a ring where no bacterial growth was evident. Fleming concluded that the mould was killing bacteria by producing a chemical, 'penicillin', but he was unable to isolate the penicillin. This much is common knowledge and Fleming's appreciation of the importance of his initial observation justifies his reputation as the discoverer of penicillin, even though antibacterial moulds were known and possibly used as early as the nineteenth century.

Chain and Florey reinvestigated the bactericidal properties of *Penicillium notatum* at Oxford in 1938 and isolated the active constituent known as penicillin N (**1**). A collossal Anglo-American research programme was initiated during World War II with the objective of producing usable quantities of this drug. It was discovered that *Penicillium chrysogenum*, which was obtained from a mouldy melon, also produced penicillins and mutant strains were developed which gave improved yields of penicillin. When grown in corn-steep liquor *Penicillium chrysogenum* gave penicillin G (**2**) and this fermentation was developed into a viable process for large scale penicillin G production.

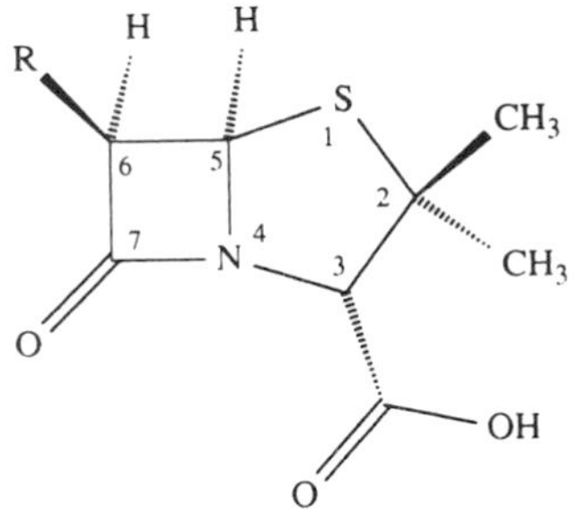

	R
1	$HO_2CCH(NH_2)(CH_2)_3CONH$-
2	$PhCH_2CONH$-
3	H
4	NH_2

The availability of penicillin G (benzyl penicillin) dramatically improved the outlook for sufferers of diphtheria (caused by *Corynebacterium diphtheriae*), tetanus (*Clostridium tetani*), gonorrhoea (*Neisseria gonorrhoea*), pneumonia (*Streptococcus pneumoniae*), and many other life-threatening diseases. Certain bacteria are less

* Species are named scientifically in two parts; the first part has a capital letter and is the genus, whilst the second part is peculiar to the species. For example *Canis familiaris* is the dog and *Canis lupus* is the wolf.

susceptible, and in general those which do not take Gram's stain (for example *Pseudomonas*) are difficult to kill with penicillin. Additionally it was soon noticed that some Gram-positive bacteria, which were sensitive to penicillin, mutated into strains which were penicillin-resistant. Consequently alternative bactericidal compounds were urgently required to sustain the initial striking improvement in prognosis for patients with common infections.

It is remarkable that sea water does not become dangerously burdened with bacteria since rivers and sewage which are heavily contaminated with micro-organisms discharge into the sea. Believing that this could be due to an antibiotic (a substance produced by micro-organisms to kill or inhibit the growth of competing organisms) Professor Giuseppe Brotzu collected micro-organisms from sea water near a sewage outlet in Sardinia in 1945. From these he isolated a fungus of the *Cephalosporium* genus which secreted substances capable of inhibiting bacterial growth. One of these substances was cephalosporin C (**5**) and showed small but definite activity against both Gram-negative and penicillin-resistant bacteria. Other cephalosporins with better activity were synthesized and novel analogues, such as the cephamycins, which show strong activity against Gram-negative and penicillin-resistant strains of bacteria, were discovered.

The penicillins and cephalosporins are called classical β-lactam antibiotics, but more distant structural analogues with antibiotic properties continue to be found. These include oxacephems and carbapenems in which the sulphur is replaced by oxygen or carbon, and nocardicins and monobactams (or sulfazecins) which are monocyclic.

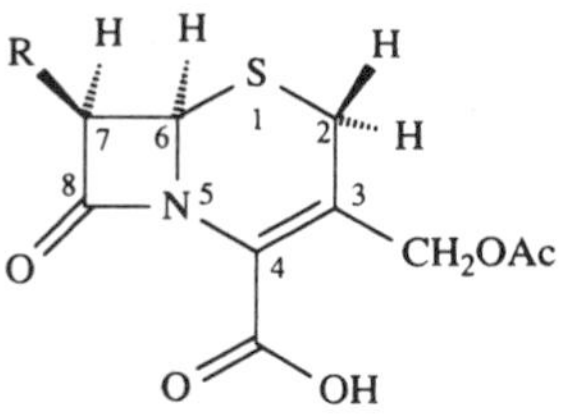

	R
5	$HO_2CCH(NH_2)(CH_2)_3CONH$-
6	H

8.1 STRUCTURE AND NOMENCLATURE

The systematic name for penicillin G (**2**) is [2*S*-[2α,5α,6β)]-3,3-dimethyl-7-oxo-6-phenylacetylamino-4-thia-1-azabicyclo[3,2,0]heptane-2-carboxylic acid. This name has not proved popular in the scientific literature. Instead semi-systematic nomenclature is used in which common basic structures have trivial names and any extra substituents are specified by name and position number. Penicillins and compounds which differ from them at position-6 only are named as a derivatives of penicillanic acid (**3**) and are numbered from sulphur as shown; thus compound (**4**) is 6-aminopenicillanic acid. Where there are other modifications the basic bicyclic β-lactam-thiazolidine structure is referred to as a penam in which the 'a' indicates that the rings are saturated and changes to an 'e' (penem) when a double bond is included. Prefixes are used to indicate replacement of one of the atoms in the ring structure but the numbering of the framework is not changed; for example the naturally occurring, potent antibiotic thienamycin (**7**), discovered in 1976, is a 1-carbapenem.

The nomenclature of cephalosporins follows a similar pattern in which close analogues of cephalosporin C are named as derivatives of cephalosporanic acid (**6**). The β-lactam-dihydrothiazine system is termed a cephem and the lowest-numbered position of the double bond may be specified by prefixes; thus cephalosporin C is a Δ^3-cephalosporin or 3-cephem. Other prefixes are added to describe nuclear analogues such as moxalactam (**8**) which is a 1-oxacephem.

In the 1970s a group of monocyclic β-lactam compounds, nocardicins (**9**), were found as components of fermentation broth of *Nocardia uniformis*. The most potent, nocardicin A (**10**), has strong activity against many Gram-negative bacteria but no activity against Gram-positive species. They may be named as derivatives of nocardicinic acid such as 3-aminonocardicinic acid (**11**).

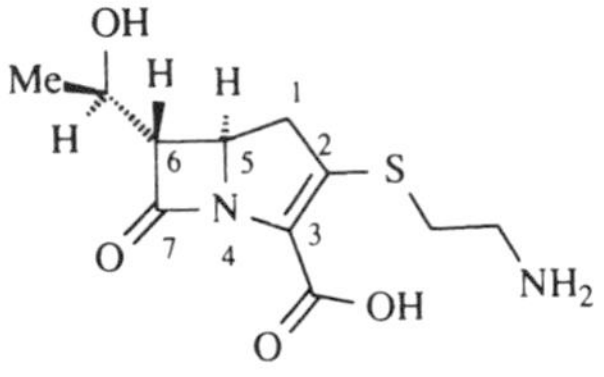

Thienamycin **7**

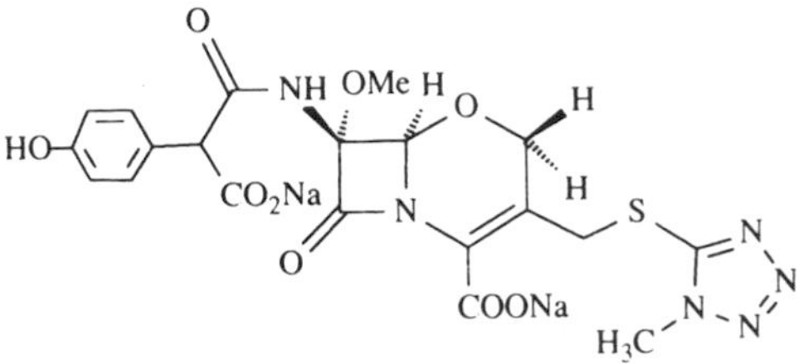

Moxalactam **8**

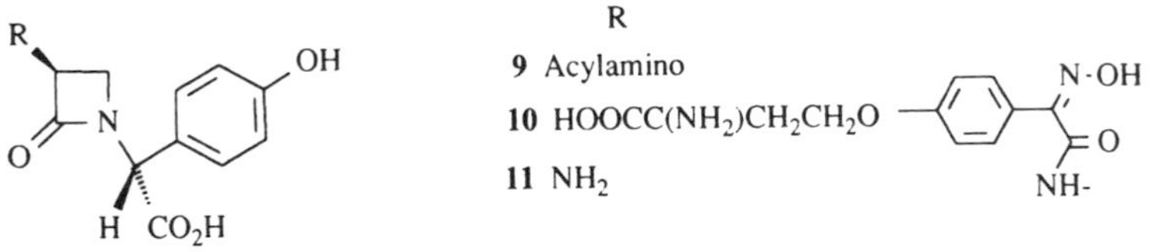

R
9 Acylamino
10 $HOOCC(NH_2)CH_2CH_2O$
11 NH_2

Nocardicins

It was discovered in 1981 that some bacteria produce monobactams (**12**), properly named 3-acylamino-2-oxoazetidine-1-sulphonic acids. Naturally occurring monobactams are (3*S*)-enatiomers and synthetic (3*R*)-enantiomers have been found to have no antimicrobial activity.

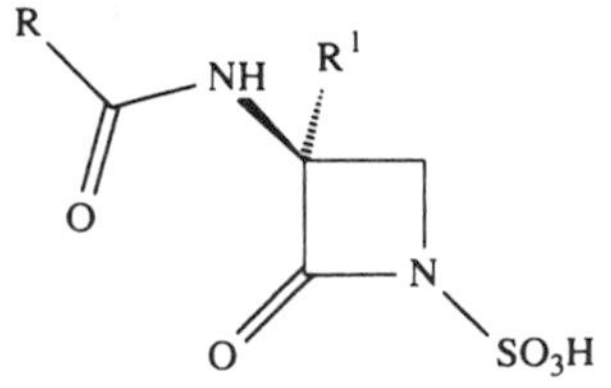

Monobactams **12** R^1 = H or OMe

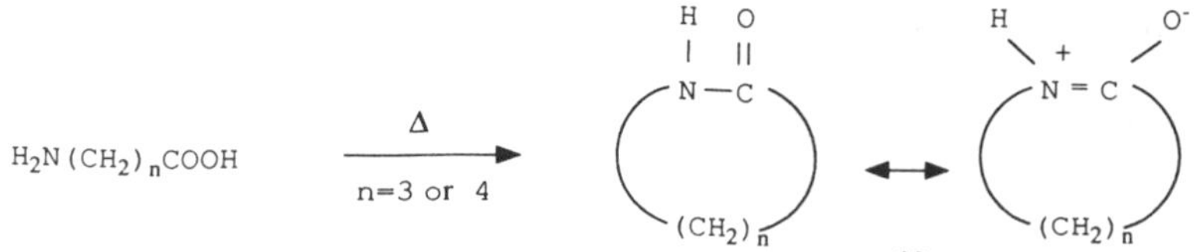

$H_2N(CH_2)_nCOOH \xrightarrow[n=3 \text{ or } 4]{\Delta}$

13

Scheme 8.1 *Formation of and amide conjugation in lactams*

8.2 PROPERTIES OF β-LACTAMS

Lactams are cyclic amides and those having the thermodynamically favoured five- or six-membered rings (**13**, n = 3 or 4] may be produced (Scheme 8.1) by heating γ- or δ-amino acids. Heating a β-amino acid does not result in formation of a β-lactam (**13**, n = 2) because the four-membered ring is strained and so less favoured thermodynamically. The strain inherent in β-lactams prepared by other means is apparent from their three-dimensional shapes, carbonyl stretching frequencies, and high reactivities. This high reactivity of β-lactam rings is crucial for the antibiotic properties of penicillins and cephalosporins.

8.2.1 Geometries and Carbonyl Stretching Frequencies of β-Lactams

In amides, conjugation leads to thermodynamic stabilization and this is most efficient when both the nitrogen and the carbonyl carbon atoms are trigonal (*i.e.* flat with 120° bond angles) and coplanar thus permitting the maximum orbital overlap. In β-lactams bond angles of 120° at carbon and nitrogen would lead to huge strain in the four-membered ring and although simple β-lactams are planar these atoms are not trigonal and so the delocalization energy is smaller than for acyclic amides. This difference has been used to explain many physical and biological properties of β-lactams but simple ring strain is probably more important.

Structures of biologically active penicillins and cephalosporins obtained by *X*-ray crystallography show that their lactam nitrogens are pyramidal. Some closely related, but biologically inactive cephalosporin analogues, such as Δ^2-cephalosporins (**14**), have a near-planar nitrogen. It has been suggested that a pyramidal nitrogen indicates poor conjugation and that this is essential for biological activity. However, in penicillins and cephalosporins a pyramidal nitrogen is not an essential condition for biological activity and there is little evidence for a correlation between planarity and conjugation.

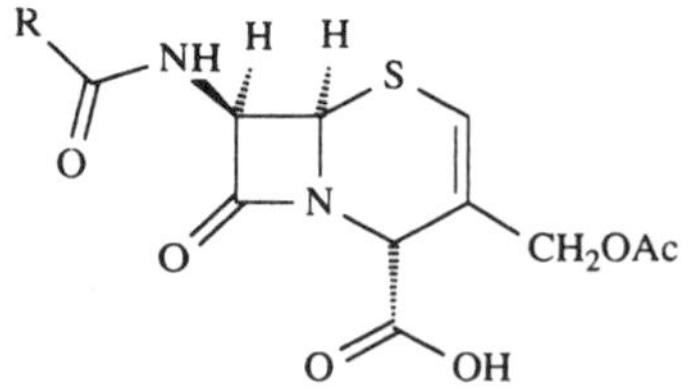

14

Table 8.1 *Typical carbonyl stretching frequencies*

Carbonyl compound	*Carbonyl stretching frequency*/cm^{-1}
Four-membered ring ketones	1775
Acyclic amides	1675
β-Lactams (*e.g.* nocardicins)	1745
Δ^2-Cephalosporins	1770
Penicillins and Δ^3-cephalosporins	1780
Penems	1800
Clavulanic acid (**85**)	1800

Another estimate of the extent of conjugation in amides may be obtained from their carbonyl stretching frequencies. A single bond between two atoms is more easily stretched than a double bond and so has a lower frequency for natural oscillations. In general as the bond-order between two atoms increases, so the infrared absorption frequency decreases. Amide conjugation reduces the double bond character in a carbonyl group and so shorter stretching frequencies (*i.e.* lower wavenumbers, cm^{-1}) are found in amides than in ketones. The carbonyl stretching frequencies shown in Table 8.1 suggest that amide conjugation is poor in β-lactams and negligible in penicillins and Δ^3-cephalosporins. However, these data were not measured under identical conditions and current thinking is that, although lactam carbonyl stretching frequencies appear to depend on the extent of conjugation, the differences are more probably due to changes in bond lengths or force constants.

8.2.2 Reactivity of β-Lactams

The nitrogen of amides is normally unreactive as a nucleophile because the lone pair on nitrogen is in conjugation with the carbonyl double bond and so not available for nucleophilic attack. In β-lactams conjugation is less efficient and so the lone pair on nitrogen is readily donated to an electrophile as illustrated (Scheme 8.2) by the facile alkylation of the β-lactam with an alkyl tosylate. However, penicillins and cephalosporins themselves show little evidence of increased nucleophilicity of the lactam nitrogen and their basicities (pK_a −5) are actually less than for typical amides. This may be due to the proximity of electron-withdrawing groups and elevation of nucleophilicity by inhibition of conjugation does not seem important.

The formation of the eight-membered ring illustrated in Scheme 8.2 indicates, in comparison with acyclic amides, the high susceptibility of a β-lactam carbonyl to nucleophilic attack. This increased reactivity of β-lactams, also apparent from their ease of alkaline hydrolysis, is thought to be primarily due to the release of ring strain accompanying conversion of the sp^2 carbonyl carbon into an sp^3 carbon.

The reactivities of penicillins are even higher than those of simple β-lactams because fusion of the five-membered thiazolidine ring results in extra ring strain,

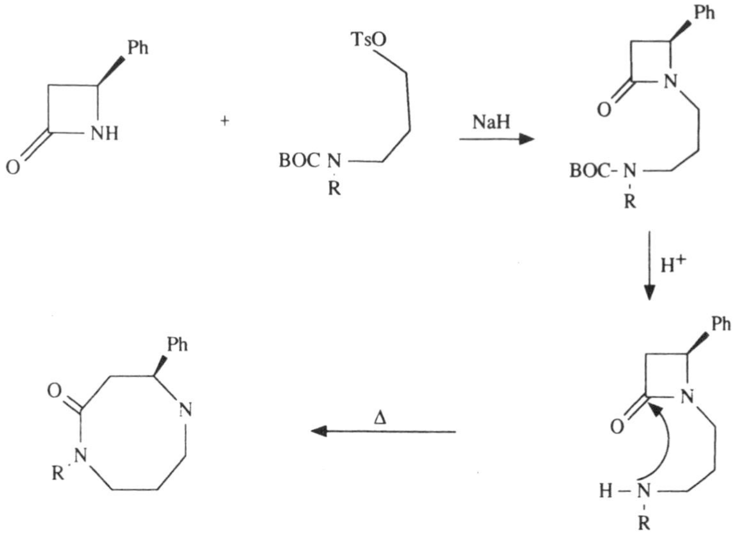

Scheme 8.2 *Reactivity of the β-lactam amido group*

a pyramidal nitrogen, and possibly reduced conjugation. In cephalosporins extra reactivity is due chiefly to the presence of the 3,4-double bond which conjugates with the nitrogen lone pair, further weakening resonance stabilization in the β-lactam. More importantly, this double bond allows nucleophilic attack at the β-lactam carbonyl to be accompanied by displacement of the allylic acetoxy group at the 3-position. This is discussed further in Section 8.6.3.

One may expect, therefore, that as the ease of basic hydrolysis of β-lactams increases, the C=O stretching frequency will increase, the nitrogen will become more pyramidal and more nucleophilic, and, as explained in Section 8.2.3, the biological activity may increase. This view has been demonstrated to be too simplistic but remains a useful guide.

8.2.3 Mechanism of Bactericidal Action

The details of the mechanisms of bactericidal action of β-lactam antibiotics are extremely complex, but the bare essentials are straightforward. Penicillins and cephalosporins are lethal to bacteria but innocuous to mammals because their toxicity results from interference with the biosynthesis of rigid bacterial cell walls which have no mammalian counterpart. There is a high osmotic pressure in a growing bacterial cell due to the build up of nutrients; in the absence of a strong cell wall this causes rupture of the delicate cell membrane, loss of the cell contents and death.

The bacterial cell wall is synthesized by cross-linking peptidoglycan chains and the enzymes which catalyse this process mistake β-lactam antibiotics for the last two or three amino acid residues of such peptidoglycan chains. Once the antibiotic is at the enzymic active site, nucleophilic groups on the enzyme (such as OH) attack the carbonyl of the β-lactam to give a product or products which

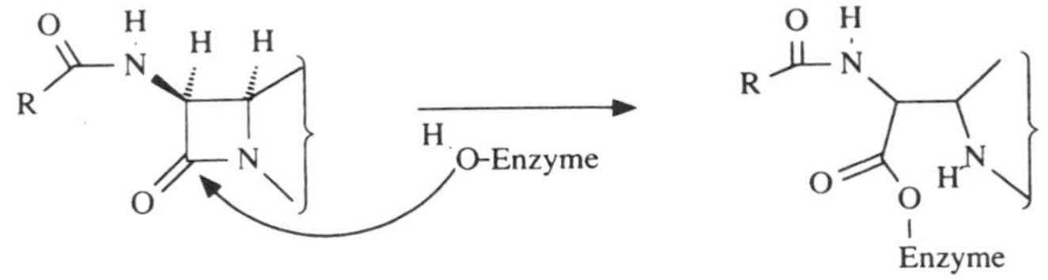

Scheme 8.3 *Irreversible bacterial enzyme inhibition by a β-lactam*

are no longer capable of effective catalysis causing irreversible enzyme inhibition (Scheme 8.3). Effective β-lactam antibiotics must be sufficiently stable to reach the bacterial enzyme without decomposition but then reactive enough to be cleaved by reaction with a nucleophilic site on the enzyme causing mutual deactivation.

To prevent suicide, micro-organisms which produce β-lactam antibiotics avoid substantial antibiotic production during their growing phases. These organisms also have various defence mechanisms, such as production of β-lactam cleaving enzymes (β-lactamases), which assist survival in this self-created poisonous environment. β-Lactamases are also responsible for deactivating penicillin in Gram-negative and penicillin-resistant bacteria.

8.3 ORGANOSULPHUR CHEMISTRY

Sulphur occurs directly below oxygen in Group VI of the Periodic Table and organosulphur compounds are often very similar in properties to the analogous oxygen-containing organic compounds. The valence shell of sulphur is $(3s)^2(3p)^4$ but the natures of the *p*-orbitals are such that strong π-bonding is virtually impossible. Promotion of electrons into the normally vacant *d*-orbitals permits better π-bonding and allows sulphur to have oxidation states of II, IV, or VI which is, of course, a major difference from oxygen. A single valence bond structure is a poor representation of such bonding and sulphur–oxygen double bonds are often indicated by an arrow intended to convey a largely polar bond with some double bond character, as might alternatively be indicated by the two canonical forms (**15**). While this type of π-bonding between sulphur and oxygen is present in many stable molecules, similar bonding between sulphur and carbon is inefficient and such structures are often unstable. Thioacetone, $Me_2C{=}S$, for example, spontaneously trimerizes to a trithiane.

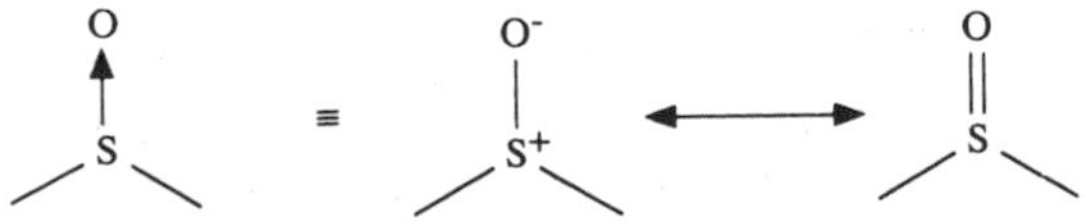

(**15**)

Sulphur is less electronegative than oxygen as indicated by the chemical shift of protons in methyl groups next to oxygen (δ 3.5–4.5) as opposed to those in methyl groups adjacent to sulphur (δ 2.5–3.0). However, C—H groups next to sulphur are usually more acidic than those next to oxygen since sulphur is better

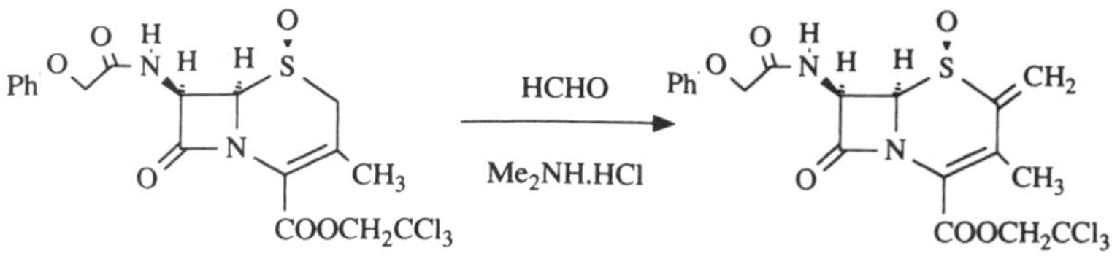

$$\xrightarrow[\text{Me}_2\text{NH.HCl}]{\text{HCHO}}$$

$COOCH_2CCl_3$ → $COOCH_2CCl_3$

Scheme 8.4 *Alkylation at C2 of a cephalosporin sulphoxide*

able to stabilize an adjacent negative charge by allowing overlap with its vacant *d*-orbitals. Since position-2 of cephalosporin sulphoxides is made acidic by both the adjacent sulphur and by the allylic double bond it may be alkylated, for example by reaction (Scheme 8.4) with formaldehyde and a secondary amine (Mannich reaction) to give a C2 methylene derivative.

The brief survey of organosulphur compounds which follows concentrates on those reactions which can not be readily predicted from the behaviour of analogous oxygen compounds.

8.3.1 Thiols, RSH

The most remarkable property of thiols is their odour. All the lower thiols (R = Me, Et, Pr, Bu, or Ph) are volatile and have strong, often nauseating odours meriting the warning, normally present on their bottle labels, 'stench!'.

Thiols may be prepared by reaction of sodium hydrosulphide, NaSH, with an alkyl halide or with an alkyl sulphonate as shown (**16**) for ethanethiol:

$$\text{EtX} + \text{NaSH} \longrightarrow \text{EtSH} + \text{NaX} \qquad \textbf{(16)}$$

$$\text{X} = \text{halide or } \text{R}^1\text{OSO}_2$$

Elimination is an important side-reaction in the preparation of alcohols by hydrolysis of alkyl halides with sodium hydroxide, but even tertiary alkyl halides may be employed in the above thiol synthesis since the hydrosulphide ion is more nucleophilic and much less basic than hydroxide.

Thiols are more acidic than alcohols and are easily alkylated by reaction with a weak base and an alkyl halide to give a dialkyl sulphide. Ethanethiol thus gives (**17**) ethyl isobutyl sulphide (isobutylthioethane) on reaction with sodium hydroxide and 1-bromo-2-methylpropane:

$$\text{EtSH} \xrightarrow{\text{NaOH}} \text{EtS}^- \xrightarrow{\text{Me}_2\text{CHCH}_2\text{Br}} \text{Me}_2\text{CHCH}_2\text{SEt} \qquad \textbf{(17)}$$

The older name for thiols is mercaptans, which is derived from their ability to react with salts of mercury to give (**18**) crystalline mercaptides:

$$2\text{RSH} + \text{HgO} \longrightarrow \text{H}_2\text{O} + (\text{RS})_2\text{Hg} \qquad \textbf{(18)}$$

Salts of other heavy metals such as lead react similarly.

Gentle oxidation of a thiol gives (**19**) a disulphide (*cf.* cystine) whilst more powerfully oxidizing conditions give a sulphonic acid such as ethanesulphonic acid from ethanethiol:

$$2EtSH \xrightarrow{I_2} EtSSEt \xrightarrow{HNO_3} EtSO_3H \qquad (19)$$

8.3.2 Sulphides, Sulphoxides, Sulphones, and Sulphonium Salts

Dialkyl sulphides (thioethers) may be prepared by alkylation (**17**) of thiols and are desulphurized to alkanes by hydrogenation (**20**) over Raney nickel:

$$RSR^1 + H_2 \xrightarrow{Ni} RH + R^1H \qquad (20)$$

They may also be alkylated (**21**) to give trialkylsulphonium salts:

$$Me_2S + R^1X \longrightarrow Me_2S^+R^1X^- \qquad (21)$$

Oxidizing agents convert dialkyl sulphides first into sulphoxides and then sulphones. Dimethyl sulphide thus gives (**22**) dimethyl sulphoxide (methylsulphinylmethane) with one mole of hydrogen peroxide and dimethyl sulphone when two moles are used:

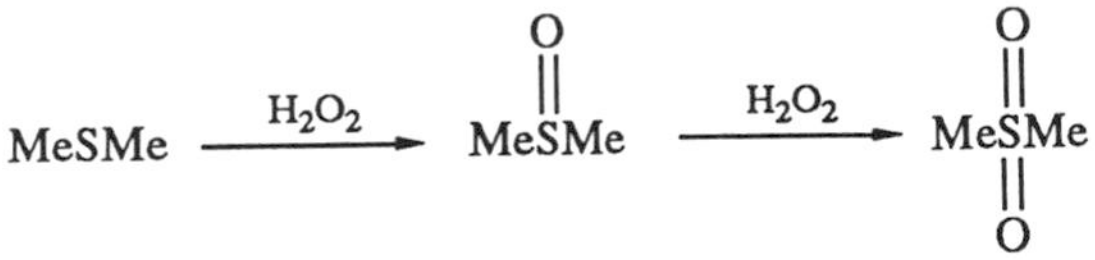

$$MeSMe \xrightarrow{H_2O_2} MeS(=O)Me \xrightarrow{H_2O_2} MeS(=O)_2Me \qquad (22)$$

Dimethyl sulphide may be regenerated from the sulphoxide by treatment with a deoxygenating reagent such as triphenylphosphine.

Sulphoxides, $R_2CHS(O)R$, are much less acidic than ketones since enethiolates, $R_2C{=}S(O^-)R$, are less stable (poor π-bonding between C and S) than enolates, $R_2C{=}C(O^-)R$, and do little to stabilize the anions, $R_2C^-S(O)R$.

Sulphoxides having two different alkyl groups (**23**) and sulphonium salts (**24**) having three different alkyl groups may be resolved into enantiomers because the sulphur is a centre of asymmetry, with the lone pair of electrons forming the fourth substituent on a distorted tetrahedron.

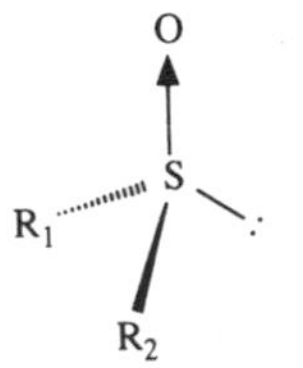

O, S, R_1, R_2

23

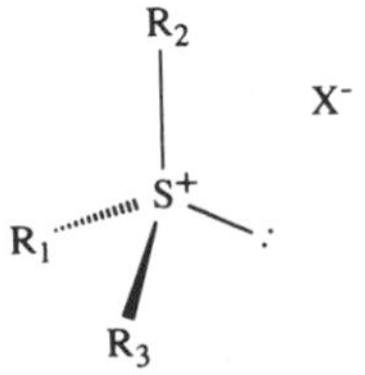

R_2, S^+, R_1, R_3, X^-

24

8.3.3 Organic Oxyacids of Sulphur

The oxidation state of sulphur in sulphenic, sulphinic, and sulphonic acids increases in alphabetical order while their associated reactivities decrease.

8.3.3.1 Sulphenic Acids, RSOH. Sulphenic acids are encountered as fleeting intermediates in the thermal decomposition of sulphoxides. Thermolysis of t-butyl sulphoxide, for example (**25**), gives t-butanesulphenic acid with elimination of butene.

$$Me_3C\overset{O}{S}{-}CMe_2(CH_2{-}H) \longrightarrow Me_3CSOH + CH_2{=}C(Me)_2 \qquad (25)$$

Sulphenic acids react readily with double and triple bonds, as illustrated by the reaction (**26**) of t-butanesulphenic acid with acrylic acid, and such reactions are used to 'trap' sulphenic acids.

$$Bu^tSOH + CH_2{=}CHCO_2H \longrightarrow Bu^t\overset{O\uparrow}{S}CH_2CH_2CO_2H \qquad (26)$$

In the absence of an alkene or alkyne the sulphenic acid forms (**27**) its anhydride (a thiolsulphinate) or disproportionates (**28**) giving a sulphinic acid.

$$RSOH \longrightarrow H_2O + R\overset{O\uparrow}{S}SR \qquad (27)$$

$$RSOH \longrightarrow RSH + RSO_2H \qquad (28)$$

8.3.3.2 Sulphinic Acids, RSO_2H. Although sulphinic acids are thermally unstable, disproportionating to the sulphinyl sulphone and a sulphonic acid (**29**), they are readily isolable compounds commonly prepared by the reduction (**30**) of sulphonyl halides.

$$3RSO_2H \longrightarrow RS(O_2)SR + RSO_3H + H_2O \qquad (29)$$

$$p\text{-}CH_3C_6H_4SO_2Cl \xrightarrow[\text{iii, } H^+]{\text{i, } Zn/H_2O \quad \text{ii, } Na_2CO_3} p\text{-}CH_3C_6H_4SO_2H \qquad (30)$$

They add to multiple bonds in a similar manner to sulphenic acids but more commonly sulphinate ion acts as a nucleophile such as in their alkylation which normally occurs on sulphur, giving (**31**) a sulphone.

$$RSO_2^- + R^1X \longrightarrow RS(O_2)R^1 + X^- \qquad (31)$$

Hydrogen peroxide and many other oxidizing agents convert sulphinic acids into the corresponding sulphonic acids.

8.3.3.3 Sulphonic Acids, RSO_3H. Sulphonic acids are stable acids which are as strong as typical mineral acids. Aryl sulphonic acids are prepared by sulphonation of arenes using fuming sulphuric acid, but alkyl sulphonic acids are prepared by alkylation (**32**) of sulphite ion:

$$RBr + HOSO_2Na \longrightarrow RSO_3Na \longrightarrow RSO_3H \qquad (32)$$

This method has even been used to prepare tertiary alkyl sulphonic acids, albeit in low yield, and the intermediate sodium sulphonates find use as detergents.

Alkanesulphonic acids may be reacted with phosphorus pentachloride to give alkanesulphonyl chlorides. These are commonly used to convert alcohols into their alkanesulphonates which readily undergo nucleophilic displacements (**33**).

$$MeSO_2H \xrightarrow{PCl_5} MeSO_2Cl \xrightarrow{R^1OH} MeSO_2R^1 \xrightarrow{Nu^-} R^1Nu + MeSO_2^- \qquad (33)$$

Reaction of sulphonyl chlorides with amines gives sulphonamides such as *p*-aminobenzenesulphonamide (sulphanilamide, **34**) which was widely used as an antibacterial drug prior to the availability of β-lactam antibiotics and is still used medicinally in eye-drops.

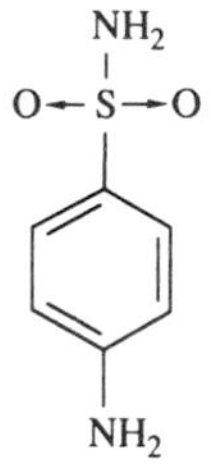

NH_2
O←S→O
NH_2

34

8.4 MICROBIOLOGICAL PENICILLIN PRODUCTION

8.4.1 Biosynthesis

Penicillin N would result from cyclization of the tripeptide δ-(D-α-aminoadipyl)-L-cysteinyl-D-valine (**35**) and so it appears to be derived from cysteine, valine and α-aminoadipic acid with the latter two amino acids in their unnatural D-configurations.

In fact only L-amino acids are used as precursors in the biosynthetic pathway (Scheme 8.5). First L-α-aminoadipic acid condenses with L-cysteine to give the dipeptide in which cysteine is joined to the side-chain of aminoadipic acid. Then

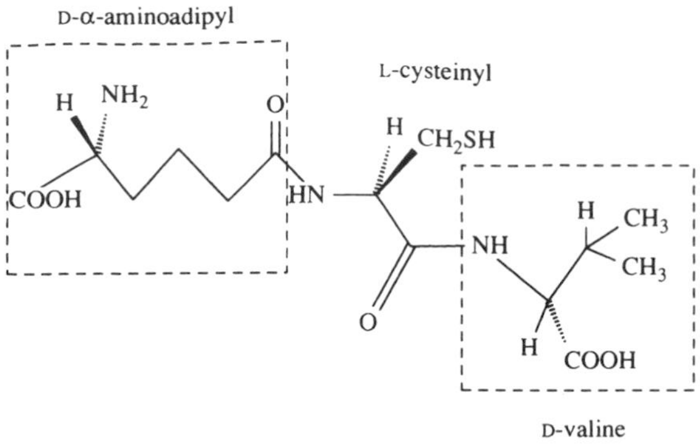

D-α-aminoadipyl

L-cysteinyl

D-valine

Tripeptide derived from penicillin N **35**

L-valine couples *with inversion of configuration* giving δ-(L-α-aminoadipyl)-L-cysteinyl-D-valine (**36**). This tripeptide has been isolated from the mycelium (vegetative filaments) of *Penicillium chrysogenum* and also from *Cephalosporium acremonium*.

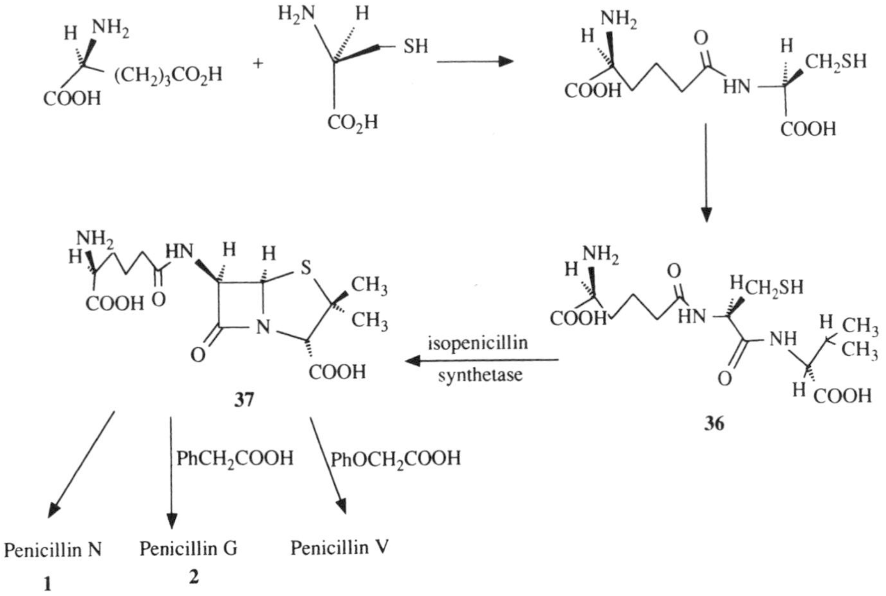

isopenicillin synthetase

37

36

$PhCH_2COOH$

$PhOCH_2COOH$

Penicillin N **1**

Penicillin G **2**

Penicillin V

Scheme 8.5 *Biosynthesis of penicillins*

The tripeptide then cyclizes to isopenicillin N (**37**) catalysed by the enzyme isopenicillin N synthetase. The four-membered ring, or lactam, is probably the first ring to form but the intermediate remains bound to the enzyme until the second ring is joined. The double cyclization of one molecule of tripeptide results in the removal of four hydrogen atoms and requires one molecule of dioxygen.

Finally the nature of the side-chain is altered. Penicillin N is produced from isopenicillin N by inversion of the configuration in the amino acid side-chain. Alternatively the side-chain may be replaced by a phenylacetamido group to give penicillin G or by a phenoxyacetamido group (from phenoxyacetic acid) to give penicillin V.

8.4.2 Industrial Penicillin G Production

The commercial production of penicillins uses a high-yielding strain of the fungus (or mould, the names are virtually interchangeable) *Penicillium chrysogenum* which is taken from a master stock and grown in a series of inoculum development steps. When the desired characteristics are reached the mycelium pellets (fungal growths) are transferred to the production stage. A typical fermentation would use either corn-steep liquor (40 g l^{-1}) or an aqueous solution of glucose (5 g l^{-1}) and lactose (30 g l^{-1}) with inorganic nutrients such as ammonium sulphate. In the presence of phenylacetic acid (0.5 g l^{-1}) penicillin G is produced but addition of other α-substituted acetic acids to the fermentation medium gives penicillins bearing the corresponding α-substituted acetamido side-chain. The broth is kept at a pH between 6.8 and 7.4 and is well aerated to facilitate oxygen uptake and evolution of carbon dioxide. The concentration of carbohydrate falls steeply at first but is then maintained by feeding with glucose.

When the final fermented mixture reaches optimum penicillin levels (up to 2%) the mycelium is separated on a rotary vacuum filter and the filtered broth is cooled to 0 °C. The penicillin is converted into the free acid by addition of dilute sulphuric or phosphoric acid and then rapidly extracted into butyl acetate or amyl acetate. Addition of sodium or potassium acetate to the organic phase converts the penicillin into its sodium or potassium salt, which crystallizes and may be filtered off.

8.5 STRUCTURAL MODIFICATIONS OF PENICILLINS

Achieving a chosen chemical modification of an oligopeptide is difficult because of the presence of other reactive groups; in penicillins the problems are accentuated by the requirement for reaction conditions that do not destroy the β-lactam ring. Although selective reaction at one functional group may sometimes be achieved by the use of sufficiently mild or discriminating reaction conditions, here we will consider protection of reactive groups to allow reaction at only the desired site. Protection of the sulphur is normally accomplished by conversion into a sulphoxide. Nitrogen-protection, where necessary, generally employs the same amine-protecting groups as used for amino acids, but a large variety of methods are used for carboxylate-protection.

8.5.1 Protection of the Carboxylic Acid Group

Of the acid-protecting groups employed in peptide synthesis, benzyl, *p*-nitrobenzyl, and t-butyl esters have been used successfully for protection of penicillins. It is more common, however, to find deployment of acid-protecting groups which have been developed specially for use with β-lactams and methods for attachment and removal of the more common ones are described below. The use of

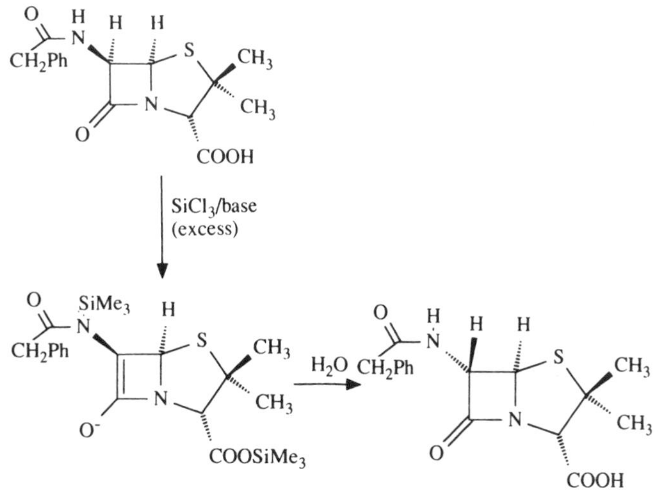
COOH
SiCl3/base
(excess)
COOSiMe3
H2O
COOH

38

Scheme 8.6 *Epimerizations of benzylpenicillin*

these protecting groups is now routine and many examples are given in the following pages, but careful technique is necessary. For example, use of excess reagents in the silylation of benzylpenicillin (Scheme 8.6) can lead to epimerization at the C6 position to give, after hydrolysis, 6-epi-benzylpenicillin (**38**). Epimerization at C6 of penicillins or C7 of cephalosporins results in virtual destruction of biological activity.

8.5.1.1 Trichloroethyl (TCE) Esters. The normal way to make a TCE ester (Scheme 8.7) is to react the acid with trichloroethanol in the presence of the condensing agent (see Section 4.5.4.1) dicyclohexylcarbodi-imide.

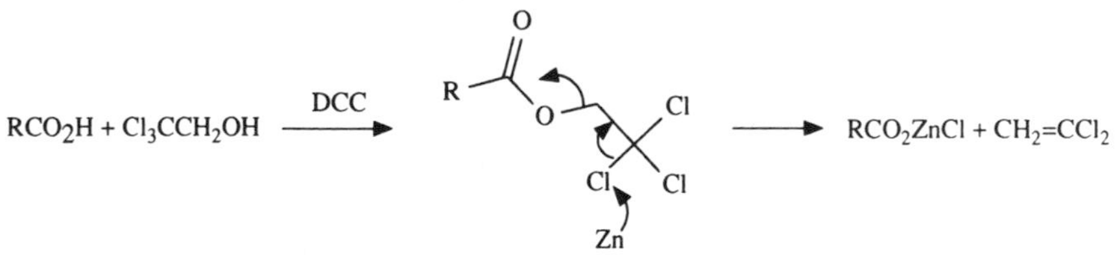
RCO2H + Cl3CCH2OH
DCC
RCO2ZnCl + CH2=CCl2
Zn

Scheme 8.7 *Trichloroethyl esters; formation and mechanism of zinc-induced removal*

The TCE ester may be selectively removed by an elimination induced by zinc as shown in Scheme 8.7. Trifluoroacetic acid (TFA) is often used as solvent but simply warming a TCE ester in methanol with zinc usually regenerates the free acid and constitutes a very mild deprotection procedure.

8.5.1.2 Diphenylmethyl Esters. Diphenylmethyl esters are usually produced by reaction (**39**) of the carboxylic acid with diphenyldiazomethane in diethyl ether.

This mild method of attachment is complemented by mild removal processes such as treatment with aluminium chloride in anisole (phenyl methyl ether) or TFA in anisole.

$$RCOOH + Ph_2CN_2 \longrightarrow RCOOCHPh_2 \xrightarrow{H^+} RCOOH + Ph_2CHOH \quad (39)$$

This facile acidic hydrolysis proceeds via an $A_{AC}1$ pathway, encouraged by the relative stability of the intermediate diphenylmethyl carbocation (Ph_2CH^+) in which there is extensive delocalization of the positive charge.

8.5.1.3 Silyl Esters. Trimethylsilyl and dimethylchlorosilyl esters are prepared (see Scheme 8.8 for example) from acids using trimethylchlorosilane or dimethyldichlorosilane respectively with *N*,*N*-dimethylaniline as base. These esters are unstable in hydroxylic solvents and so the acid may be regenerated by treatment with water or aqueous methanol.

8.5.2 Side-chain Modification in Penicillins

Penicillin G is an extremely valuable drug but causes allergic responses, is unstable in acid, and has little activity against Gram-negative and penicillin resistant-bacteria. An early indication that modification of the side-chain (the acylamino group at C6) could be useful was shown by the activity of penicillin N against Gram-negative bacteria such as *Salmonella typhi* which causes typhoid.

8.5.2.1 Enzymic Removal of the Side-chain from Penicillins. Penicillin G and penicillin V are easily prepared by microbiological fermentation and merely require side-chain removal followed by reacylation to give semi-synthetic penicillins. An enzymic method for side-chain removal was sought since simple basic hydrolysis of penicillin G causes β-lactam cleavage and results in yields of less than 0.5% of the desired 6-aminopenicillanic acid (6-APA, **4**). It was found that the benzyl side-chain may be removed selectively by incubating aqueous penicillin G (up to 6% solution) with the enzyme penicillin G acylase. Immobilizing the enzyme by attachment to a solid support facilitates its separation for re-use. Purification of the fermented mixture continues with extraction with an organic solvent to remove phenylacetic acid and penicillin G followed by low temperature evaporation of the aqueous solution which induces crystallization of 6-APA in 90% yield.

8.5.2.2 Chemical Removal of the Side-chain from Penicillins. A chemical procedure for side-chain removal (Scheme 8.8) was developed subsequent to the enzymic one and, although multi-step, it may be conducted without isolation of intermediates and is high-yielding. Penicillin G is first converted into its dimethylchlorosilyl ester and then the one remaining active hydrogen, the N—H of the side chain, is functionalized by reaction with phosphorus pentachloride at -40 °C. This gives the imino chloride (**40**), which is converted into the imino ether (**41**) by addition of n-butanol at -40 °C. The imino ether and the silyl ester functions are then hydrolysed simultaneously by pouring into cold water thus producing 6-APA. Provided the reaction with butanol occurs at or below -40 °C, overall yields of better than 90% are obtained.

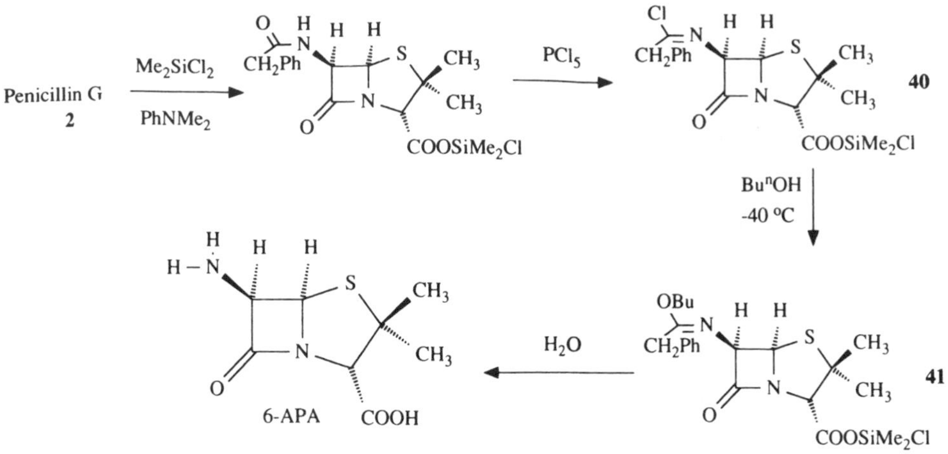

Scheme 8.8 *Chemical removal of side-chain from benzylpenicillin*

8.5.2.3 Semi-synthetic Penicillins. A wide variety of acyl chlorides in pyridine have been used to acylate 6-APA to give the semi-synthetic penicillins (Scheme 8.9) whose structure–activity relationships have been deduced so enabling the development of a sizable armoury of useful drugs (Table 8.2 gives a few examples).

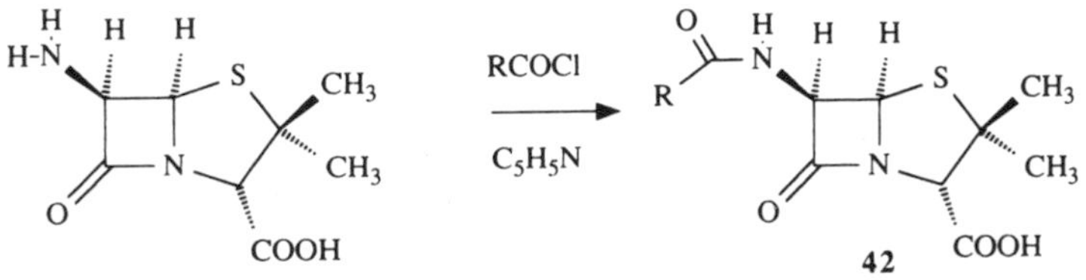

Scheme 8.9 *Synthesis of semi-synthetic penicillins from 6-APA*

Table 8.2 *Structures and antimicrobial activities of natural and semi-synthetic penicillins*

Penicillin (**42**)	*Side-chain* (*R*)	*Oral use*	*Sensitive bacteria*	
			Gram-negative	*Penicillin-resistant*
Penicillin G	$PhCH_2$	–	–	–
Penicillin V	$PhOCH_2$	+	–	–
Penicillin N	D-$NH_2CH(COOH)(CH_2)_3$	?	+	?
Methicillin	2,6-dimethoxyphenyl	–	+	+
Ampicillin	D-$PhCH(NH_2)$	+	+	–
Amoxycillin	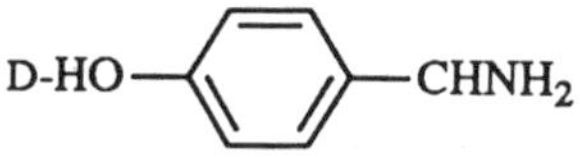 D-HO–C_6H_4–$CHNH_2$	+	+	–
Pivampicillin ($CH_2OCH_2COCH_3$ ester)	D-$PhCH(NH_2)$	+	+	–
Carbenicillin	$PhCH(COOH)$	–	+	–
Cloxacillin	3-methyl-5-(2-chlorophenyl)isoxazolyl	+	+	+

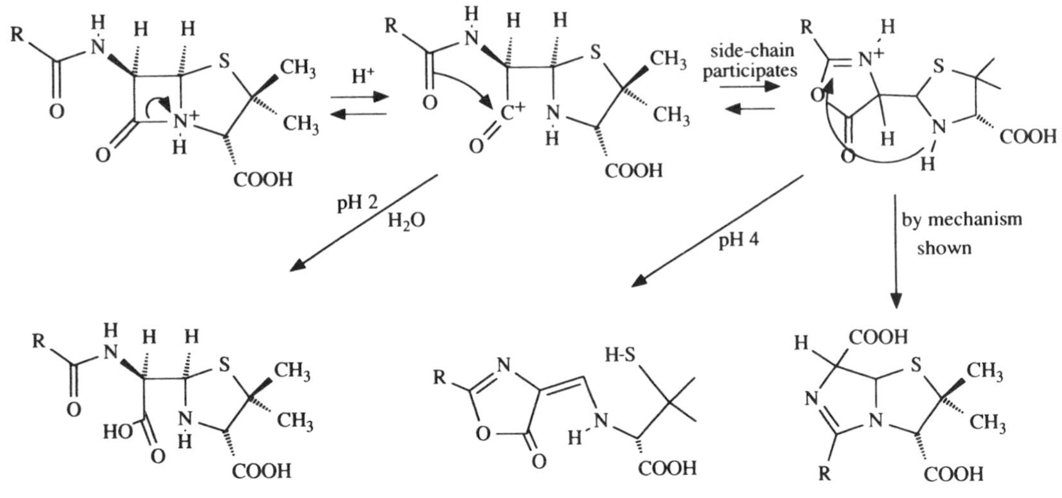

A Penicilloic Acid (**43**)

A Pencillenic Acid (**44**)

A Penillic Acid (**45**)

Scheme 8.10 *Decomposition of penicillins in acid*

The decomposition of penicillin G (Scheme 8.10) in aqueous acid to give a penicilloic acid (**43**), a penicillenic acid (**44**), or a penillic acid (**45**) depending on pH is accelerated by anchimeric assistance from the side-chain. The penicillenic acid (**44**) formed after administration of penicillin G may react with amino groups of proteins to give antigenic substances implicated as factors in penicillin allergy. Changing the side-chain may reduce acid lability but has little effect on the allergenicity of penicillins.

Penicillin G (but not penicillin V) is too unstable under the acid conditions in the stomach for oral use, necessitating administration by injection. Oral activity is bestowed on penicillins such as ampicillin and cloxacillin having either an electron-withdrawing side-chain, which may inhibit anchimeric assistance of β-lactam cleavage, or with a side-chain bearing a basic amino group which may be protonated in preference to the lactam nitrogen. In ampicillin, carbenicillin, and amoxycillin, polar amino or carboxyl side-chain groups improve water-solubility and assist cell penetration particularly into Gram-negative bacteria. Side-chains having the D-configuration are used since the L-isomers tend to be weak antibiotics and are readily cleaved by amidases. Absorption of these derivatives through the gut wall tends to be inefficient but may be improved when the dipolar carboxyl group is transiently masked as in pivampicillin, which being an acyloxymethyl ester is rapidly hydrolysed in the body to its component acid, ampicillin.

Another problem amenable to side-chain modification is the inactivation of penicillin G and penicillin V by β-lactamases (enzymes which cleave β-lactam rings) produced by Gram-negative bacteria such as *Escherichia coli*, which causes gastro-enteritis, and penicillin-resistant bacteria (*e.g. Staphylococcus aureus*), responsible for many skin and soft tissue infections. The steric bulk of the side-chains of methicillin and cloxacillin inhibits the approach of β-lactamase enzymes and is thus responsible for their activities against penicillin-resistant bacterial strains. The resistance of Gram-negative bacteria is largely due to production of different β-lactamases and the activity of penicillins active against them is not so easily assigned to essential molecular characteristics. Those penicillins having activity against Gram-negative organisms tend to have weak activity towards penicillin-resistant bacterial strains and towards Gram-positive bacteria.

8.5.3 Ring-expansion of Penicillins

The discovery that cephalosporin C is active against Gram-negative bacteria, and has inherent resistance to both acid hydrolysis and to β-lactamase activity, initiated research into cephalosporin production. Use of the readily available penicillins as starting materials requires expansion of the five-membered thiazolidine ring to a six-membered dihydrothiazine ring without destruction of the labile β-lactam. Selective ring-opening of the thiazolidine has been achieved at each of the four possible bonds but 1,2-bond cleavage is used almost exclusively. Before considering this in detail we shall look at a less practical but interesting earlier method.

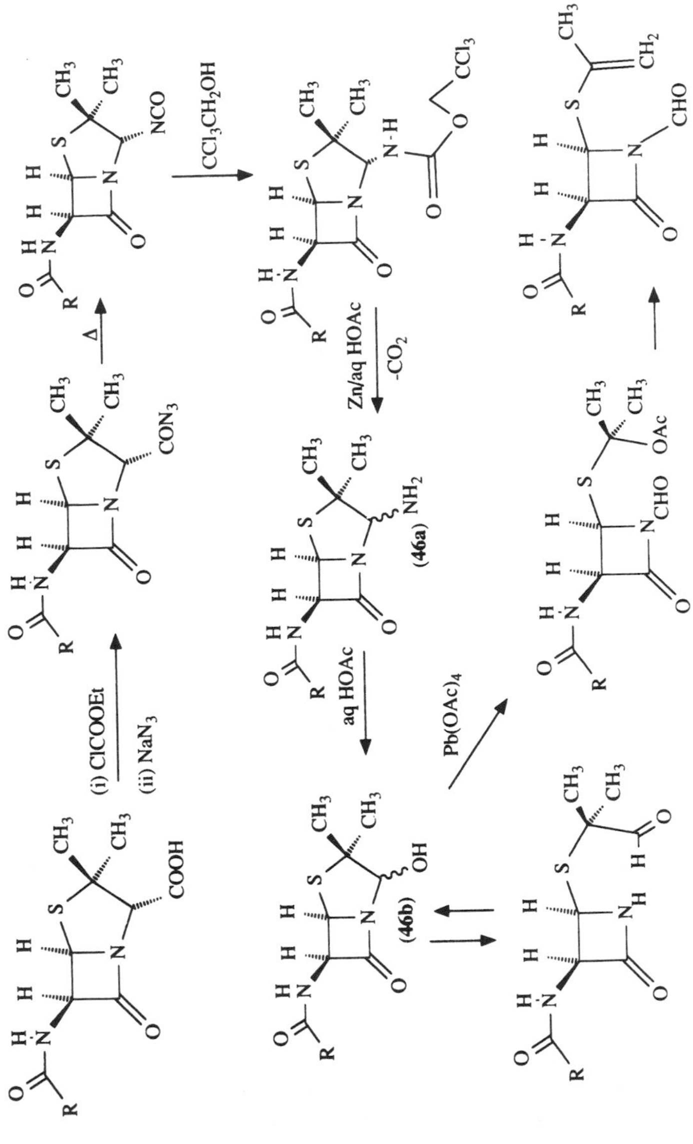

Scheme 8.11 *Curtius rearrangement to cleave 2,3-bond in penicillin*

8.5.3.1 Penam Ring Opening Using the Curtius Rearrangement. The Curtius thermal rearrangement of acid azides ($RCON_3$) yielding first isocyanates (RNCO) and then amines (RNH_2) or urethanes ($RNCOR^1$) by reaction with water or alcohols proceeds under mild conditions compatible with preservation of a β-lactam. Penicillins may be converted into their acid azides by activation using methyl chloroformate in triethylamine followed by addition of sodium azide. Thermal rearrangement of the penicillin azide in trichloroethanol produces (Scheme 8.11) first a reactive isocyanate and then the trichloroethylurethane. Removal of the trichloroethyl group using zinc in aqueous acetic acid gives the unstable *gem*-diamine intermediate (**46a**) which hydrolyses to an α-hydroxyamine (**46b**) (in equilibrium with the ring-opened aldehyde) which may be reacted with lead tetra-acetate (*cf.* cleavage of 1,2-diols, Section 2.3.9.4) to give the *N*-formyl β-lactam derivative. Being an acetate of a tertiary alcohol this readily eliminates acetic acid to give a vinyl sulphide which may serve as a precursor for cephalosporin synthesis.

8.5.3.2 Cephalosporins from Penicillin Sulphoxides. Conversion of penicillins into cephalosporins during biosynthesis involves oxidation of a methyl group at the 2-position followed by its incorporation into the sulphur-containing ring. The chemical equivalent is to subject a penicillin sulphoxide to an acid-catalysed thermal elimination to give an unsaturated sulphenic acid in which a 2-methyl group is dehydrogenated to an alkene and then incorporated into a six-membered ring.

The synthesis of penicillin sulphoxides is simple and stereospecific as illustrated by preparation of the sulphoxide (**47**) in excellent yield on treatment of the trichloroethyl ester of penicillin G with sodium periodate. Only the sulphoxide with the *S*-configuration at sulphur (oxygen on the upper face) is obtained. This stereospecificity is believed to be due to hydrogen bonding of the side-chain to the reagent and is also observed when the sulphoxides are prepared by reaction of penicillin esters with peracids. The sulphoxide derivatives are more stable to acidic and to basic conditions than the parent penicillins and are widely used synthetic intermediates.

Heating the penicillin sulphoxide (**47**, Scheme 8.12) in acetic anhydride gives the expected sulphenic acid acetate (**48**) and this rapidly cyclizes with loss of acetate ion to give the thiiranium ion intermediate (**49**). Nucleophilic attack of acetate ion at one of the two carbons of the thiiranium ion gives either the penam (**52**) or the cepham (**51**), while abstraction of a proton leads to the deacetoxycephalosporin ester (**50**). The latter product is favoured by conditions which allow thermodynamic control of product formation and the protected deacetoxycephalosporin (**50**) is formed in 60% yield when the sulphoxide (**47**) is heated with 5% acetic anhydride in dimethylformamide at 130 °C. Removal of the trichloroethyl protecting group gives a deacetoxycephalosporin, which on side-chain modification gives useful antibiotics such as the orally active cephalexin. Alternatively further chemical reactions, described below, may convert the deacetoxycephalosporin ester (**55**, see Scheme 8.14) into a cephalosporanic acid.

An acid-protecting group is used in the above thermal elimination since the

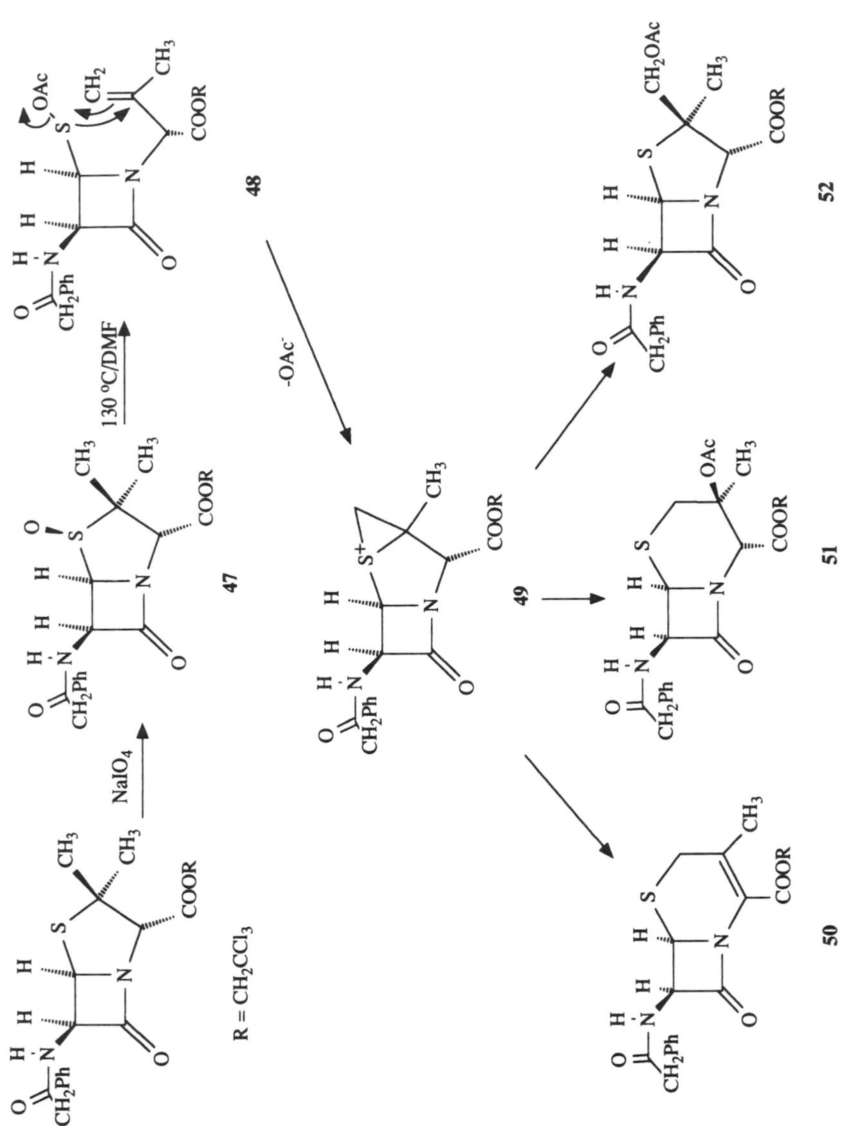

Scheme 8.12 *Conversion of a benzylpenicillin ester into a deacetoxycephalosporin ester*

free acid might decarboxylate or cause β-lactam cleavage on heating. This necessitates two extra steps in the production of deacetoxycephalosporins from penicillins but this inconvenience has been minimized by use of the trimethylsilyl ester group for protection which permits a two-stage sequence (Scheme 8.13) from penicillin G. Penicillin G is first oxidized to its sulphoxide which is heated with 2-picoline (2-methylpyridine) and excess trimethylsilyl chloride. This treatment esterifies the acid group and provides the acidic catalyst for the desired rearrangement. Pouring the reaction mixture into water hydrolyses the silyl protecting group to give the deacetoxycephalosporin in 70% overall yield.

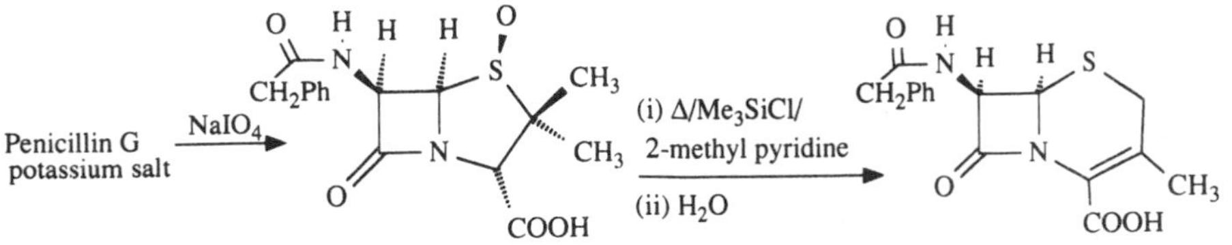

Penicillin G potassium salt $\xrightarrow{NaIO_4}$ [sulphoxide] $\xrightarrow[\text{(ii) } H_2O]{\text{(i) } \Delta/Me_3SiCl/\text{2-methyl pyridine}}$ [deacetoxycephalosporin]

Scheme 8.13 *One-pot conversion of penicillin G sulphoxide into a deacetoxycephalosporin*

Conversion of a deacetoxycephalosporin ester into a cephalosporin requires functionalization of the 3-methyl group. First (Scheme 8.14) the sulphoxide is prepared in order to protect the sulphur and to direct the subsequent bromination to the desired position. This bromination is accomplished using *N*-bromosuccinimide (NBS) with a free radical initiator (light) and it is selective for the allylic position. Such allylic brominations proceed via a free-radical chain reaction (**53**) in which a very low static concentration of Br_2 is maintained by reaction of the NBS with HBr formed during the bromination.

$$Br\bullet + CH_3CR{=}CR \longrightarrow \bullet CH_2CR{=}CR + HBr$$

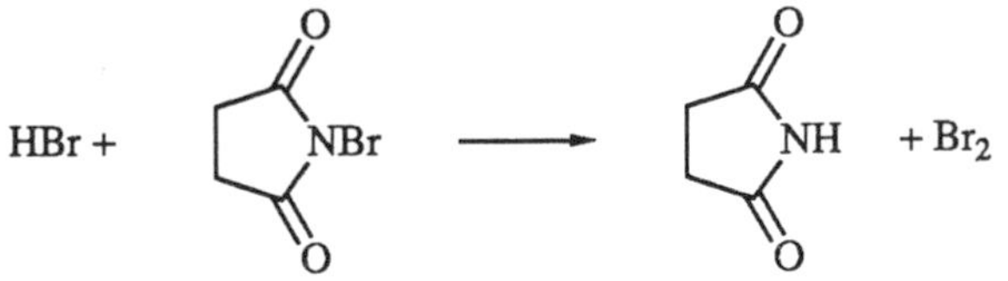

$$HBr + \text{NBS} \longrightarrow \text{succinimide (NH)} + Br_2 \qquad \textbf{(53)}$$

$$\bullet CH_2CR{=}CR + Br_2 \longrightarrow BrCH_2CR{=}CR + Br\bullet$$

Under these conditions addition of bromine to carbon–carbon double bonds can not compete successfully and so the reaction is highly regioselective.

Allyl bromides are reactive alkylating agents and so nucleophilic substitution may readily convert this product (**54**) into its acetate, which on reduction of the sulphoxide and removal of the trichloroethyl group (R) yields the cephalosporanic acid (**55**).

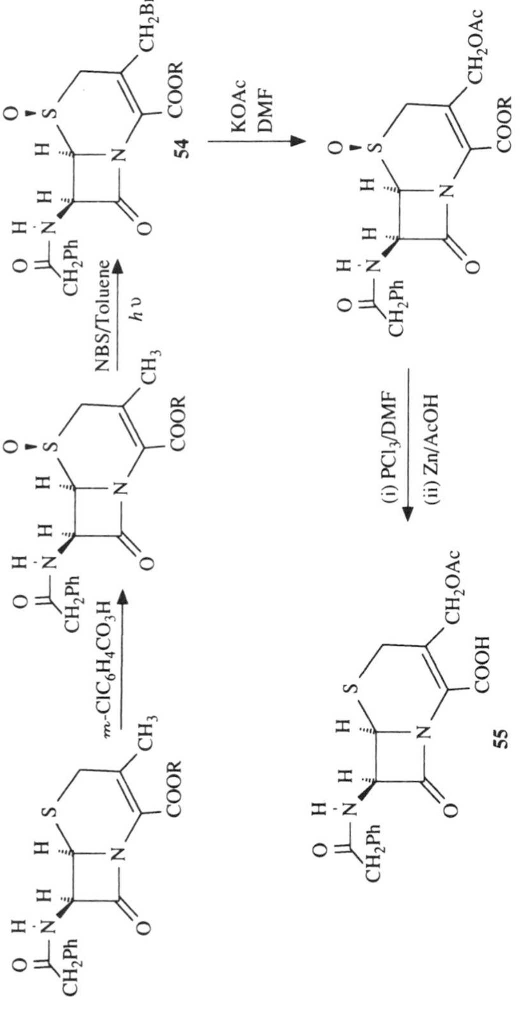
m-ClC6H4CO3H
NBS/Toluene
hν
54
KOAc
DMF
(i) PCl3/DMF
(ii) Zn/AcOH
55
CH2Ph
COOR
CH3
CH2Br
CH2OAc
COOH

Scheme 8.14 *Synthesis of a cephalosporin from a deacetoxycephalosporin*

8.5.3.3 3-Heterosubstituted Cephems from Penicillin Sulphoxides. Cephalosporin analogues bearing an electron-withdrawing heteroatom at C3 should have increased reactivity due to a greater electrophilic character in the β-lactam carbonyl and hence improved biological activity. This expectation was confirmed in the 1970s when several 3-substituted cephems, such as the orally absorbed broad-spectrum antibiotic cefaclor (Table 8.3, p. 242), were found to possess potent antibiotic properties. A highly successful approach to their synthesis is to heat a penicillin sulphoxide in a trialkyl phosphite. Trivalent phosphorus compounds are deoxygenating reagents and so the intermediate sulphenic acid produced on thermolysis is reduced to a thiol. Being both a powerful nucleophile and *cis* to the side-chain, the SH group attacks the carbonyl of the 7-amido group to give the cyclic dihydrothiazole derivative (**56**, Scheme 8.15).

Ozonolysis of this methylene compound (**56**) is performed at low temperature to avoid the danger of overoxidation to a sulphoxide and gives the methyl ketone which exists principally in the enol form (**57**) which is protected as the enamine (**58**) by conversion into a mesylate followed by nucleophilic substitution by morpholine. The methyl group is allylic and so may be readily halogenated using bromine in pyridine. Careful hydrolysis of the dihydrothiazole ring liberates the thiol which is alkylated by reaction with the bromomethyl group to form a six-membered ring. This treatment also hydrolyses the enamine so giving the 3-hydroxy-cephem (**59**). 3-Halogeno- and 3-methoxy-cephalosporin analogues are accessible from this 3-hydroxy-cephem by reaction with phosphorus trihalides or diazomethane respectively. Although this route to 3-heterosubstituted cephems appears lengthy it may be followed by sequential reactions in one pot and is therefore feasible for industrial use.

8.5.3.4 Thermolysis of a 6-Epipenicillin Sulphoxide. The course of thermolysis of a penicillin sulphoxide in a trialkyl phosphite or phosphine (Scheme 8.15) is fundamentally altered if the relative configuration at position-6 is changed: heating the 6-epipenicillin (**60**) in triphenylphosphine gives (Scheme 8.16) the dihydro-oxazole (**61**) which contains no sulphur. Presumably a sulphenic acid is formed which may be deoxygenated to a thiol but it does not cyclize onto the side-chain since a highly strained *trans* ring-junction would result. However, the side-chain is well positioned for nucleophilic displacement of the *trans*-related sulphur-containing group, so leading to the dihydro-oxazole.

Recyclization with incorporation of oxygen is important for the synthesis of the expensive oxacephem antibiotic moxalactam (latamoxef). For this purpose (Scheme 8.16) the product (**61**) is first subjected to allylic chlorination to give a CH_2Cl group in place of the methyl and then hydrolysed to CH_2OH. Treatment with osmium tetroxide adds two OH groups across the double bond and then addition of boron trifluoride catalyses ring-opening of the dihydro-oxazole which is accompanied by attack of the CH_2OH group to give the desired oxazine. The product is converted into an oxacephem by elimination of the remaining hydroxyl group using thionyl chloride in pyridine. A further seven steps, shown in Scheme 8.23, are necessary to synthesize moxalactam.

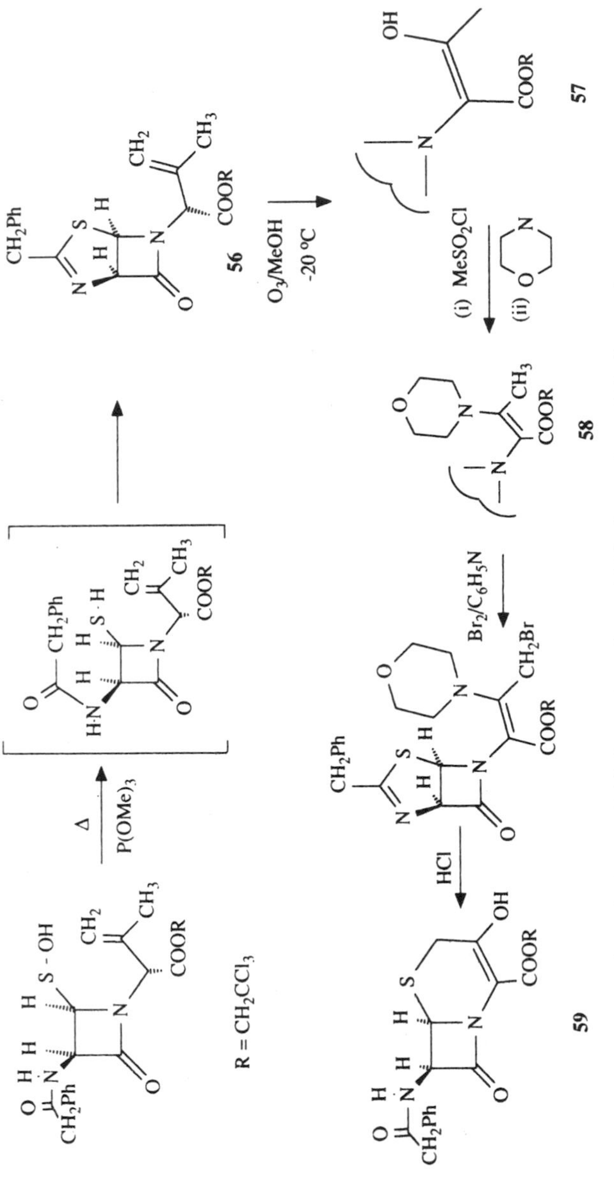

Scheme 8.15 *Conversion of a penicillin G ester into a 3-hydroxycephem*

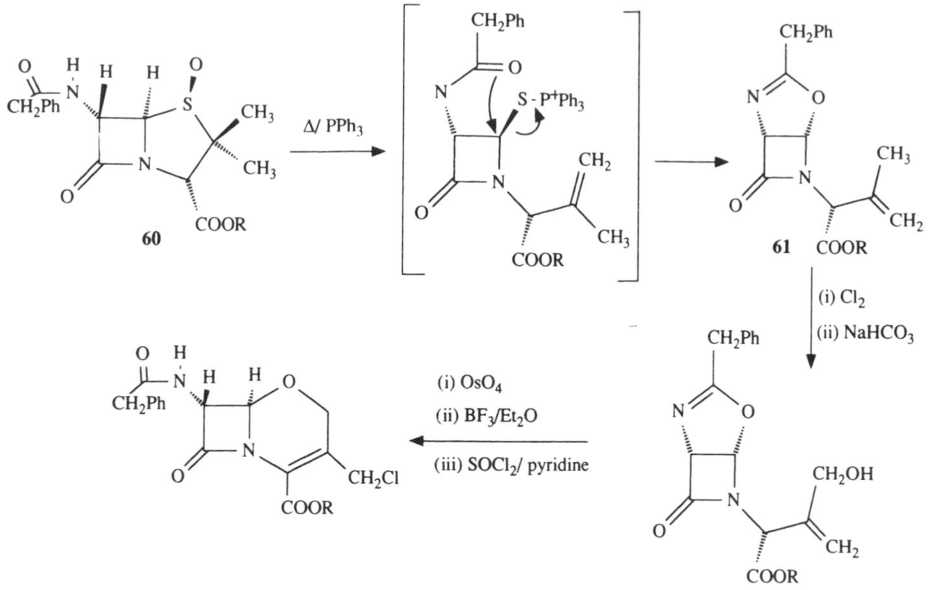

Scheme 8.16 *Synthesis of an oxacephem from a 6-epipenicillin*

8.6 SYNTHESIS OF CEPHALOSPORIN DERIVATIVES

The most important positions for modification of cephalosporins in the synthesis of antibiotics are the side-chain, to give derivatives analogous to the semi-synthetic penicillins, and at C3 by substitution of the 3-acetoxy group. Cephalosporins have more resistance to acidic hydrolysis than have penicillins but still require gentle reaction conditions to avoid destruction. Protection of functional groups during chemical modification is therefore routine and is accomplished by the same methods as described for penicillins. As with penicillins, excess reagents should be avoided to minimize side reactions such as the base-catalysed (acetic anhydride in pyridine is particularly effective) rearrangement of 3-cephems to 2-cephems. Cephalosporin starting materials originate from thermolysis of penicillin sulphoxides or from fermentation.

8.6.1 Microbiological Production of Cephalosporins

Cephalosporin C is produced by a fermentation which, although costly, still appears to be favoured commercially over the chemical route from penicillin G. High-yielding strains of the fungus *Cephalosporium acremonium* (now reclassified as *Acremonium chrysogenum*) give cephalosporin C in submerged fermentations under conditions similar to those described for microbial penicillin production. The biosynthetic pathway (Scheme 8.17) is via penicillin N, which is initially formed in the same manner (Section 8.4.1) as by *Penicillium chrysogenum*. The C2 methyl group on the upper face of penicillin N is then incorporated into the five-membered ring with removal of two hydrogens, giving a five-membered dihydrothiazine ring. This deacetoxycephalosporin is enzymically oxidized to deacetylcephalosporin C (**62**) in which the oxygen in the CH_2OH group comes from molecular oxygen. Highly efficient aeration is thus required for optimum cephalosporin production. The biosynthetic pathway to cephalosporin C is completed by addition of an acetyl group which, in *C. acremonium*, is catalysed by an acetyl transferase.

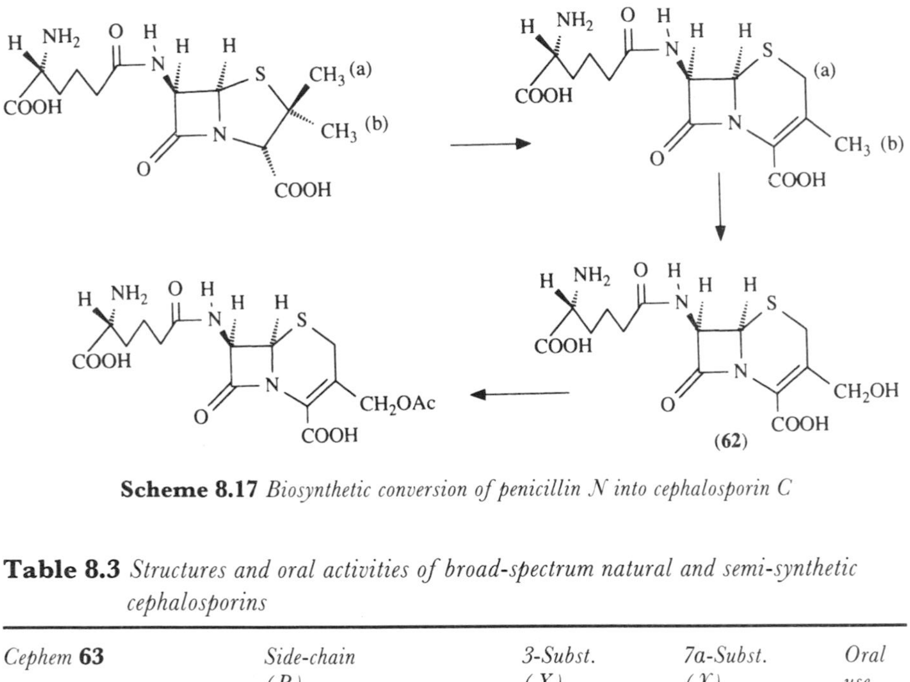

Scheme 8.17 *Biosynthetic conversion of penicillin N into cephalosporin C*

Table 8.3 *Structures and oral activities of broad-spectrum natural and semi-synthetic cephalosporins*

Cephem **63**	*Side-chain (R)*	*3-Subst. (X)*	*7a-Subst. (Y)*	*Oral use*
Cephalexin	$PhCH_2(NH_2)$	CH_3	H	+
Cephalothin	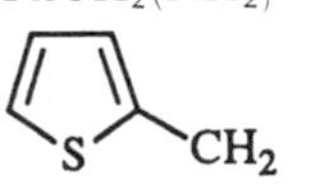 S CH_2	CH_2OAc	H	–
Cefoxitin	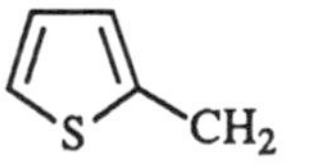 S CH_2	CH_2OCONH_2	OMe	–
Cephamycin C	$NH_2CH(COOH)(CH_2)_3$	CH_2OCONH_2	OMe	+
Ceftazidime	**64a**, R′ = Me	pyridinium	H	–
Cefotaxime	**64a**, R′ = CMe_2COOH	CH_2OAc	H	–
Cefaclor	$PhCH(NH_2)$	Cl	H	+

8.6.2 Side-chain Removal and Replacement

Cephalosporin C is not used clinically because high concentrations are required for bactericidal action but the synthesis and testing of many analogues has led to the development of dozens of useful drugs. The most obvious modification is to change the side-chain and to this end methods for deacylating cephalosporin C to 7-aminocephalosporanic acid (7-ACA) were sought. No microbiological or enzymatic method was found and so a chemical route similar to that described for conversion of penicillin G into 6-APA (Section 8.5.2.2) is used. First cephalosporin C is esterified (Scheme 8.18) using dimethylsilyl dichloride and then the side-chain amido group is converted into an imino chloride which is hydrolysed to 7-aminocephalosporanic acid via the t-butyl imino ether in 92% overall yield.

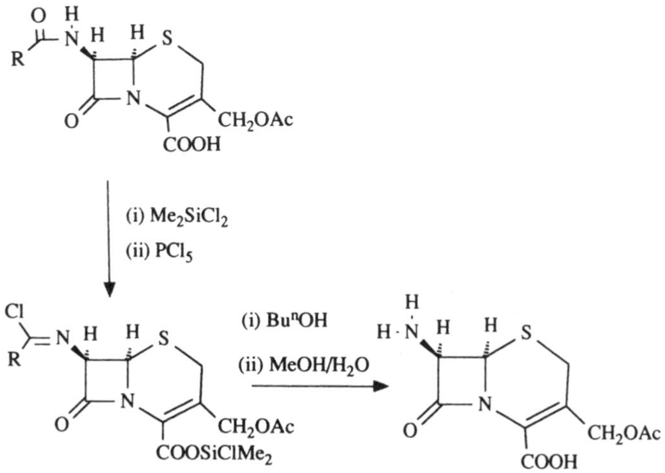

(i) Me_2SiCl_2
(ii) PCl_5

(i) Bu^nOH
(ii) $MeOH/H_2O$

Scheme 8.18 *Side-chain removal from cephalosporins*

Reacylation to give semi-synthetic cephalosporins (Table 8.3) having modified side-chains is accomplished by reaction of 7-ACA with the selected acyl halide and an organic base such as dimethylaniline. Cephalothin, the first modified cephalosporin to be marketed, has a thiophene-2-acetamido side-chain and is active against β-lactamase-producing staphylococci and some Gram-negative bacteria such as *E. coli*. Although resistant to acid hydrolysis it is not absorbed from the gut and so is not given orally. Cephaloglycin, which has the same side-chain as ampicillin [R = D-PhCH(NH_2)], exhibits activity against Gram-negative and penicillin-resistant bacteria and is orally active which makes it particularly valuable in urinary tract infections.

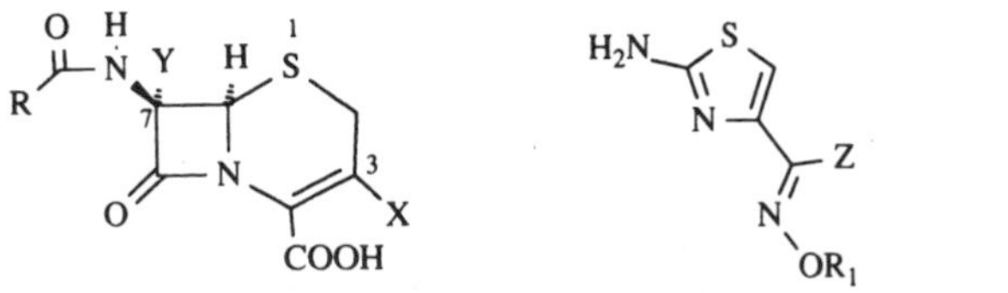

(63)

(64a) Z is attachment point

(64b) Z =COOH

8.6.2.1 7-Oximinoacylcephalosporins. Simple side-chain replacement may also give the oximinocephalosporin group of antibiotics, exemplified by cefotaxime (Table 8.3). Being particularly stable to β-lactamases they have good activity against Gram-negative bacteria especially if the the oxime is *O*-substituted with a bulky alkyl group. The geometry of the oxime in the crucial oximinoacylamino side-chain of these derivatives was found to be important for biological activity: those possessing the *syn*-geometry, as shown in a side-chain precursor (**64b**), being between three and fifteen times more active against a range of Gram-negative and Gram-positive micro-organisms than those with an *anti*-geometry.

Fortunately the stereoselective synthesis of the type of oxime required (**64b**) is fairly easy since metallic anions such as magnesium favour the formation of the *syn*-oxime by complexation between the oxime's OH group and the carboxylate. A relatively slow isomerization accompanies *O*-alkylation of the oxime but retention of the *syn*-geometry is observed in rapid alkylations using reactive electrophilic reagents. Activation for coupling with 7-ACA by forming the acyl chloride has been successfully developed for industrial scale production of ceftazidime but coupling using dicyclohexylcarbodi-imide with hydroxybenzotriazole is preferred for small scale work to avoid the danger of acid-catalysed isomerization.

8.6.3 Modifications of Cephalosporins at C3

The dihydrothiazine ring of cephalosporins is more reactive than the thiazolidine ring in penicillins and offers a wide choice of possible modifications. Replacement of the 3-acetoxy group is the most widely exploited and this potential was indicated (Scheme 8.19) at an early stage of development when a solution of cephalosporin C in pyridine acetate buffer produced the pyridinium analogue (**66**). Pyridine is not a powerful nucleophile but the displacement is promoted by the stability of the intermediate allylic carbocation (**65**). Good

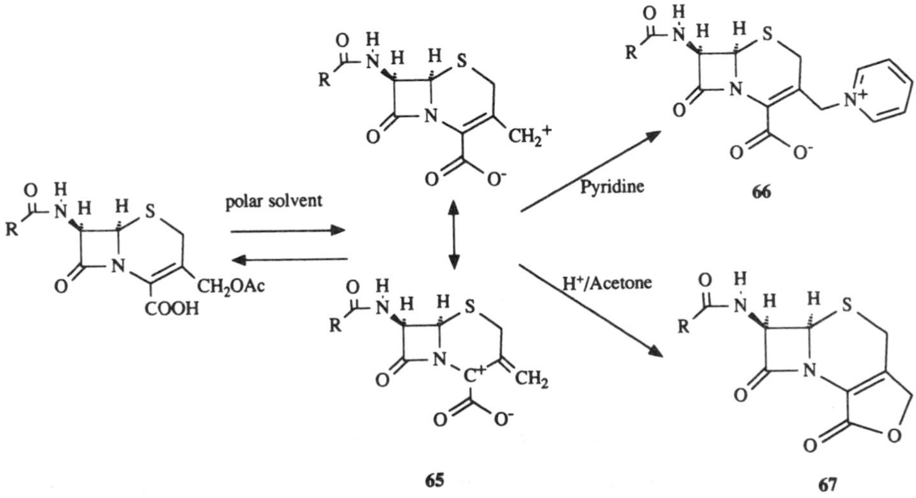

Scheme 8.19 *Replacement of the C3 acetoxy group*

yields of the pyridinium betaine may be obtained by reacting the cephalosporin with pyridine containing aqueous sodium iodide (to promote the nucleophilic substitution).

Sulphur or nitrogen nucleophiles easily displace the acetoxy group to give 3-substituted cephalosporins and reaction of cephalosporins with dilute hydrochloric acid in acetone causes intramolecular substitution to give the lactone (**67**). Substitution with an oxygen nucleophile is difficult since they are generally sufficiently basic to cause β-lactam cleavage, but Δ^2-cephalosporins may be substituted at position-3 using an oxygen-nucleophile and the resulting derivatives may be converted into Δ^3-isomers via 1-sulphoxides. Alternatively 3-bromomethylcephems (Section 8.5.3.2) are sufficiently reactive to give the 3-alkoxymethyl-cephalosporins by direct nucleophilic displacement.

Commercially 3-alkoxymethyl- and 3-acyloxymethyl-cephalosporins are prepared from cephalosporins by enzymatic hydrolysis of the acetoxymethyl group to a hydroxymethyl group on fermentation with a yeast such as *Rhodosporidium toruloides* followed by alkylation or acylation. The metabolically stable 3-carbamoyloxymethyl group of cefuroxime may be introduced, for example (Scheme 8.20), by reaction of the 3-hydroxymethyl derivative (**68**) with trichloroacetyl isocyanate. The protected carbamate which results is hydrolysed to sodium cefuroxime (**69**) using sodium 2-ethylhexanoate in alcohol.

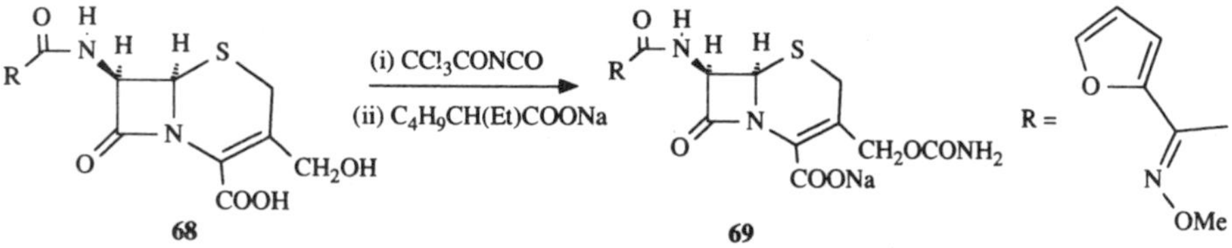

O H
R N H H S
O N
CH$_2$OH
COOH
68
(i) CCl$_3$CONCO
(ii) C$_4$H$_9$CH(Et)COONa
O H
R N H H S
O N
CH$_2$OCONH$_2$
COONa
69
R =
O
N
OMe

Scheme 8.20 *Synthesis of cefuroxime*

8.6.3.1 Deacetoxycephalosporins and 3-Hetero-substituted Cephalosporins. We have already seen that deacetoxycephalosporins may be produced from the thermal elimination of penicillin sulphoxides. They may also be produced (Scheme 8.21) by reduction of a cephalosporanic acid to a 3-methylene-cephalosporin (**70**) followed by isomerization. Chromium(II) acetate, hydrogen with palladium catalyst, or even electrolysis has been used for the reduction and the isomerization may be accomplished by silylation followed by hydrolysis.

Alternatively the 3-methylene-cephalosporin may by ozonized at −80 °C to give a 3-hydroxy-cephalosporin, which may readily be converted into 3-chloro-, 3-bromo-, or 3-methoxy-cephalosporins (**71**) by reaction with phosphorus trichloride, phosphorus tribromide, or diazomethane respectively.

8.6.4 7α-Methoxy-cephalosporin Derivatives

In 1971 it was reported that the antibiotic cephamycin C (Table 8.3), a natural product from a strain of *Streptomyces clavuligerus*, was highly active against β-lac-

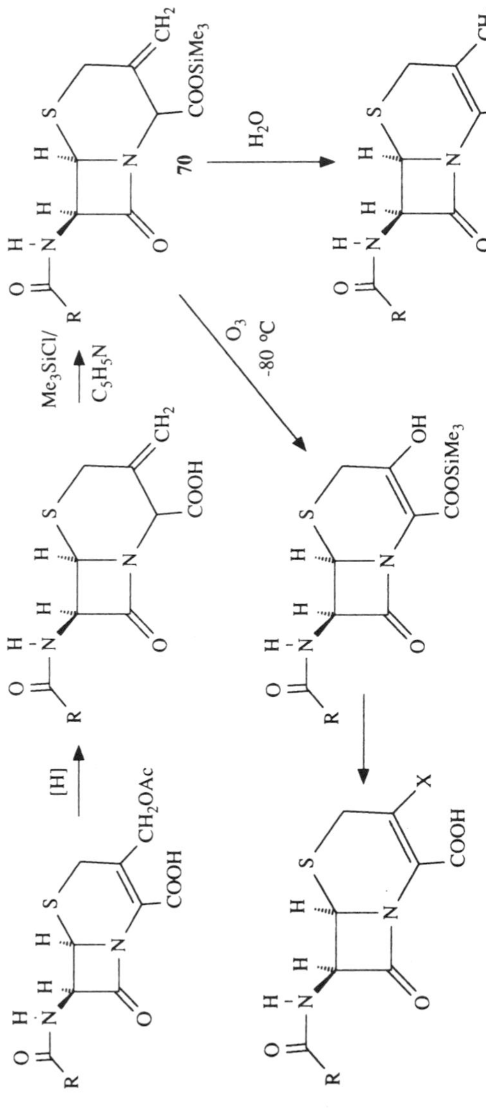
[H]
Me3SiCl/
C5H5N
70
H2O
O3
-80 °C
CH2OAc
COOH
CH2
COOSiMe3
CH3
OH
X

71 X = Cl, Br or OMe

Scheme 8.21 *Conversion of cephalosporins into 3-hetero-substituted derivatives*

tamase-producing micro-organisms. A 7α-methoxy substituent confers this remarkable level of stability and the observation led to the development of antibiotics such as cefoxitin and moxalactam. At first sight introduction of a 7α-methoxy substituent into a cephalosporin is a daunting task but it has proved surprisingly simple (Scheme 8.22). Methoxylation of a CH situated between NH and C=O groups such as in the methyl ester of *N*-(phenylacetyl)-2-phenylglycine [**72**, R′ = PhCH(COOPh), R = OMe] may be accomplished using t-butyl hypochlorite in methanol. The NH is first converted into NCl and then HCl is eliminated to give an acylimine from which the methoxy derivative is generated by methanolysis (Scheme 8.22).

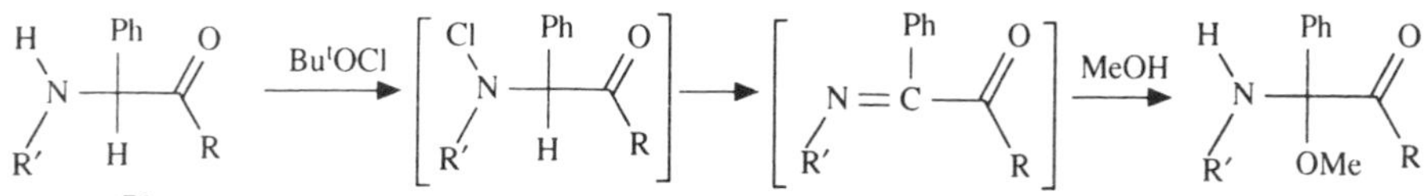

Scheme 8.22 *Methoxylation using t-butyl hypochlorite in methanol*

For the synthesis of 7α-methoxycephalosporins from a 7-acylamino-3-cephem the methanolysis step looks reasonable because the desired product results from attack of methoxide from the less hindered side but the treatment with t-butyl hypochlorite could be troublesome since chlorination at C2 or isomerization of the double bond are likely side-reactions. However, treatment of cephalosporin C with lithium methoxide and t-butyl hypochlorite at −80 °C gives a 70% yield of 7α-methoxycephalosporin C. In penicillins the sulphur is more prone to oxidation by t-butyl hypochlorite and the β-lactam is more easily hydrolysed but synthesis of 6α-methoxypenicillins may usually be accomplished similarly.

Acidic hydrolysis of the 7α-methoxy group of cephalosporins would give a 7-keto group with loss of the side-chain but this does not normally happen under conditions mild enough for preservation of the β-lactam. The range of manipulations possible is illustrated (Scheme 8.23) by the final steps in the synthesis of moxalactam (**8**).

8.7 TOTAL SYNTHESIS OF β-LACTAM ANTIBIOTICS

Although the total synthesis of penicillins and cephalosporins is unnecessary in the manufacture of β-lactam antibiotics (but see thienamycin, Section 8.7.3.2) it has been a notable scientific achievement and has provided unnatural analogues useful in the identification of structure–activity relationships. Since penicillins and cephalosporins are inherently unstable, construction of the reactive bicyclic structures is usually left until late in synthetic sequences. Two of the most famous syntheses, Sheehan's synthesis of penicillin V and Woodward's synthesis of cephalosporin C, are described below.

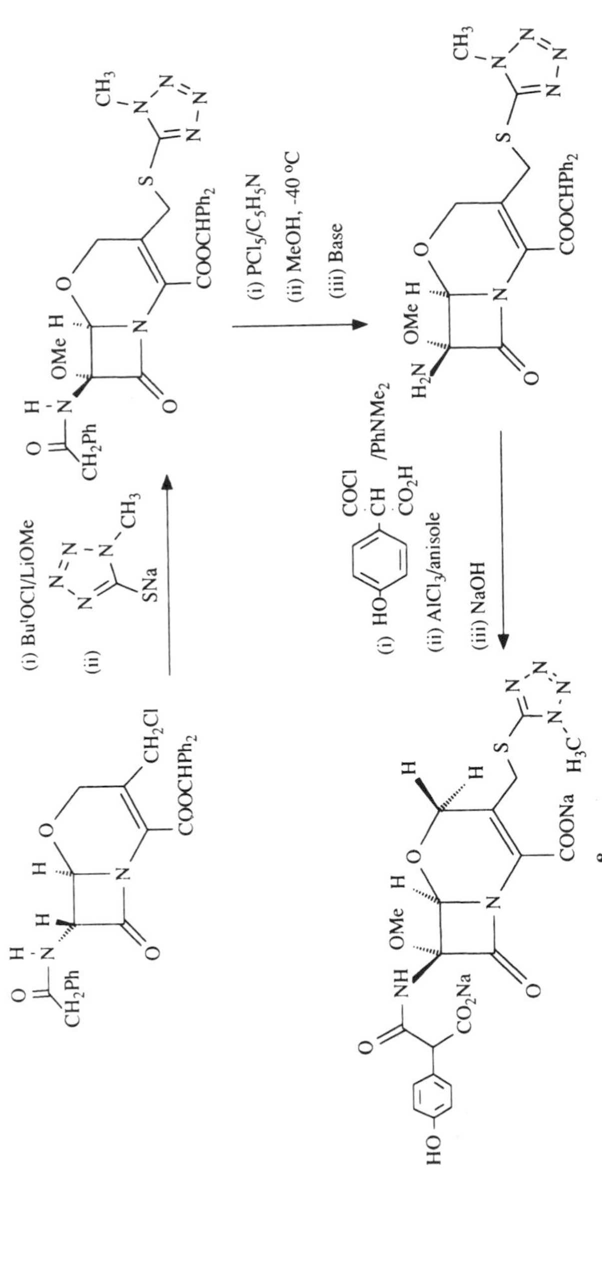

Scheme 8.23 *Final stages in the synthesis of moxalactam*

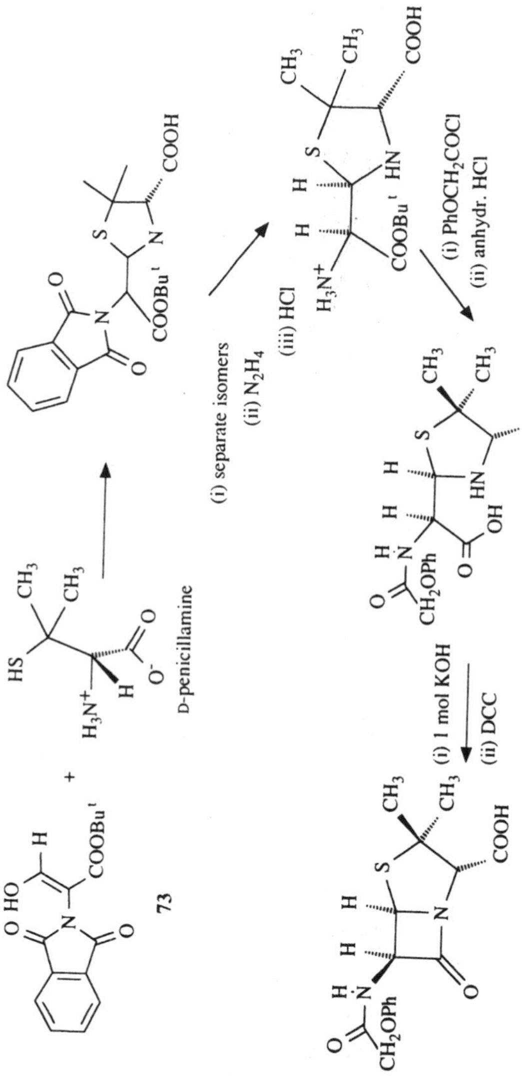

73 + D-penicillamine

(i) separate isomers
(ii) N_2H_4
(iii) HCl

(i) $PhOCH_2COCl$
(ii) anhydr. HCl

(i) 1 mol KOH
(ii) DCC

Scheme 8.24 *A total synthesis of penicillin*

8.7.1 Total Synthesis of Penicillin V

The total synthesis of penicillin V shown in Scheme 8.24 was the first (1959) to give a respectable yield and is instructive in its skilful use of protecting groups. Starting from the aminothiol D-penicillamine, prepared in a six-step synthesis, reaction with the protected formylglycine derivative (**73**) gives a thiazolidine ring. Two new chiral centres are produced but only two of the four possible isomers are formed. After separation of the desired diastereoisomer the nitrogen-protecting phthalimido group is removed by treatment with hydrazine (as in the Gabriel synthesis of amines) and the product is converted into an ammonium salt by addition of the required amount of hydrochloric acid. The side-chain is attached by reaction with phenoxyacetyl chloride and then the t-butyl ester is removed using anhydrous hydrochloric acid. The thiazolidine dicarboxylic acid is then reacted with one equivalent of potassium hydroxide to give the mono-potassium salt of the stronger acid thus introducing the simplest of acid-protecting groups, a potassium ion. The β-lactam ring may then be joined between the NH of the thiazolidine and the free carboxylate group using dicyclohexyl-carbodi-imide (DCC) giving penicillin V. Less than 10% yield was obtained in this cyclization because the amido group of the side-chain also reacts with the acid group to give an oxazolone. It has since been discovered that much better yields are obtainable when a tritylamino group is present in place of the side-chain. Alternatively the phthalimido group may be retained until after β-lactam formation and then removed under mild conditions (Scheme 8.25) compatible with preservation of the penam ring.

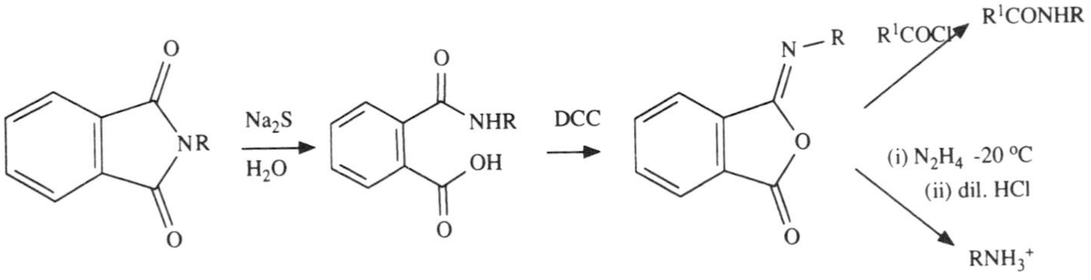

Scheme 8.25 *Mild removal of a phthalimido group*

8.7.2 Total Synthesis of Cephalosporins

The β-lactam ring may be formed early in a synthetic pathway if construction of the destabilizing second ring is delayed as in the preparation (Scheme 8.26) of cephalothin and cephalosporin C. L-Cysteine is used as starting material and requires the introduction of a nitrogen α to sulphur. Since this position is of low reactivity the thiol, amino, and carboxylate groups are first protected by reactions with acetone (giving a thiazolidine), t-butyloxycarbonyl chloride (giving the BOC derivative), and finally diazomethane (giving the methyl ester). Hydroxylation next to sulphur may then be achieved by UV irradiation in the

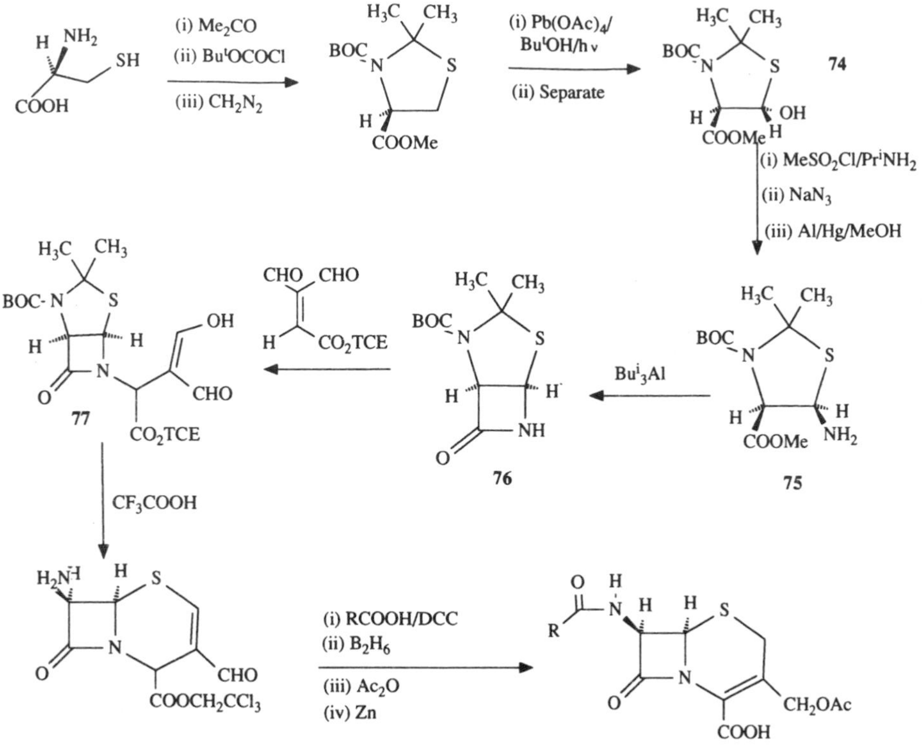

Scheme 8.26 *Total synthesis of cephalosporins*

presence of lead tetra-acetate and t-butyl alcohol. The desired alcohol (**74**) is separated from the resulting diastereomeric mixture and esterified to the mesylate which, with sodium azide, readily undergoes nucleophilic substitution (with inversion of configuration) to give the azide. Reduction using aluminium amalgam gives a *cis* amino-ester (**75**) which on treatment with the condensing agent tri-isobutylaluminium cyclizes to the β-lactam (**76**). The α,β-unsaturated dialdehyde then reacts with the electrophilic β-lactam nitrogen in a Michael-type conjugate addition to give the β-lactam derivative (**77**). Treatment with TFA releases the thiol group which acts as a nucleophile in an intramolecular conjugate addition forming the dihydrothiazine ring. The remaining steps result in attachment of the side-chain, isomerization of the double bond, reduction of the aldehyde to an alcohol, acetylation of the alcohol, and removal of the trichloroacetyl group respectively.

The β-lactam (**76**) has also been reacted with other aldehydes and the products converted into nuclear analogues of cephalosporins such as 2-thiacephems by similar routes.

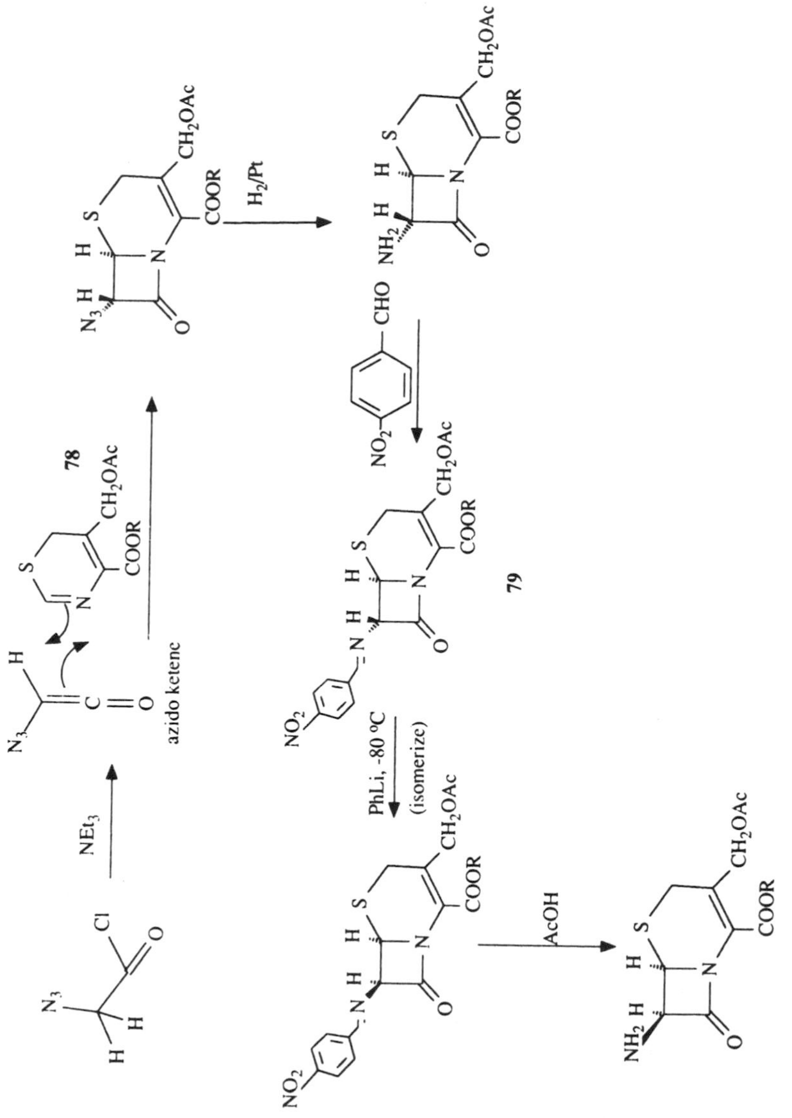

N_3CH_2COCl → (NEt_3) → azido ketene $N_3CH{=}C{=}O$ + **78** → N_3-cephem (COOR, CH_2OAc) → (H_2/Pt) → NH_2-cephem → (p-$NO_2C_6H_4CHO$) → **79** → (PhLi, -80 °C (isomerize)) → → (AcOH) → NH_2-cephem (COOR, CH_2OAc)

Scheme 8.27 *Total synthesis of an ester of (±)-7-ACA*

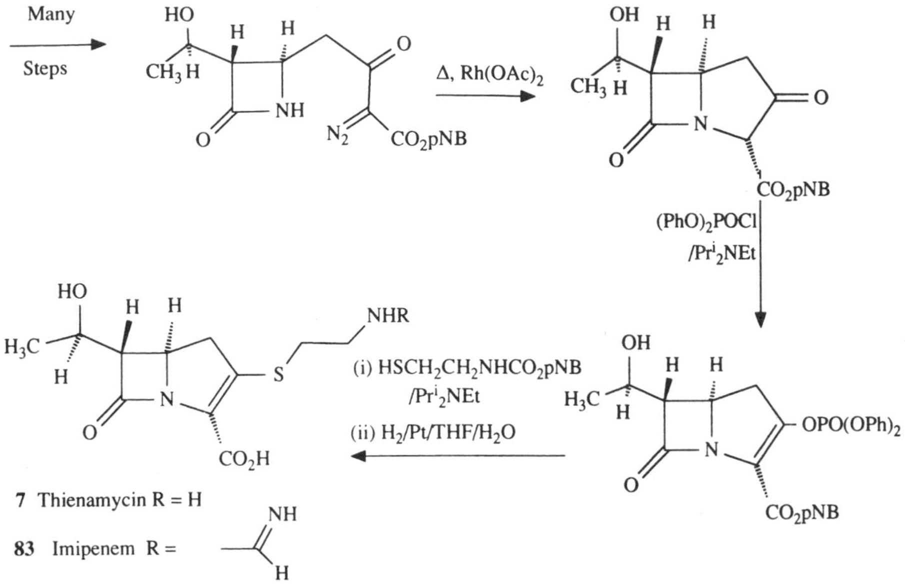

Many Steps

Δ, $Rh(OAc)_2$

$(PhO)_2POCl$ / Pr^i_2NEt

(i) $HSCH_2CH_2NHCO_2pNB$ / Pr^i_2NEt

(ii) $H_2/Pt/THF/H_2O$

7 Thienamycin R = H

83 Imipenem R = $CH{=}NH$

Scheme 8.30 *Synthesis of thienamycin*

pose under normal conditions of application and so the design of novel antibacterial penem acids requires a balance between desirable reactivity of the β-lactam system and sufficient stability. Thienamycin, for example, whose primary amino group may nucleophilically attack the β-lactam of another molecule, is too short-lived in the body to be clinically useful. Its *N*-formimoyl derivative, imipenem (**83**) has reduced nucleophilicity which gives degradation resistance while retaining the wide-ranging potency and exceptional antipseudomonal power of thienamycin. The drug tienam, containing imipenem, was approved for clinical use in the UK in 1988.

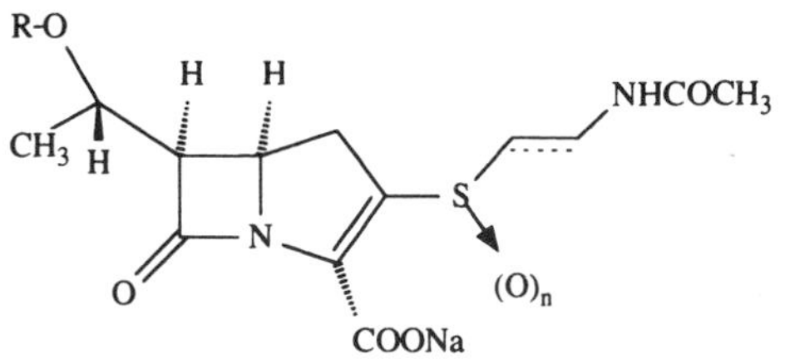

R = H, NaO_3S

n = 0,1,2

84 Olivanic acids

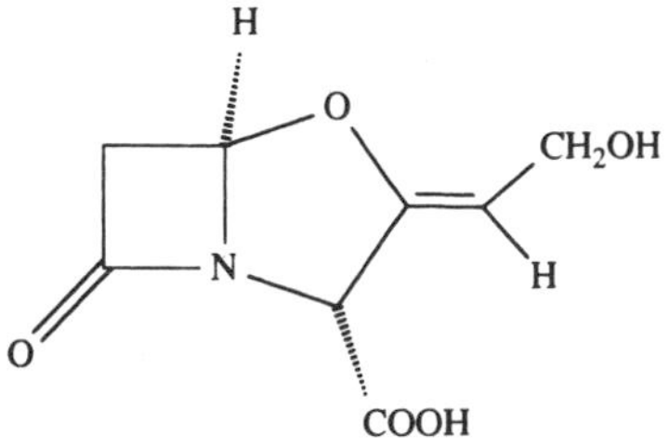

85 Clavulanic acid

Construction of such strained bicyclic systems is generally left until late in the synthetic sequences but may be accomplished in excellent yield. The total synthesis of thienamycin (Scheme 8.30) is commercially important since fermen-

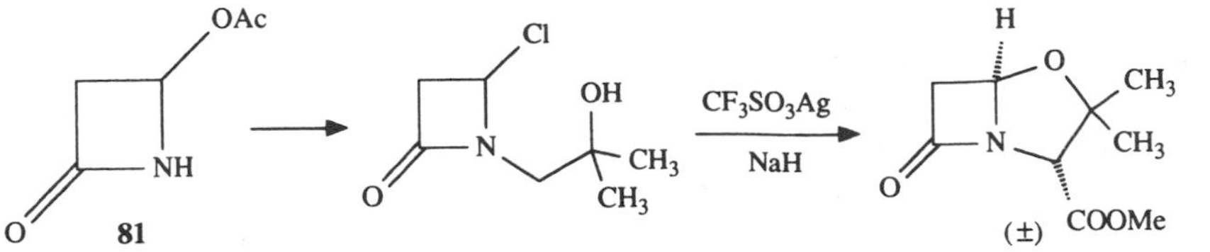

Scheme 8.31 *Synthesis of an oxapenam*

tation methods have proved uneconomic. A β-lactam ring is first produced either in racemic form by cycloaddition or optically pure by condensation of an aspartic acid derivative and then modified in a sequence of high-yielding steps ready for cyclization. The cyclization is nearly quantitative and proceeds via the insertion of a carbene, generated by thermal decomposition of a diazoketone, into the NH bond of the β-lactam. Reaction with diphenyl chlorophosphate introduces the double bond by formation of an enol phosphate and the phosphate may then be displaced by the appropriate sulphide. Finally removal of protecting groups liberates thienamycin.

Another analogue of the penems is clavulanic acid (**85**) which was isolated from *Streptomyces clavuligerus*. It has weak antibacterial activity but is a potent β-lactamase inhibitor and, in the drug augmentin, is used to extend the activity of amoxycillin to a wider range of organisms. Penicillins which are deactivated by β-lactamases give products which readily leave the enzyme's surface so allowing renewed access to the enzyme's active site and continual conversion of penicillin into products. Clavulanic acid has a β-lactam ring which is recognized by β-lactamase enzymes and gives some products which, because of the double bond in the side-chain, remain temporarily or permanently attached to the enzyme. The active site is thus blocked and such inhibitors are known as suicide substrates. The 4-acetoxy-β-lactam (**81**) has been used in the synthesis of derivatives of clavulanic acid by the route indicated in Scheme 8.31.

The β-lactam ring of carumonam (Scheme 8.32) is formed in a cycloaddition between the ketene derived from CBZ-glycine and a valine derivative. The

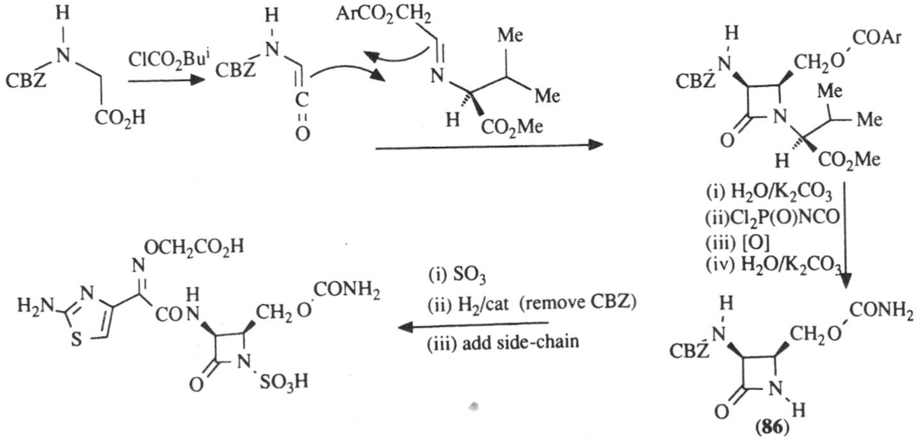

Scheme 8.32 *Outline synthesis of the monobactam carumonam*

L-valine derivative acts as a chiral template to give mainly the chiral *cis* product shown. The ester protective groups are carefully hydrolysed, the carbamoyl moiety is added, and then the chiral substituent on the lactam nitrogen removed by oxidative electrolysis to give **86**. Reaction with sulphur trioxide gives the *N*-sulphonate from which carumonam is produced by removal of the CBZ group, addition of the side-chain protected as its diphenylmethyl ester, and final deprotection using trifluoroacetic acid.

8.8 NUCLEAR MAGNETIC RESONANCE SPECTRA

Preliminary identification of synthetic or naturally occurring β-lactam antibiotics is usually based on their NMR spectra. Approximate values for proton and ^{13}C chemical shifts observed for penicillin V methyl ester and for a related cephalosporin are shown in Figure 8.1.

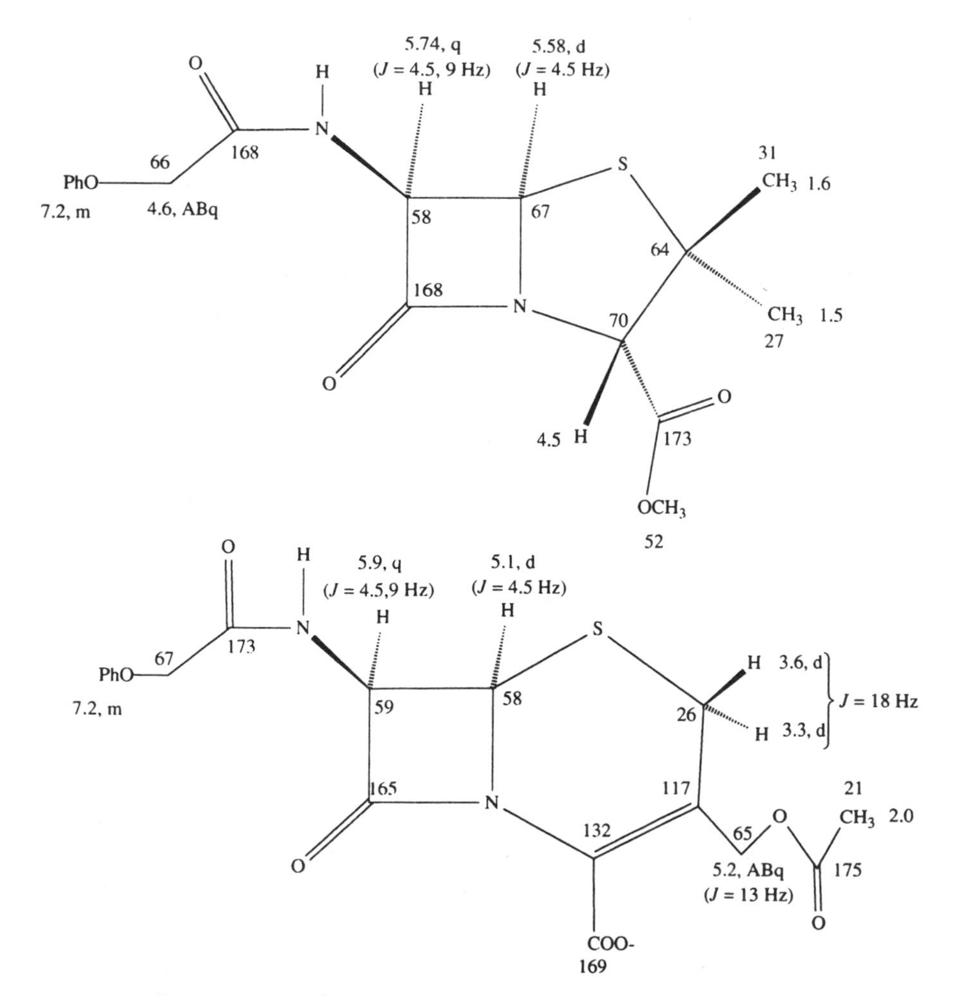

Figure 8.1 *^{13}C (integer) and ^{1}H (decimal) δ-values for a penicillin and a cephalosporin*

Some protons in penicillins and cephalosporins, which at first sight appear to be in identical environments, give separate signals owing to the lack of symmetry. The two methyl groups at C2 of penicillins, for example, are fixed on different sides of the ring and so are diastereotopic (*i.e.* not chemically equivalent) and each gives a singlet. Similarly the protons in each of the two methylene groups of cephalosporins (at C2 and in the CH_2OAc group) have different chemical shifts and are strongly coupled (J 12–19 Hz) to each other. Such methylene protons thus give a pair of doublets but since the chemical shifts are fairly close the pattern is often similar to a quartet and is referred to as ABq.

Coupling between H5 and H6 in penicillins (H6 and H7 in cephalosporins) is diagnostic of the relative configuration at these centres. A coupling constant of 4–5 Hz is found for the natural *cis*-configuration while *trans*-isomers such as 6-epipenicillin show $J = 1.5–2$ Hz. The proton on the carbon to which the side-chain is attached (H6 in penicillins) normally also couples ($J = 8–11$ Hz) to the vicinal amido proton so giving a quartet.

Bibliography

The following list of books and articles is a selection for further reading. The first mentioned in a group is usually the most basic.

General

I.L. Finar, 'Organic Chemistry', Longman, Harlow, 5th Edn., 1975, Vol. 2: Stereochemistry and the Chemistry of Natural Products.

L. Stryer, 'Biochemistry', Freeman, New York, 1988.

R.H. Thomson, 'The Chemistry of Natural Products', Blackie, Glasgow, 1985.

'Biological Compounds', ed. E. Haslam, 'Comprehensive Organic Chemistry', Vol. 5, Pergamon Press, Oxford, 1979.

Chapter 1 (Intermolecular Forces)

S.N. Vingradov and R.H. Linnell, 'Hydrogen Bonding', Van Nostrand Reinhold, New York, 1971, Chapter 1.

Chapter 2

H.S. El Khadem, 'Carbohydrate Chemistry: Monosaccharides and the Oligomers', Academic Press, San Diego, 1988.

R.J. Ferrier and P.M. Collins, 'Monosaccharide Chemistry', Penguin Books, Harmondsworth, England, 1972.

'Carbohydrate Chemistry', ed. J.F. Kennedy, Clarendon Press, Oxford, 1988.

Chapter 3

D.A. Rees, 'Polysaccharide Shapes', Chapman and Hall, London, 1977.

G.O. Aspinall, 'The Polysaccharides', Vols. 1 and 2, Academic Press, New York, 1982.

'Carbohydrate Chemistry', ed. J.F. Kennedy, Clarendon Press, Oxford, 1988.

Chapter 4

P.D. Bailey, 'An Introduction to Peptide Chemistry', Wiley, Chichester, 1990.

R.E. Dickerson and I. Geis, 'The Structure and Action of Proteins', Harper & Row, New York, 1969.

T.E. Creighton, 'Proteins', Freeman & Co. New York, 1984.

Chapter 5

'Nucleic Acids in Chemistry and Biology', ed. G.M. Blackburn and M.J. Gait, IRL Press, Oxford, 1990.

R.L.P. Adams, J.T. Knowler, and D.P. Leader, 'The Biochemistry of the Nucleic Acids', 10th Edn., Chapman and Hall, London, 1986.

Chapter 6

I.L. Finar, 'Organic Chemistry', Longman, Harlow, 5th Edn. 1975, Vol. 2, pp. 507–605.

W. Templeton, 'An Introduction to the Chemistry of the Terpenoids and Steroids', Butterworths, London, 1969.

R. Wiechert, *Angew. Chem. Int. Edn.*, 1977, **16**, 506.

L.F. Fieser and M. Fieser, 'Steroids', Chapman and Hall, London, 1959.

Chapter 7

R.F. Newton and S.M. Roberts, 'Prostaglandins and Thromboxanes', Butterworth Scientific, London, 1982.

P.H. Bentley, *Chem. Soc. Rev.*, 1973, **2**, 29.

Chapter 8

G. Lowe, in 'Comprehensive Organic Chemistry', Vol. 5, ed. E. Haslam, Pergamon Press, Oxford, pp. 289–320.

M.I. Page, *Adv. Phys. Org. Chem.*, 1987, **23**, 165.

Subject Index and Abbreviations

A page whose number is given in **bold face** is a principal reference and/or includes the structure of the indexed compound. Prefixes such as *cis-* or *o-* are ignored in the alphabetical order of the index.

Abbreviations are indexed with key page numbers but more information is often indexed under the full names. The one-letter abbreviations for amino acids are not included here, but in Table 4.1, p. 75.